AF560606

CARBOHYDRATES AND PROTEINS

CARBOHYDRATES AND PROTEINS

By

Amit Arora

DISCOVERY PUBLISHING HOUSE
NEW DELHI-110002

First Published – 2006

Reprinted – 2024

ISBN: 978-81-8356-178-5

Carbohydrate and Proteins

Published by:

DISCOVERY PUBLISHING HOUSE

4383/4B, Ansari Road, Darya Ganj
New Delhi-110 002 (India)
Phone: +91-11-23279245; 23253475; 43596065
Mobile: +91 9811179893 / +91 9871656464
E-mail: discoverybooksindia@gmail.com
orderdphbooks@gmail.com
namitwasan9@gmail.com
web: www.discoverypublishinggroup.com

Printed at:
Infinity Imaging Systems
Delhi

Preface

This book is written for B.Sc., B.Sc. (Hons.) and M.Sc. students of various universities of India. In this book my aim has been describe the fundamental principles of organic chemistry. Since I do not consider the chemistry of natural products to be fundamental chemistry but rather the application of fundamental principles. The subject matter described in this book covers much of the basic organic chemistry that is needed by a student who wish to study chemistry as a main subject at degree level. The arrangement of the subject-matter is based on homologous series and in general, descriptions of reactions are followed by discussion of their mechanisms and these includes an elementary account of the sort of evidence that led workers to suggest mechanisms that are acceptable at the present time.

We are sure you will appreciate the thought and care that has gone into making of this text. This book has everything you want to have in your advanced organic chemistry text.

The author wish to thank all those who helped in the preparation of the text and illustration, suggestion for improvement and criticism will be gratefully received and acknowledge.

Author

CONTENTS

1

CARBOHYDRATES

INTRODUCTION

Carbohydrates are substances with the general formula $C_x(H_2O)_y$, and were called carbohydrates (hydrates of carbon) because they contained hydrogen and' oxygen in the same proportion as in water. However, a number of compounds have been discovered which are carbohydrates by chemical behaviour, but do not conform to the formula $C_x(H_2O)_y$, *e.g.*, 2-deoxyribose, $C_5H_{10}O_4$. It is also important to note that all compounds conforming to the formula $C_x(H_2O)_y$ are not necessarily carbohydrates, *e.g.*, formaldehyde, CH_2O; acetic acid, $C_2H_4O_2$; etc.

All carbohydrates are polyhydroxy-aldehydes or ketones, or substances that yield these on hydrolysis. Some partially methylated sugars occur naturally, and one natural nitro-sugar is known.

NOMENCLATURE

The names of the simpler carbohydrates end in *-ose*, carbohydrates with an aldehydic structure are known as *aldoses*, and those with ketonic, *ketoses*. The number of carbon atoms in the molecule is indicated by a Greek prefix, *e.g.*, a *tetrose* contains four carbon atoms, a *pentose* five, a *hexose* six, etc.

Carbohydrates are divided into two main classes, *sugars* and *polysaccharides* (*polysaccharoses*). Sugars are crystalline substances with a sweet taste and soluble in water. Polysaccharides are more complex than the sugars, their molecular weights being far greater. Most of them are non-crystalline substances which are not sweet, and are insoluble or less soluble in water, than the sugars.

Sugars are subdivided into two major groups:

1. *Monosaccharide* (*monofacchuroses*). These are sugars which cannot be hydrolysed into smaller molecules. Their general formula is $C_nH_{2n}O_n$, (there are exceptions), where n is 2--10 (but see later), and the most important are the pentoses and hexoses.

2. *Oligosaccharides.* These yield 2-10 monosaccharide molecules on hydrolysis, *e.g.*, *disaccharides* (two monosaccharide molecules which may, or may not, be the same); *trisaccharides* (three monosaccharide molecules): etc.

Polysaccharides (*polysaccharoses*) are carbohydrates which yield a large number of mono-saccharide molecules on hydrolysis. The most widely spread polysaccharides have the general formula $(C_6H_{10}O_5)_n$, *e.g.*, starch, cellulose, etc., a group of polysaccharides which are not so widely spread in nature is the *pentosans*, $(C_5H_8O_4)_n$.

Monosaccharides

The simplest monosaccharide is glycolaldehyde, $HOCH_2CHO$. This does not contain an asymmetric carbon atom and is therefore not optically active. Since all naturally occurring sugars are optically active, it is more satisfactory to exclude glycolaldehyde from the group of sugars, and to define sugars as optically active polyhydroxy-aldehydes or ketones.

Triose, $C_3H_6O_3$, *Glyceraldehyde*, $HOCH_2CHOHCHO$, is an aldotriose. It contains one asymmetric carbon atom, and can therefore exist in two optically active forms, D- and L-; both are known. Glyceraldehyde has been chosen as the standard configuration in sugar chemistry.

Dihydroxyacetone, $HOCH_2COCH_2OH$, is not optically active, and hence, by the definition given above, is not a sugar. If the definition be rejected, *i.e.*, the proviso that a sugar is always optically active is rejected, then dihydroxyacetone will be a ketotriose.

Tetroses, $C_4H_8O_4$. The structure of an aldotetrose is $HOCH_2$ CHOHCHOHCHO. There is only one structure possible for a ketotetrose, viz., $HOCH_2COCHOHCH_2OH$. This contains one asymmetric carbon atom, and corresponds to D- and L-erythrulose.

Pentoses, $C_5H_{10}O_5$. The *aldopentoses* are an important group of sugars (monosaccharides), and their structure has been elucidated by methods similar to those used for glucose and fructose. Their structure is:

$OCHCHOHCHOHCHOHCH_2OH$.

This contains three structurally different asymmetric carbon atoms, and can therefore exist in eight optically active forms. All are known, and correspond to the D- and L-forms of *arabinose*, xylose, ribose and lyxose.

L(+)-*Arabinose*, m.p. 158°C, occurs naturally as pentosans (arabans) in various gums, *e.g.*, gum arabic. It is usually obtained by the hydrolysis of cherry-gum with dilute sulphuric acid.

D(+)-*Arabinose* occurs in certain glucosides.

O(+)-*Xylose* (wood-sugar), m.p. 145°C, occurs as entosans (xylans) in wood gums, bran and straw, from which it may be obtained by hydrolysis with dilute sulphuric acid.

D(+)-*Ribose*, m.p. 95 C, occurs in plant nucleic acids, and in liver and pancreas nucleic acids.

All the other aldopentoses are synthetic compounds.

The chemical properties of the aldopentoses are similar to those of the aldohexoses (see glucose, below); pentoses do not undergo fermentation. All the aldopentoses are converted quantitatively into *furfural* when warmed with dilute acid:

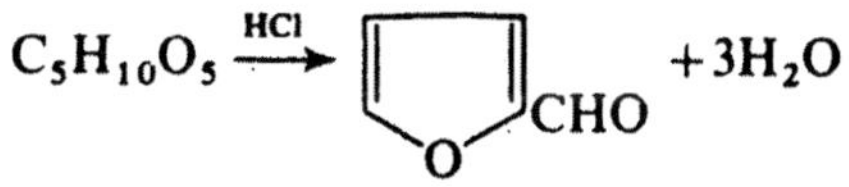

$$C_5H_{10}O_5 \xrightarrow{HCl} \text{(furan ring with CHO)} + 3H_2O$$

Fig. 1.1

A very important aldopentose is 2-deoxyribose, $HOCH_2CHOHCHOHCH_2CHO$ (the prefix 'deoxy' indicates the replacement of a hydroxyl group by hydrogen;). It occurs in nucleic acids.

Ketopentoses with the structure $HOCH_2COCHOHCHOHCH_2OH$ have been prepared. The molecule contains two structurally different asymmetric carbon atoms, and can therefore exist in four optically active forms.

All are known, and correspond to D- and L-forms of ribulose and xylulose.

Hexoses, $C_6H_{12}O_6$. The *aldohexoses* are another important group of monosaccharides, and their structure has been elucidated as follows:

(i) Analysis and molecular-weight determinations show that the molecular formula of the aldohexoses is $C_6H_{12}O_6$.

(ii) When treated with acetic anhydride, aldohexoses form the penta-acetate. This indicates the presence of five hydroxyl groups

(iii) Aldohexoses form an oxime when treated with hydroxylamine, and therefore contain a carbonyl group.

(iv) When an aldohexose is oxidised with bromine-water, a pentahydroxy-acid of formula $C_6H_{12}O_7$ is obtained. This indicates that the carbonyl group is present in an *aldehydic* group.

(v) When reduced with concentrated hydriodic acid and red phosphorus at 100°C, aldohexoses give a mixture of 2-iodohexane and n-hexane. This indicates that the six carbon atoms in an aldohexose are in a straight chain.

This conclusion is supported by ascending the series by the Kiliani reaction and reducing the product, a polyhydroxy acid, with hydriodic acid; n-heptanoic acid is produced.

The foregoing reactions show that the structure of the aldohexoses is:

$$OCHCHOHCHOHCHOHCHOHCH_2OH.$$

This contains four structurally different asymmetric carbon atoms, and can therefore exist in sixteen opticany active forms. All are known, and correspond to the D- and L-forms of glucose, *mannose*, *galactose*, *allose*, *altrose*, *gulose*, *idose* and *talose*.

Although the foregoing evidence indicates an open-chain structure, there is further evidence which shows that the sugars actually exist as *six-membered rings* in the *solid* state.

The ring structure is hexagonal (I), but it is convenient to use the planar formula (II), *e.g.*, α-D-glucose, (II), corresponds to the open-chain formula (III), and it should be noted that the way of drawing the conventional planar formula of an aldose is with the carbon chain vertical and the aldehyde group at the top.

In the D-series the hydroxyl group is always to the right on the bottom asymmetric carbon atom. The L-series is then the enantiomers of the D-series.

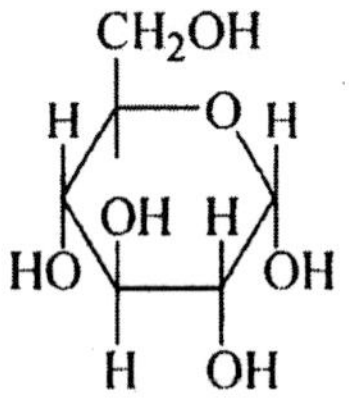
(I)

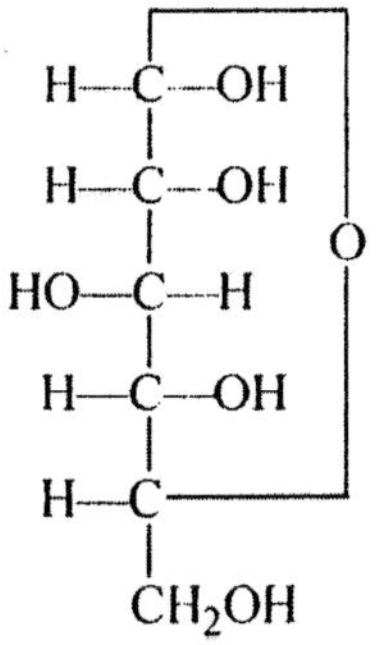
(II)

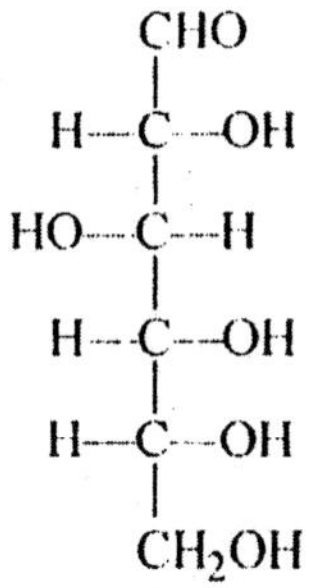
(III)

D (+)-*Glucose, dextrose* (*grape-sugar*), is found in ripe grapes, honey and most sweet fruits; it is also a normal constituent of blood, and occurs in the urine of diabetics. Commercially, pure D(+)-glucose is manufactured by heating starch with dilute hydrochloric acid under pressure:

$$(C_6H_{10}O_5)_n + nH_2O \xrightarrow{HCl} nC_6H_{12}O_6$$

D(+)-Glucose, m.p. 146°C, has a sweet taste, but is not as sweet is cane-sugar. Naturally-occurring glucose is dextrorotatory (hence name *dextrose*). It is a strong reducing agent, reducing both Fehling's solution and ammoniacal silver nitrate. When heated with sodium hydroxide, an aqueous solution of glucose turns brown. Glucose forms a cyanohydrin with hydrogen cyanide (also the Kiliani reaction), and an oxime with hydroxylamine. All of these reactions are typical of aldehydes, but heating with phenyl-hydrazine in acetic acid does not give the phenylhydrazone; an *osazone* is formed.

Oxidation of sugars. The nature of the product formed by the oxidation of aldoses depend; on the oxidising agent used. Mild oxidising reagents such as bromine-water or an alkaline solution of iodine oxidise only the aldehyde group to give *aldonic acids*. Thus, glucose give: *gluconic acid*, m.p. 131°C.

$$HOCH_2(CHOH)_4CHO + [O] \xrightarrow[(50\%)]{Br_2/H_2O} HOCH_2(CHOH)_4CO_2H$$

This type of oxidation may also be carried out with mercuric oxide in the presence of calcium carbonate (calcium salt is obtained). Aldonic acids (also as calcium salt), however, are best prepared by the electrolysis of a solution of the aldose in the presence of calcium bromide and calcium carbonate. Gluconic acid is manufactured by the oxidation of glucose by the enzyme glucose oxidase, which is present in *Aspergillus niger*, *Penicillium chrysogenum*, etc.

The method of naming aldonic acids is to add prefixes to ionic to indicate the parent aldose. *e.g.*, galactonic and arabonic acids.

Oxidation of aldoses with strong oxidising reagents, *e.g.*, nitric acid (and under suitable conditions), produces α, ω-dicarboxylic acids known as saccharic or aldaric acids, *e.g.*, glucose gives glucosaccharic (still often called saccharic acid) or glucaric acid (some oxalic acid is also obtained):

$$HOCH_2(CHOH)_4CHO \xrightarrow[HNO_3]{[O]} HO_2C(CHOH)_4CO_2H \ (40\text{–}46\%)$$

The general method of naming these acids is to indicate the parent aldose by a prefix, *e.g.*, mannosaccharic or mannaric acid, ribosaccharic or ribaric acid.

Aldaric acids readily form monolactones, but in some cases the dilactone may be produced.

Uronic acids are produced by oxidation of only the terminal —CH_2OH group in an aldose, *e.g.*, glucose forms glucuronic acid:

$$HOCH_2(CHOH)_4CHO \xrightarrow{[O]} HO_2C(CHOH)_4CHO$$

This oxidation may be carried out in several ways. A general method is to protect the aldehyde group, usually as the isopropylidene derivative, which is then oxidised with oxygen in the presence of Pt-C as catalyst, *e.g.*, glucuronic acid from glucose (cyclic structures are used). (III) is 1,2:5,6-di-isopropylideneglucofuranose and (IV) is 1,2-isopropylidene-glucofuranose.

CHOH
|
CHOH
|
CHOH
|
CHOH
|
CH
|
CH_2OH

$\xrightarrow[HCl]{Me_2CO}$

CHO\
| CMe_2
CHO/
|
CHOH
|
CH
|
CHO\
| CMe_2
CH_2O/

$\xrightarrow{AcOH}$

$$\begin{array}{l} CHO\diagdown \\ |\quad\quad CMe_2 \\ CHO\diagup \\ | \\ CHOH \\ | \\ CH \\ | \\ CHOH \\ | \\ CH_2OH \end{array} \xrightarrow[\text{(ii) HCl}]{\text{(i) } O_2/Pt\text{-}C} \begin{array}{l} CHO \\ | \\ CHOH \\ | \\ CHOH \\ | \\ CHOH \\ | \\ CHOH \\ | \\ CH_2H \end{array}$$

Acetone usually condenses with *cis*-5-hydroxyl groups on adjacent carbon atoms and hence, to form the di-isopropylidene derivative, the ring changes size (the OH groups on C-3 and C-4 are *trans*. see also mutarotation.

Aldoses (and ketoses) are oxidised by periodic acid with complete breakdown of the carbon chain.

Reduction of sugars. As with oxidation, the product of reduction depends on the nature of the reagent.

Sodium amalgam, sodium borohydride, high-pressure catalytic hydrogenation (Ni—H_2), or electrolytic reduction in acid solution all convert aldoses into the corresponding alcohols, *e.g.*, glucose is converted into *sorbitol.*

$$HOCH_2(CHOH)_4CHO \xrightarrow{e;H^+} HOCH_2(CHOH)_4CH_2OH \quad (50\%)$$

Reduction of glucose with concentrated hydriodic acid and red phosphorus at 100°C produces 2-iodohexane; prolonged heating finally gives n-hexane.

Sugar derivatives. The sugars form a variety of derivatives, *e.g.*, ethers, esters, isopropylidene compounds (see above), etc. Because of the easy removal of the iso-propylidene group (dilute acid), these derivatives are very useful for protecting alcoholic groups (they are stable to alkali).

Aldohexoses, when heated with concentrated hydrochloric acid, first give hydroxymethyl-furfural, which then decomposes into laevulic acid and other products.

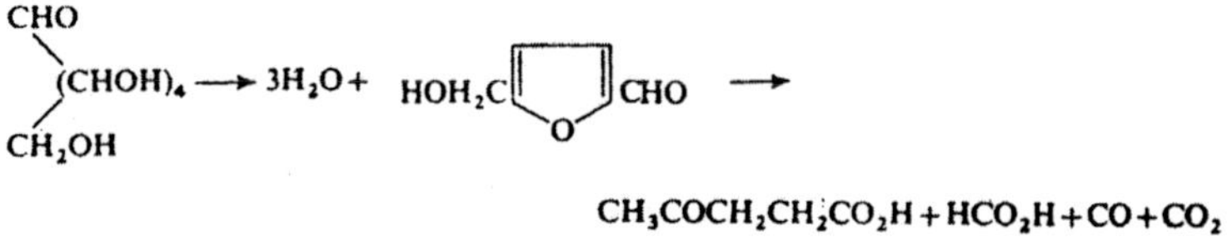

$$\begin{matrix} CHO \\ (CHOH)_4 \\ CH_2OH \end{matrix} \longrightarrow 3H_2O + HOH_2C\text{-}C_4H_2O\text{-}CHO \longrightarrow$$

$$CH_3COCH_2CH_2CO_2H + HCO_2H + CO + CO_2$$

Fig. 1.2

Glucose is fermented by yeast to ethanol and forms glucosates with various metallic hydroxides, *e.g.*, with calcium hydroxide it forms calcium glucosate, $C_6H_{12}O_6.CaO$; the structure of these glucosates is uncertain.

Aldohexoses (and aldopentoses) have.a hydroxyl group at C-l in the ring structure. This does not behave as an alcoholic group, *e.g.*, glucose combines with monohydric alcohols, in the presence of hydrochloric acid, to form glucosides.

D(+)-*Mannose*, m.p. 132°C, may be obtained by carefully oxidising with nitric acid:

$$HOCH_2(CHOH)_4CH_2OH + [O] \xrightarrow{HNO_3} HOCH_2(CHOH)_4CHO + H_2O$$

Commercially, mannose is prepared by hydrolysing the polysaccharide seminine with boiling dilute sulphuric acid. Seminine (a *mannan*) occurs in many plants, particularly in the shell of the ivory nut which is used as the starting material for D(+)-mannose.

D(+)-*Galactose* (m.p. of monohydrate, 118°C) is found in several polysaccharides (galactans), and is combined with glucose in the disaccharide *lactose*. Galactose may be prepared by hydrolysing lactose and separating it from glucose by fractional crystallisation from water, in which glucose is more soluble than galactose.

Rhanmoase, $CH_3(CHOH)_4CHO$, is a 6-deoxyhexose: it occurs in several glycosides.

Ketohexoses. The only important ketohexose is D(–)-fructose, m.p. 102°C, the structure of which has been elucidated as follows:

(i) Analysis and molecular-weight determinations show that the molecular formula of fructose is $C_6H_2O_6$...

(ii) When treated with acetic anhydride, fructose forms the penta-acetate. This indicate the presence of five hydroxyl groups.

(iii) Fructose forms an oxime when treated with hydroxylamine, and therefore contains a carbonyl group.

(iv) When oxidised with nitric acid, fructose is converted into a mixture of trihydroxyglutaric, tartaric and glycollic acids. Since a mixture of acids each containing fewer carbon atoms than fructose is obtained, the carbonyl group in fructose must be present in a *ketonic* group.

(v) Fructose may be reduced to a hexahydric alcohol, *sorbitol*, which, on reduction with hydriodic acid and red phosphorus at 100°C, gives a mixture of 2-iodohexane and n-hexane. The formation of the latter two compounds indicates that the six carbon atoms in fructose are in a straight chain.

(vi) On ascending the series by the *Kiliani reaction*, and reducing the product with hydriodic acid, 2-methylhexanoic acid is obtained. This shows that the ketonic group in fructose is adjacent to one of the terminal carbon atoms.

The foregoing reactions show that the structure of fructose is:

$$HOCH_2CHOHCHOHCHOHCOCH_2OH$$

This contains three structurally different asymmetric carbon atoms, and can therefore exist in eight optically active forms. Of these the following six are known: D(–)- and L(+)-*fructose*, D(+)- and L(–)-*sorbose*, D(+)-*tagatose* and L(–)-*psicose*.

Naturally-occurring fructose (*fruit-sugar*) is laevorotatory, and is therefore also known as laevulose (cf. dextrose). D(–)-Fructose is found in fruits and honey, and occurs combined with glucose in the disaccharide *cane-sugar*, from which it may be obtained by hydrolysis and fractional crystallisation. Fructose is prepared commercially by hydrolysis of *inulin*, a polysaccharide which occurs in dahlia tubers and Jerusalem artichokes:

$$(C_6H_{10}O_5)_n + nH_2O \xrightarrow{HCl} nC_6H_{12}O_6$$

The chemical reactions of fructose are similar to those of glucose, any differences being due to the fact that the former is a ketone and the latter an aldehyde. Thus, the most noticeable difference is the behaviour towards oxidising agents; fructose is not oxidised by bromine-water or an alkaline solution of iodine, and oxidation with nitric acid gives a mixture of acids, each containing less than six carbon atoms.

Reaction of glucose and fructose with phenylhydrazine. Fischer (1884, 1887) showed that both aldoses and ketones react with excess of phenylhydrazine to form *osazones* (these contain two phenylhydrazino groups), aniline and ammonia. The overall equation may be written:

$$\begin{matrix} CHO \\ | \\ CHO \\ \vdots \end{matrix} \xrightarrow{3PhNHNH_2} \begin{matrix} CHO{=}NNHPh \\ | \\ CHO{=}NNHPh \\ \vdots \end{matrix} + PhNH_2 + NH_3$$

$$\xleftarrow{3PhNHNH_2} \begin{matrix} CH_2OH \\ | \\ CO \\ \vdots \end{matrix}$$

Fischer also suggested a mechanism for this reaction, but it has been rejected in favour of that proposed by Weygand (1940). According to this mechanism, the first step is the formation of a phenylhydrazone, which then undergoes the *Amadori rearrangement* (1926). Weygand had originally proposed two mechanisms, but later (1958-1963) concluded that both were operating. According to Shemyakin *et al.* (1965), however, only one of these mechanisms is operating, viz.,

Glucose

$$\begin{matrix} CHO \\ | \\ CHOH \\ \vdots \end{matrix} \xrightarrow{PhNHNH_2} \begin{matrix} CH{=}NNHPh \\ |\;\;H \\ C \\ \vdots\;\;OH \end{matrix} \rightleftharpoons \begin{matrix} CH{-}NH \\ \|\;\;\;\;NHPh \\ C\;\;\;\;\;H \\ \vdots\;\;O \end{matrix}$$

glucose phenylhydrazone

$$\xrightarrow{-PhNH_2} \begin{matrix} CH{=}NH \\ | \\ C{=}O \\ \vdots \end{matrix} \xrightarrow[(-H_2O)]{-PhNHNH_2}$$

$$\begin{matrix} CH{=}NH \\ | \\ C{=}NNHPh \\ \vdots \end{matrix} \xrightarrow{-PhNHNH_2} \begin{matrix} CH{=}NNHPh \\ | \\ C{=}NNHPh \\ \vdots \end{matrix} + NH_3$$

glucosazone

Fructose

$$\begin{matrix} CH_2OH \\ | \\ C{=}O \\ \vdots \end{matrix} \xrightarrow{PhNHNH_2} \begin{matrix} CH{-}OH \\ |\;\;H \\ C{=}NNHPh \\ \vdots \end{matrix} \rightleftharpoons \begin{matrix} CH{-}O{-}H \\ \|\;\;\;\;\;\;\;\;\vdots \\ \|\;\;\;\;NHPh \\ C{-}NH \\ | \end{matrix}$$

fructose phenylhydrazone

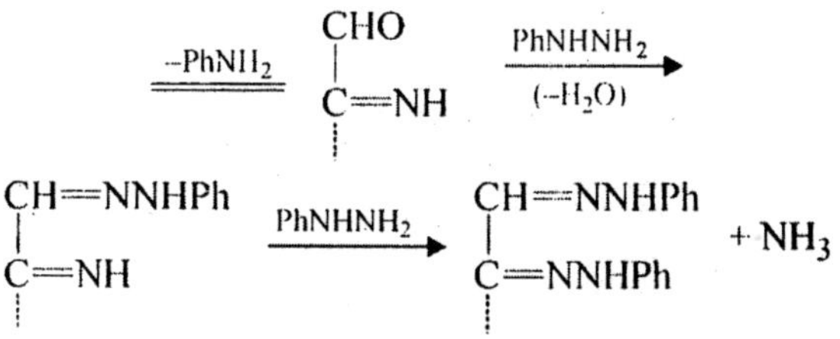

$$\xrightleftharpoons{-PhNH_2} \begin{array}{l} CHO \\ | \\ C{=}NH \\ \vdots \end{array} \xrightarrow[(-H_2O)]{PhNHNH_2}$$

$$\begin{array}{l} CH{=}NNHPh \\ | \\ C{=}NH \\ \vdots \end{array} \xrightarrow{PhNHNH_2} \begin{array}{l} CH{=}NNHPh \\ | \\ C{=}NNHPh \\ \vdots \end{array} + NH_3$$

fructosazone

Weygand's mechanism does not explain why only the first two carbon atoms should be involved. At first sight, it might have been expected that the Amadori rearrangement could carry on down the chain to produce a hexaphenylhydrazone. The failure to undergo further reaction has been explained by stabilisation of the osazone by chelation, *i.e.*, phenylosazones have *ring* structures (Fieser and Fieser, 1944). Two alternatives (1) and (II) were proposed.

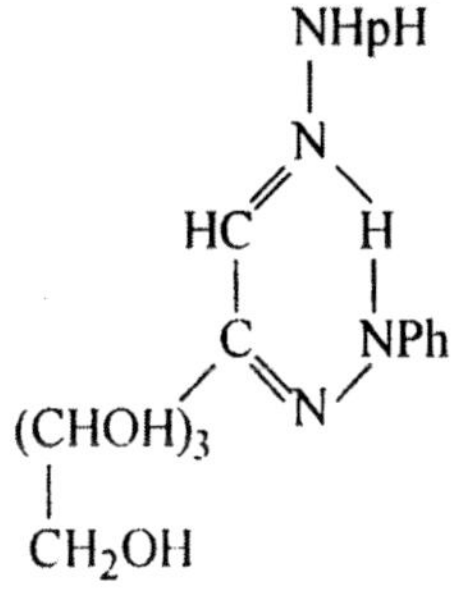

NHpH – N; HC; H; C; NPh; N; (CHOH)$_3$; CH$_2$OH

(I)

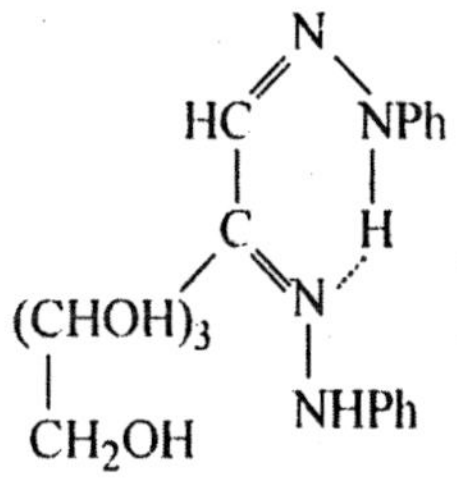

N; HC; NPh; C; H; N; NHPh; (CHOH)$_3$; CH$_2$OH

(II)

NHPh; N; CH; H; O; CH; NPh; N; CH—(CHOH)$_2$; CH$_2$OH

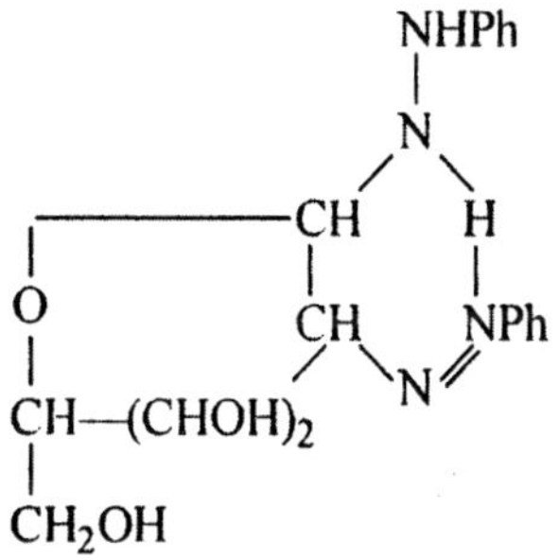

(III)

Mester (1955, 1965) believed that (1) was the correct structure, but Blair *et al.* (1969), from their chemical and i.r. and u.v. spectra studies, believe that neither is correct, and have proposed (III), the phenylazo structure (at C-2) for the compound in the *solid state*, and that *in solution*, (III) is in equilibrium with the open-chain form.

Osazones are yellow crystalline solids and are used to characterise the sugars. They may be hydrolysed with hydrochloric acid, both phenylhydrazino groups being eliminated; the dicarbonyl compound formed is known as an *osone*, *e.g.*, *glucosazone* forms *glucosone*;

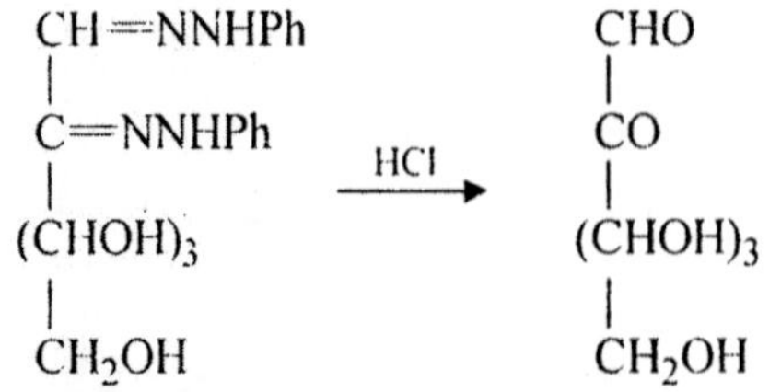

$$\begin{array}{c} CH{=}NNHPh \\ | \\ C{=}NNHPh \\ | \\ (CHOH)_3 \\ | \\ CH_2OH \end{array} \xrightarrow{HCl} \begin{array}{c} CHO \\ | \\ CO \\ | \\ (CHOH)_3 \\ | \\ CH_2OH \end{array}$$

A more convenient method of obtaining the osone is to add benzaldehyde to a solution of the osazone; benzaldehyde phenylhydrazone is precipitated, leaving the osone in solution. Osones react with phenylhydrazine in the cold to form osazones.

Epunerisation. Aldoses which produce the same osazones must have identical configurations on all their asymmetric carbon atoms except the *alpha* (since only the aldehyde group and α-carbon atom are involved in osazone formation). Such sugars are known as *epimers*. Fischer (1890) changed an aldose into its epimervia the aldonic acid.

Thealdonic acid was heated with pyridine (or quinoline), whereupon it was converted into an equilibrium mixture of the original acid and its epimer. These were separated, and the epimeric acid lactone reduced to an aldose (the Kiliani reaction below).

The mechanism of epimerisation is not certain, but it is probably similar to racemisation, *e.g.*, epimerisation of glucose into toannose:

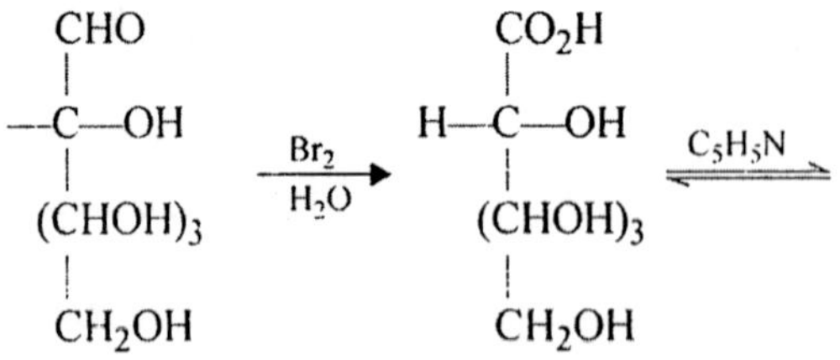

$$\underset{\text{D(+)-glucose}}{\begin{array}{c} CHO \\ | \\ -C-OH \\ | \\ (CHOH)_3 \\ | \\ CH_2OH \end{array}} \xrightarrow[H_2O]{Br_2} \underset{\text{gluconic acid}}{\begin{array}{c} CO_2H \\ | \\ H-C-OH \\ | \\ (CHOH)_3 \\ | \\ CH_2OH \end{array}} \xrightleftharpoons{C_5H_5N}$$

$$\underset{\text{}}{\begin{array}{c} \text{C}(=\text{O})\text{O}^- \\ | \\ \text{H—C—OH} \\ | \\ (\text{CHOH})_3 \\ | \\ \text{CH}_2\text{OH} \end{array}} \underset{+\text{H}^+}{\overset{-\text{H}^+}{\rightleftharpoons}} \begin{array}{c} \text{C}(=\text{O})\text{O}^- \\ | \\ ^-\text{C—OH} \\ | \\ (\text{CHOH})_3 \\ | \\ \text{CH}_2\text{OH} \end{array} \longleftrightarrow$$

$$\begin{array}{c} \text{C}(\text{O}^-)\text{O}^- \\ \| \\ \text{C—OH} \\ | \\ (\text{CHOH})_3 \\ | \\ \text{CH}_2\text{OH} \end{array} \underset{-\text{H}^+}{\overset{+\text{H}^+}{\rightleftharpoons}} \underset{\text{mannonic acid anion}}{\begin{array}{c} \text{C}(=\text{O})\text{O}^- \\ | \\ \text{HO—C—OH} \\ | \\ (\text{CHOH})_3 \\ | \\ \text{CH}_2\text{OH} \end{array}} \xrightarrow[\text{(ii) lactone reduced}]{\text{(i) HCl}} \underset{\text{D(+)-mannose}}{\begin{array}{c} \text{CHO} \\ | \\ \text{HO—C—H} \\ | \\ (\text{CHOH})_3 \\ | \\ \text{CH}_2\text{OH} \end{array}}$$

This change of configuration of one asymmetric carbon atom in a compound containing two - or more asymmetric carbon atoms is known as *epimerisation.*

Method of ascending the sugar series. An aldose may be converted into its next higher aldose, *e.g.*, an aldopentose into an aldohexose, by means of the *Kiliani reaction* (1886).

Theoretically, two lactones are possible, since two cyanohydrins may be formed when hydrogen cyanide adds on to the aldopentoge (a new asymmetric carbon is produced), viz.,

$$\begin{array}{c} \text{CHO} \\ | \\ \text{CHOH} \\ | \\ \text{CHOH} \\ | \\ \text{CHOH} \\ | \\ \text{CH}_2\text{OH} \end{array} \xrightarrow[\text{aq.}]{\text{HCN}} \begin{array}{c} \text{CN} \\ | \\ \text{CHOH} \\ | \\ \text{CHOH} \\ | \\ \text{CHOH} \\ | \\ \text{CHOH} \\ | \\ \text{CH}_2\text{OH} \end{array} \xrightarrow[\text{(ii) H2SO4}]{\text{(i) Ba(OH)}_2} \begin{array}{c} \text{CO}_2\text{H} \\ | \\ \text{CHOH} \\ | \\ \text{CHOH} \\ | \\ \text{CHOH} \\ | \\ \text{CHOH} \\ | \\ \text{CH}_2\text{OH} \end{array}$$

$$\xrightarrow{\text{heat}} \begin{array}{l} \text{CO}\text{———}\\ | \\ \text{CHOH} \\ | \qquad\quad \text{O} \\ \text{CHOH} \\ | \\ \text{CH}\text{———} \\ | \\ \text{CHOH} \\ | \\ \text{CH}_2\text{OH} \end{array} \xrightarrow[\text{H}_2\text{SO}_4]{\text{Na/Hg}} \begin{array}{l} \text{CHO} \\ | \\ \text{CHOH} \\ | \\ \text{CHOH} \\ | \\ \text{CHOH} \\ | \\ \text{CHOH} \\ | \\ \text{CH}_2\text{OH} \end{array}$$

γ-lactone

$$\begin{array}{l}\text{CHO}\\ \vdots\end{array} \xrightarrow{\text{HCN}} \text{H—}\overset{\text{CN}}{\underset{\vdots}{\text{C}}}\text{—OH} + \text{OH—}\overset{\text{CN}}{\underset{\vdots}{\text{C}}}\text{—H}$$

Thus, two epimeric aldohexoses should be obtained. In practice, one cyanohydrin predominates because the asymmetry present in the molecule produces diastereotopic faces at the double bond. Hence the final product will be mainly one aldohexose, and very little of its epimer.

1,4-Aldonolactones (γ-lactones) may also be reduced to the aldoses by sodium borohydride or lithium aluminium hydride in a mixture of tetrahydrofuran and pyridine.

There are now many other methods available for stepping *et al* the sugars. Sowden *et al.* (1947) used nitromethane as follows, and the proportions of the epimers formed by this method are different from those produced by the Kiliani reaction.

$$\begin{array}{l}\text{CHO} + \text{CH}_3\text{NO}_2\\ \vdots\end{array} \xrightarrow{\text{CH}_3\text{ONa}} \begin{array}{l} \text{CH}{=}\text{NO}_2\text{Na} \\ | \\ \text{CHOH} \\ \vdots \end{array}$$

$$\xrightarrow{\text{H}_2\text{SO}_4} \begin{array}{c} \text{CHO} \\ | \\ \text{H—C—OH} \\ \vdots \end{array} + \begin{array}{c} \text{CHO} \\ | \\ \text{HO—C—H} \\ \vdots \end{array}$$

Sowden (1950) has also stepped up an aldose to a *ketose* containing *two* additional carbon atoms using the above method except that 2-nitroethanol is used instead of nitromethane.

$$\begin{array}{c}CHO\\ \vdots\end{array} + \begin{array}{c}CH_2NO_2\\ |\\ CH_2OH\end{array} \xrightarrow{CH_3ONa} \begin{array}{c}CH_2OH\\ |\\ C{=}NO_2Na\\ |\\ CHOH\\ \vdots\end{array}$$

$$\xrightarrow{H_2SO_4} \begin{array}{c}CH_2OH\\ |\\ CO\\ |\\ H{-}C{-}OH\\ \vdots\end{array} + \begin{array}{c}CH_2OH\\ |\\ CO\\ |\\ HO{-}C{-}OH\\ \vdots\end{array}$$

On the other hand, Wolfrom et at. (1946) have stepped up an aldose to a ketose with one more carbon atom by a modified Arndt-Eistert reaction.

$$\begin{array}{c}CHO\\ |\\ (CHOH)_3\\ |\\ CH_2OH\end{array} \xrightarrow[H_2O]{Br_2} \begin{array}{c}CO_2H\\ |\\ (CHOH)_3\\ |\\ CH_2OH\end{array} \xrightarrow[(ii)\ SOCl_2]{(i)\ Ac_2O} \begin{array}{c}COCl\\ |\\ (CHOAc)_3\\ |\\ CH_2OAc\end{array}$$

$$\xrightarrow{CH_2N_2} \begin{array}{c}CHN_2\\ |\\ COCl\\ |\\ (CHOAc)_3\\ |\\ CH_2OAc\end{array} \xrightarrow[(ii)\ Ba(OH)_2]{(i)\ AcOH} \begin{array}{c}CH_2OH\\ |\\ COCl\\ |\\ (CHOAc)_3\\ |\\ CH_2OAc\end{array}$$

More recently, Kochetkov et at. (1965) have stepped up an aldose to an aldose containing two additional carbon atoms, *e.g.*, the aldose is condensed with carbethoxymethylenephosphorane (see Wittig reaction), and the product is then treated as follows:

$$RCHOHCHO + Ph_3P = CHCO_2Et$$

$$\longrightarrow Ph_3PH + RCHOHCH = CHCO_2Et \xrightarrow[HClO_4]{OsO_4}$$

$$[RCHOHCHOHCHOHCO_2H] \longrightarrow RCH(CHOH)_2CO \ (\text{with O bridging CH and CO})$$

$$\xrightarrow[\text{or } NaBH_4]{Na{-}Hg} RCHOHCHOHCHO$$

Method of descending the sugar series. There are various methods of converting a sugar into its next lower sugar. All of these methods start with the aldose, and hence, in order to convert a ketose into the lower aldose, it is first necessary to transform it into an aldose (see later).

Wohls method (1893) starts with the aldose oxime, which is then treated as shown (for an aldohexose).

$$\begin{array}{c} CHO \\ | \\ CHO \\ | \\ (CHOH)_3 \\ | \\ CHO \end{array} \xrightarrow{NH_2OH} \begin{array}{c} CH{=}NOH \\ | \\ CHOH \\ | \\ (CHOH)_3 \\ | \\ CH_2OH \end{array} \xrightarrow{(CH_3CO)_2O} \begin{array}{c} CN \\ | \\ CHOCOCH_3 \\ | \\ (CHOCOCH_3) \\ | \\ CH_2OCOCH_3 \end{array}$$

$$\xrightarrow{(CH_3CO)_2O} \begin{array}{c} CHO \\ | \\ (CHOH)_3 \\ | \\ CH_2OH \end{array} + AgCN\ (55\%)$$

Zemplen (1917) modified Wohl's method by using a solution of sodium methoxide in chloroform instead of an aqueous solution ofammoniacal silver nitrate, to remove hydrogen cyanide and the acetyl groups, and thereby increased the yield of pentose to 60-70 per cent.

Weygand *et al.* (1950) have treated the oxime with l-fluoro-2,4-dinitrobenzene in aqueous sodium hydrogen carbonate; the products are the lower aldose (yield: 50-60%), hydrogen cyanide and 2,4-dinitrophenol.

Ruff's method (1898). The aldose is oxidised (by bromine-water) to the corresponding aldonic acid; when the calcium salt of this acid is treated with Fenton's reagent it is converted into the lower aldose (*cf.* oxidation of α-hydroxyacids:

$$\begin{array}{c} CHO \\ | \\ CHO \\ | \\ (CHOH)_3 \\ | \\ CH_2OH \end{array} \xrightarrow{Br_2/H_2O} \begin{array}{c} CO_2H \\ | \\ CHOH \\ | \\ (CHOH)_3 \\ | \\ CH_2OH \end{array} \xrightarrow[H_2O_2/Fe^{2+}]{\text{Ca salt}} \begin{array}{c} CHOH \\ | \\ (CHOH)_3 \\ | \\ CH_2OH \end{array} + CO_2\ (25\%)$$

Berezovski *et al.* (1949) have shown that calcium gluconate can be oxidised with hydrogen peroxide in the presence of ferric sulphate and barium acetate to give D-arabinose (44%).

Weennan's reaction (1913). This is the reaction whereby an α-hydroxy- or α-methoxy-amide is degraded by means of a cold solution of sodium hypochlorite (cf. Hofmann amine preparation)

α-*Hydroxyamides*

$$\begin{array}{l}CONH_2\\|\\CHOH\\|\\R\end{array} \xrightarrow{NaOH/NaOCl} \left[\begin{array}{l}NCO\\|\\CHOH\\|\\R\end{array}\right] \xrightarrow{NaOH} \begin{array}{l}CHO + NaNCO\\|\\R\end{array}$$

α-*Methoxyamides*

$$\begin{array}{l}CONH_2\\|\\CHOCH_3\\|\\R\end{array} \xrightarrow{NaOH/NaOCl} \left[\begin{array}{l}NCO\\|\\CHOCH_3\\|\\R\end{array}\right]$$

$$\xrightarrow{NaOH} \begin{array}{l}CHO + CO_2 + NH_3 + CH_3OH\\|\\R\end{array}$$

More recent methods of descending the sugars are also available. Thus, Macdonald *et al.* (1953) have stepped down an aldose as follows:

$$\begin{array}{l}CHO\\|\\CHOH\\|\\R\end{array} \xrightarrow{C_2H_5SH} \underset{\text{mercaptal}}{\begin{array}{l}CH(SC_2H_5)_2\\|\\CHOH\\|\\R\end{array}} \xrightarrow{C_2H_5CO_3H} \underset{\text{disulphone}}{\begin{array}{l}CH(SO_2C_2H_5)_2\\|\\CHOH\\|\\R\end{array}}$$

$$\xrightarrow{NH_4OH} \begin{array}{l}CHOH + CH_2(SO_2C_2H_5)_2\\|\\R\end{array}$$

On the other hand, Perlin (1954) has shown-that aldoses may be stepped down by direct oxidation with lead tetra-acetate or sodium bismuthate, *e.g.*, D-mannose gives D-arabinose (35%).

Conversion of an aldose into a ketose. The aldose is converted into its osazone, which is then hydrolysed with hydrochloric acid to the osone. On reduction with zinc and acetic acid, the osone is converted into the ketose (an aldehyde group is reduced more readily than a ketonic group):

$$\begin{array}{l} CH{=}NNHC_6H_5 \\ | \\ C{=}NNHC_6H_5 \\ \vdots \end{array} \xrightarrow{HCl} \begin{array}{l} CHO \\ | \\ CO \\ \vdots \end{array} \xrightarrow{e;H^+} \begin{array}{l} CH_2OH \\ | \\ CO \\ \vdots \end{array}$$

Conversion of a ketose into an aldose. The ketose is reduced (preferably by catalytic reduction, to the corresponding polyhydric alcohol, which is then oxidised to a monocarboxylic acid (only one of the terminal CH_2OH groups being oxidised). On warming, the acid is converted into the γ-lactone which, on reduction with sodium amalgam in faintly acid solution, is converted into the aldose:

$$\begin{array}{l} CH_2OH \\ | \\ CO \\ | \\ CHOH \\ | \\ CHOH \\ | \\ CHOH \\ | \\ CH_2OH \end{array} \xrightarrow{H_2/Ni} \begin{array}{l} CH_2OH \\ | \\ CHOH \\ | \\ CHOH \\ | \\ CHOH \\ | \\ CHOH \\ | \\ CH_2OH \end{array} \xrightarrow[HNO_3]{[O]} \begin{array}{l} CH_2OH \\ | \\ CHOH \\ | \\ CHOH \\ | \\ CHOH \\ | \\ CHOH \\ | \\ CH_2OH \end{array}$$

$$\xrightarrow{heat} \begin{array}{l} CO\text{---} \\ | \\ CHOH \quad O \\ | \\ CHOH \\ | \\ CH\text{---} \\ | \\ CHOH \\ | \\ CH_2OH \end{array} \xrightarrow{e;\,H^+} \begin{array}{l} CHO \\ | \\ CHOH \\ | \\ CHOH \\ | \\ CH \\ | \\ CHOH \\ | \\ CH_2OH \end{array}$$

Theoretically, two polyhydric alcohols may be formed on reduction of the ketose, due to the formation of a new asymmetric carbon atom:

$$\begin{array}{l} CH_2OH \\ | \\ CO \\ \vdots \end{array} \xrightarrow{H_2/Ni} \begin{array}{c} CH_2OH \\ | \\ H\text{—}C\text{—}OH \\ \vdots \end{array} + \begin{array}{c} CH_2OH \\ | \\ HO\text{—}C\text{—}OH \\ \vdots \end{array}$$

In practice, however, one predominates (*cf.* the Kiliani reaction). Furthermore, when these two alcohols are oxidised, oxidation may take place at *either* end of the chain, and hence the final product will be a mixture *of four* aldoses, but these will not be present to the same extent.

Lobry de Bruyn-van Ekenstein rearrangement (1890). When warmed with *concentrated* alkali, sugars first turn yellow, then brown and finally resinify *(cf.* aldehydes). In the presence *of dilute* alkali or amines, sugars undergo rearrangement; *e.g.*, a dilute solution of glucose, in the presence of sodium hydroxide, is converted into an almost optically inactive

solution from which have been isolated D(+)-glucose, D(+)-mannose, D(—)-fructose. The same mixture is obtained if the starting material is D(—)-fructose or D(+)-mannose. The details of the mechanism are uncertain, but it is accepted that the rearrangement occurs through an enediol:

CH=O | H—C—OH | D(+)-glucose $\xrightleftharpoons{OH^-}$ CH=O ‖ H—C—OH | enediol ⇌ CHO | HO—C—H | D(+)-mannose ; ⇌ CHO | CO | D(−)-mannose

Since the Lobry de Bruyn-van Ekenstein rearrangement takes place in alkaline media, it is hest to carry out reactions with the sugars in neutral or acid media.

Cyclic Structures of Monosaccharides

When a monosaccharide is dissolved in water, the optical rotatory power of the solution gradually changes until it reaches a constant value, *e.g.*, a freshly prepared solution of glucose has a specific rotation of +111°; when this solution is allowed to stand, the specific rotation falls to + 52.5°, and remains constant at this value. The final stage can breached more rapidly either by heating the solution or by adding some catalyst which may be an acid or a base. This change in value of the specific rotation is known as *mutarotation*. All *reducing* sugars (except some ketoses) undergo mutarotation. To account for mutarotation, Tollens (1883) suggested an oxide ring structure for D(4-)-glucose, whereby *two* forms would be produced since, in the formation of the ring, another asymmetric carbon atom (which can exist in *two* configurations) is produced *(cf.* the Kiliani reaction). Tollens assumed that a five-membered ring (the furanose form) was produced, (I) and (II).

The difficulty of this suggestion was that there was no experimental evidence for the existence of these two forms. Tanret (1895), however, isolated two isomeric forms of D(+)-glucose, thus apparently verifying Tollens' supposition (see later). The two forms are called α- and β-D (-l-)-glucofuranose; (I) is the α-form, and (II) the β-.

```
   ┌──────────┐                                      ┌──────────┐
 H─C─OH       │          CHO                    HO─C─H        │
   │          │           │                        │          │
 H─C─OH       O        H─C─OH                   H─C─OH        O
   │          │           │                        │          │
HO─C─H        │    ⇌  HO─C─H        ⇌           HO─C─H        │
   │          │           │                        │          │
 H─C──────────┘        H─C─OH                   H─C───────────┘
   │                      │                        │
 H─C─OH                H─C─OH                   H─C─OH
   │                      │                        │
  CH2OH                  CH2OH                    CH2OH
  (I)                                              (II)
```

Ring formation of a sugar is really hemiacetal formation, one alcohol group of the sugar forming a hemiacetal with the aldehyde group of the *same* molecule, thus producing a ring structure. This equilibrium between the open and ring forms is an example of ring-chair tautomerism. The carbon atoms in the chain tend to form a loop, and this results in the OH on C_4, or C_5 being very close to the CHO group. With ketoses, it is the OH group on C_5 or C_6 that is close to the CO group. In both aldoses and ketoses, therefore, the geometry of the molecule is very favourable for hemi-acetal formation.

```
   ┌──────────┐                                      ┌──────────┐
 H─C─OH       │          CHO                    HO─C─H        │
   │          │           │                        │          │
 H─C─OH       O        H─C─OH                   H─C─OH        O
   │          │           │                        │          │
HO─C─H        │        HO─C─H                   HO─C─H        │
   │          │    ⇌      │         ⇌              │          │
 H─C─OH       │        H─C─OH                   H─C           │
   │          │           │                        │          │
 H─C──────────┘        H─C─OH                   H─C───────────┘
   │                      │                        │
  CH2OH                  CH2OH                    CH2OH
α-D(+)-glucopyranose                        α-D(+)-glucopyranose
```

Later work by Haworth, Hirst and their co-workers (1926 onwards) has shown that glucose (and other sugars) exists, not as a five-membered ring but as a six-membered ring, the two forms being α- and β-D(+)-glucopyranose. Such pairs ofstereoisomeric ringlforms of any sugar are known as *anomers*.

The widely accepted view is that monosaccharides in solution exist mainly as an equilibrium mixture of the α- and β-anomeric pyranoses, a small amount of the open-chain form together with small amounts of the α- and β-anomeric furanoses. The presence of the furanose forms has been inferred from the fact that monosaccharides undergo some reactions which lead to the formation of furanose derivatives (isopropylidene derivatives).

The mechanism of mutarotation has been the subject of a great deal of research, but it is still uncertain. A widely accepted mechanism involves a concerted attack by base and acid (water is an amphiprotic solvent) to produce the *open-chain structure*, and then this recloses to the anomeric form:

H O—H B

C

O H—A ⇌ C $+ Br^+ + A^-$

CH CHOH

α-D-

B H—O H

⇌ C

O H—A

CH

The ordinary form of D(+)-glucose is the α-anomer. m.p. 146°C, specific rotation +111°, and may be prepared by crystallisation of glucose from cold aqueous solution. The β-anomer, m.p. 148-150°C, specific rotation + 19.2°, may be obtained by crystallising glucose from hot saturated aqueous solution. Both anomers show mutarotation, the final value of the specific rotation being 52.5°. This corresponds to about 36 per cent of the α-anomer and 64 per cent of the β. It is therefore assumed that whenever a sugar is formed in solution, it immediately changes into a mixture of the two pyranose anomers. the open chain isomer (and furanoses) being present in extremely mall amount. The cyclic structure of the sugars accounts for the following facts: (i) the existence of two isomers, *e.g.*, α- and β-glucose; (ii) mutarotation; (iii) glucose and other aldoses do not give certain characteristic reactions of aldehydes, *e.g.*, Schiff's reaction, do not form a bisulphite or an aldehyde-ammonia compound.

Haworth (1926) proposed a six-membered ring formula (hexagonal formula) based on the pyran ring which is almost planar. The *pyranose* structure is applicable to nearly all the sugars, and is supported by X-ray crystal analysis (of the sugars). Thus α-D(+)-*glucopyranose* is (a).

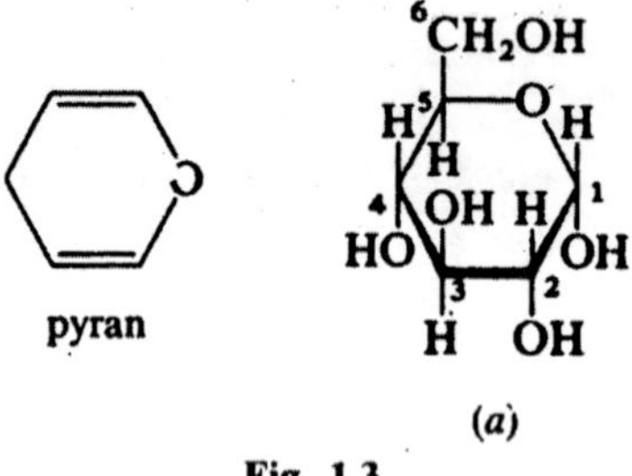

Fig. 1.3

Reeves (1950) has shown that the conformation of α-D(+)-glucopyranose is (b) and that of the β-anomer is (c). 'Both have the chair form, but in the former the glycosidic hydroxyl is axial and in the latter equatorial. As is explained on, it is a general rule that the conformation which has the greatest number of large groups in equatorial orientation is the most stable form. Thus the β-anomer will be more stable than the α-, and so will predominate in the equilibrium mixture.

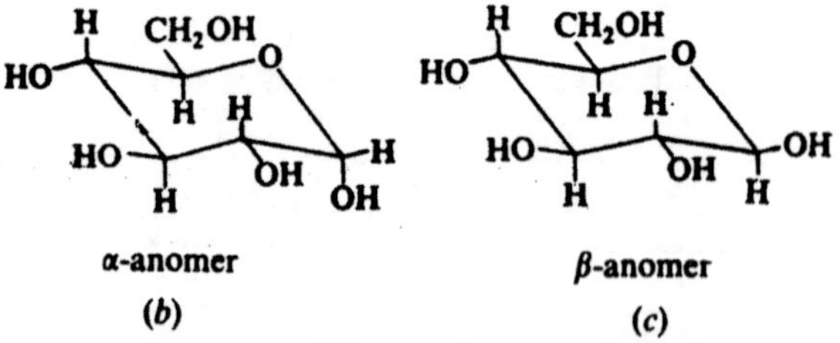

α-anomer (*b*) *β*-anomer (*c*)

Fig. 1.4

It has also been shown (Irvine, Haworth *et al.*) that glucose, fructose, etc. can exist as five membered rings, which may be regarded as derivatives of *furan*. So far, the *furanose* sugar (corresponding to Tollens' suggestion) have not been isolated in the free state, but some of their derivatives have been prepared. Thus α-D(+)-glucofuranose would (if it existed) have perspective (*i.e.*, pentagonal) formula (III); methyl α-D(+)-glucofuranoside (a known compound) has formula (IV):

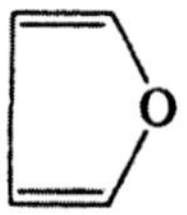

furan 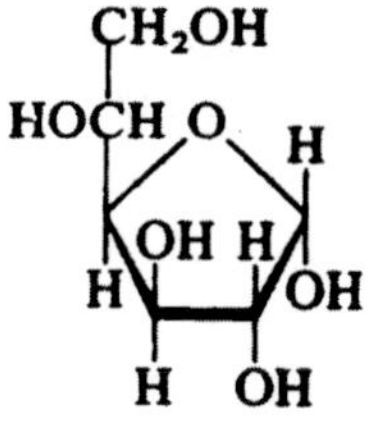 (III) 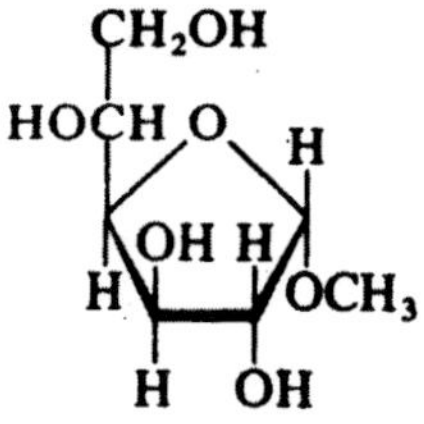 (IV)

Fig. 1.5

Pentoses also normally exist in the pyranose form, and derivatives of the furanose form have been prepared.

Conversion of the plane-diagrams into the perspective formulae may be done as follows. (V) is α-D-glucopyranose, and if the H on C_5 is interchanged with the group CH_2OH, then a Walden inversion has been effected; and if the H is now interchanged with the point of attachment of the oxygen ring, another Walden inversion is effected, and so the original configuration of (V) is retained; thus we now have (VI) (with no change in configuration). Since all horizontal bonds indicate groups lying above the plane of the paper, and vertical bonds groups lying behind this plane, then by twisting (VI) so that the oxide ring is perpendicular to the plane of the paper, and placing the oxygen atom as shown, (VII) is obtained.

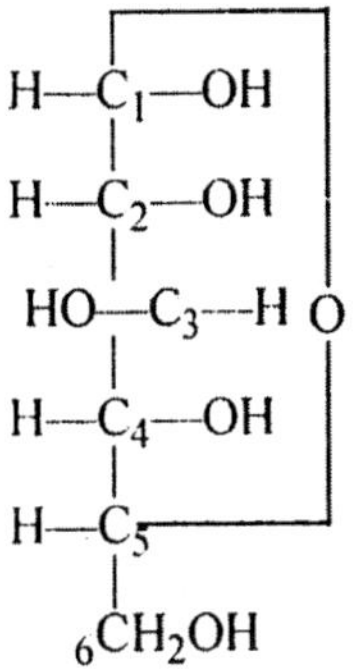

(V) 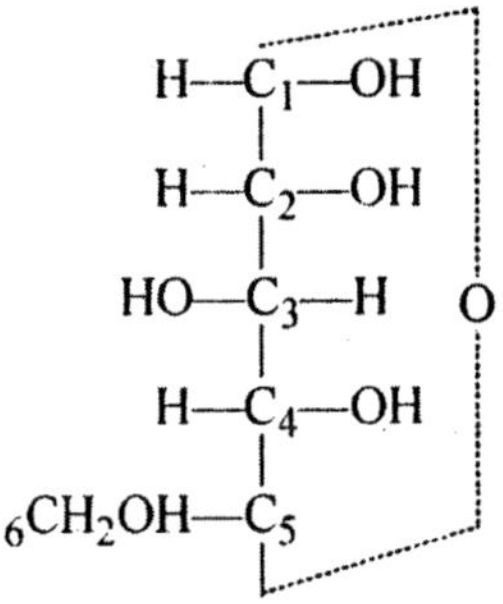 (VI) 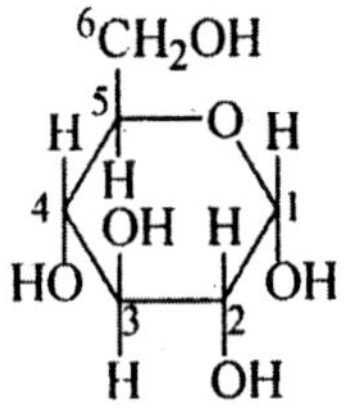 (VII)

Thus, to change from (VI) to (VII), first draw the hexagon (as shown in (VII}), and then place all the groups on the left-hand side in (VI) above the plane of the ring in (VII), and all those on the right-hand side in (VI) below, the plane of the ring in (VII).

In a similar way, perspective formulae may be obtained -for the furanose sugars, *e.g.*, methyl β-D(+)-fructofuranoside (in this case, *i.e.*, for ketoses, treat the CH_2OH group adjacent to the keto group as H in the aldoses).

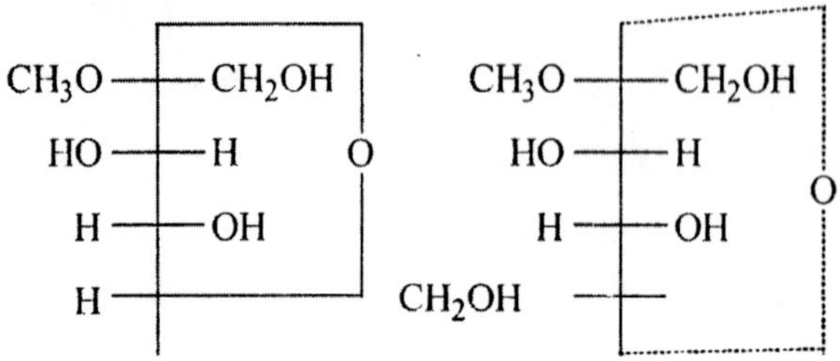

CH_3O — CH_2OH
HO — H   O
H — OH
H —

CH_3O — CH_2OH
HO — H
O
H — OH
CH_2OH —

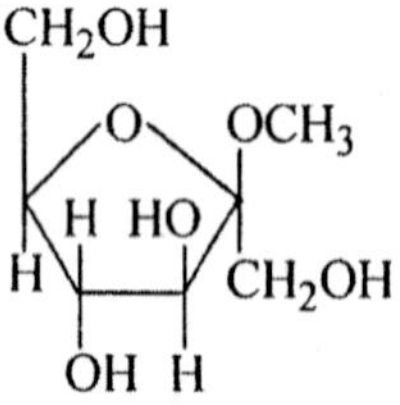

CH_2OH
O OCH_3
H HO
H CH_2OH
OH H

NMR spectroscopic studies of monosaccharides and their derivatives have led to several generalisations (for the meanings of *axial* and *equatorial*).

(i) Anomeric protons (*i.e.*, H on C-1) occur at lower field than any others (C-1 is linked to *two* oxygen atoms).

(ii) Axial hydrogens appear upfield with respect to equatorial hydrogens.

(iii) For *acetylated pyranosides*, the methyl group on axial hydroxy'i groups occurs downfield with respect to those on equatorial hydroxyl groups.

(iv) For *methylated* pyranosides, the methyl group on axial hydroxyl groups occurs upfield with respect to those on equatorial hydroxyl groups.

These generalisations are useful, but may lead to wrong conclusions in rings which have been deformed by, *e.g.*, steric effects.

Just as simple hemiacetals react with another molecule of an alcohol to form acetals, so can the hemiacetal form of a sugar react with a molecule of an alcohol to form the acetal derivative, which is known under the generic name of *glycoside*; those of glucose are known as *glucosides*, of *fructose*, fructosides, etc. For example, ethyl α-D(+)-glucopyranoside is prepared by refluxing glucose in excess of ethanol in the presence of a small amount of hydrochloric acid. These glycosides are stable compounds, and do not undergo many of the reactions of the sugars, *e.g.*, they show no reducing properties, they do not mutarotate, etc. The non-sugar part of a glycoside is known as the *aglycon*, and in most of the glycosides which occur naturally the aglycon is a phenolic compound, *e.g.*, the aglycon in *salicin* is *salicyl alcohol*, in *indican*, *indoxyl*.

Synthesis of the Monosaccharides

Photosynthesis is the most important example of biosynthesis and represents the processes whereby plants containing the pigment chlorophyll absorb light energy and utilise it to convert atmospheric carbon dioxide, in the presence of water, to carbohydrates. Photosynthesis has been shown to involving two separate types of reactions. The first involves a photochemical process in which light energy absorbed by chlorophyll, is utilised to form activated compounds. The second involves the reduction of carbon dioxide by the active molecules produced in the first process; the products are oxygen and carbohydrates (and some other compounds). Since this process can proceed in the presence or absence of light, it is referred to as the dark reactions, the first process is distinguished from this as the *light* reactions. Both the dark and light reactions require the presence of various enzymes. The overal equation for photosynthesis may be written as:

$$6O_2 + 6H_2O \longrightarrow 6(CH_2O) + 6O_2$$

Calvin *et al.* using $^{14}CO_2$ as tracer, have worked out the pathway of carbon dioxide in photosynthesis.

In the laboratory, the sugars have been synthesised in various ways:

(i) By the aldol condensation of formaldehyde in the presence of calcium hydroxide; the product is a mixture of compounds among which is a number of hexoses. This hexose mixture, know as *formose*, has been shown to contain DL-fructose (Butlerow, 1861; Loew, 1886).

(ii) By the aldol condensation of glycolaldehyde in the presence of sodium hydroxide; the product formose (E. Fischer, 1887):

$$3HOCH_2CHO \xrightarrow{NaOH} C_6H_{12}O_6$$

(iii) When oxidised with nitric acid or bromine-water, *glycerol yields* a product known as *glycerose*, which contains, among other compounds, glyceraldehyde and dihydroxyacetone. These, in the presence of barium hydroxide, are converted into a mixture of α- and β-*acrose* (E. Fischer, 1887); α-acrose is DL-fructose and /?-acrose is DL-sorbose:

$$HOCH_2CHOHCHO + HOCH_2COCH_2OH$$

$$\longrightarrow HOCH_2CO(CHOH)_3CH_2OH$$

Starting with α-acrose, Fischer isolated DL-fructosazone (DL-glucosazone), and making use of reduction, oxidation and epimerisation, he converted DL-fructosazone into D(–)-fructose, D(+)-glucose, D(+)-mannose and other aldohexoses.

(iv) When hydrolysed with barium hydroxide solution, dibromoacraldehyde forms a mixture a- and 8-acrose (E. Fischer, 1887):

$$BrCH_2CHBrCHO \xrightarrow{Ba(OH)_2} HOCH_2CHOHCHO \xrightarrow{\text{dim erises}}$$

$$HOCH_2CO(CHOH)_3CH_2OH$$

Aldotrioscs. Glyceraldehyde is the only aldotrios

CHO H—┼—OH CH_2OH	CHC HO—┼—H CH_2OH
D(+)-glyceraldehyde	L(–)-glyceraldehyde

Aldotetroses, $HOCH_2CHOHCHOHCHO$. This structure contains two asymmetric carbon atoms, and so there are four optical isomers (two pairs of enantiomers). All are known, and correspond to D- and L-threose and D- and L-erythrose.

D(+)-Glyceraldehyde may be stepped up by the Kiliani reaction to give D(–)-threose and D(–)-erythrose. The question now is: which is which ?

CHO
H—C—OH
CH_2OH

D(−)-erythrose (I)	D(−)-threose (II)
CHO	CHO
H—C—OH	HO—C—H
H—C—OH	H—C—OH
CH_2OH	CH_2OH

On oxidation, D-erythrose forms mese-tartaric acid. Therefore D-erythrose must be (I), and consequently (II) must be D-threose.

Aldopentoses, $HOCH_2CHOHCHOHCHOHCHO$. This structure contains three asymmetric carbon atoms, and so there are eight optical isomers (four pairs of enantiomers). All are known. D-Erythrose, when stepped up by the Kiliani reaction, gives D(−)-ribose and D(−)-arabinose. Similarly, D-threose gives D(+)-xylose and D(−)-lyxose.

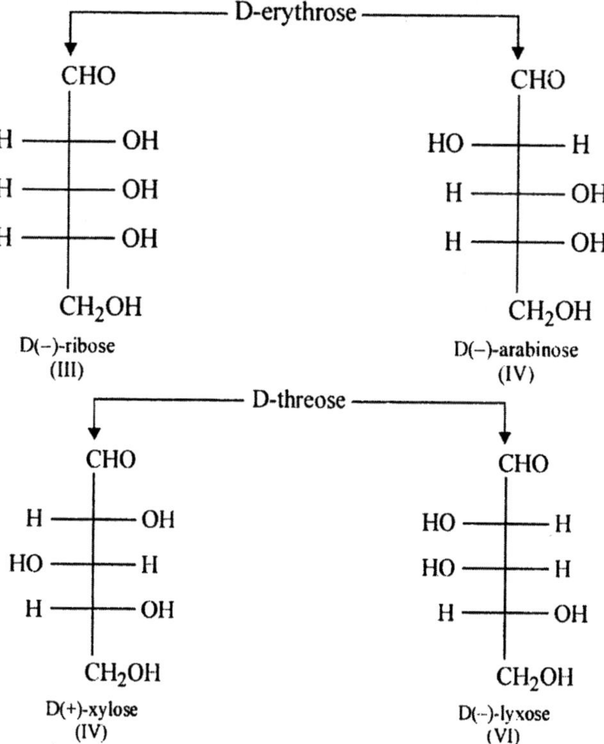

D(−)-ribose (III)

D(−)-arabinose (IV)

D(+)-xylose (IV)

D(−)-lyxose (VI)

(III) and (IV) must be ribose and arabinose, but which is which? On oxidation with nitric acid, arabinose gives an optically active dicarboxylic acid (a trihydroxyglutaric acid), whereas ribose gives an optically inactive dicarboxylic acid. When the terminal groups (*i.e.*, CHO and CH_2OH) of (III) are oxidised to carboxyl groups, the molecule produced possesses a plane of symmetry, and so this acid is inactive. The dicarboxylic acid produced from (IV), however, has no plane (or any other element) of symmetry, and so is optically active. Thus (III) is D-ribose and (IV) is D-arabinose.

(V) and (VI) must be xylose and lyxose, but which? The former, on oxidation, gives an optically inactive dicarboxylic acid, whereas the latter gives an optically active dicarboxylic acid. Therefore (V) is D-xylose and (VI) is D-lyxose.

Aldohexoscs, $HOCH_2CHOHCHOHCHOHCHOHCHO$. This structure contains four asymmetric carbon atoms, and so there are sixteen optical isomers (eight pairs of enantiomers). All are known, and may be prepared by stepping up the aldopentoses. D-ribose gives D(+)-allose and D(+)-altrose; D-arabinose gives D(+)-glucose and D(+)-mannose; D-xylose gives D(–)-gulose and D(–)-idose, and D-lyxose gives D(+)-galactose and D(+)-talose.

(VII) and (VIII) must be allose and altruse, but which is which? On oxidation with nitric acid, the former gives an optically inactive (allaric) and the latter an optically active (talaric) dicarboxylic acid. Therefore (VII) is allcse and (VIII) is altrose.

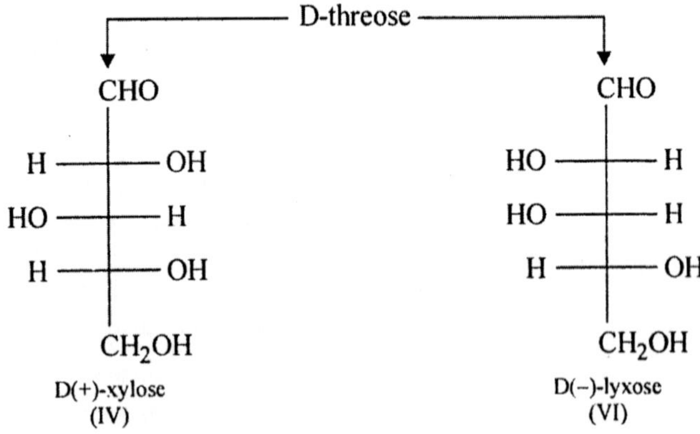

D(+)-xylose
(IV)

D(–)-lyxose
(VI)

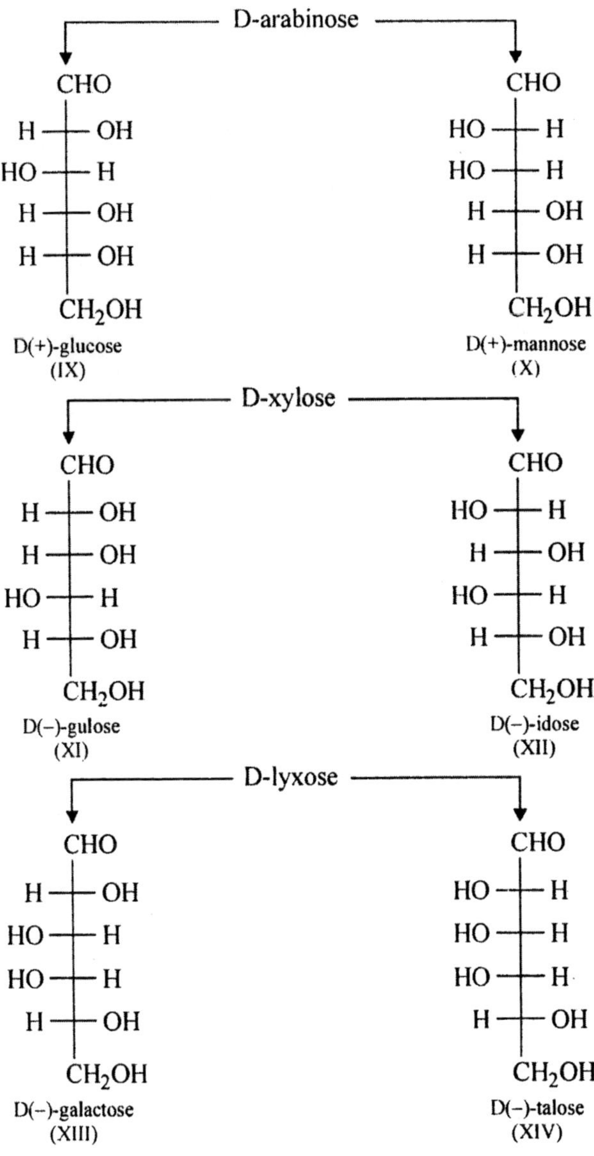

D(+)-glucose (IX)

D(+)-mannose (X)

D(−)-gulose (XI)

D(−)-idose (XII)

D(−)-galactose (XIII)

D(−)-talose (XIV)

(XIII) and (XIV) must be galactose and talose, but which is which? On oxidation with nitric acid, the former gives an optically inactive (galactaric) and the latter an optically active (talaric) dicarboxylic acid. Therefore (XIII) is galactose and (XIV) is talose.

The elucidation of the configurations of the remaining four aldohexoses is not quite so simple since, on oxidation with nitric acid, glucose and mannose *both* give optically active dicarboxylic acids, as also do gulose and idose; in all four configurations (IX, X, XI, XII), replacement of the two terminal groups (CHO and CH_2OH) by carboxyl groups leads to dicarboxylic acid whose structures have no plane (or any other element) of symmetry.

It has been found, however, that the dicarboxylic acid from glucose (glucaric acid) is the sape as that produced from gulose. Actually the two glucaric acids obtained are enantiomers, D-glucose giving D-glucaric acid and D-gulose L-glucaric acid. Since glucaric acid, $HO_2C(CHOH)_4$-CO_2H, is produced by the oxidation of the terminal groups with the rest of the molecule unaffected, it therefore follows that the 'rest of the molecule' must be the same for both glucose and gulose.

Inspection of formulae (IX, X, XI and XII) shows that only (IX) and (XI) have the 'rest of the molecule' the same; by interchanging the CHO and CH,OH groups of (IX), (XI) is obtained. Therefore (IX) must be glucose (since we know that glucose is obtained from arabinose), and (XI) must be gulose. Consequently (X) is mannose and (XII) is idose.

Ketohexoses. Fructose is a 2-ketohexose, and natural fructose is laevorotatory. Since D-glucose gives the same osazone as natural fructose, the latter must be D(–)-fructose. Further-more, since osazone formation involves only the first two carbon atoms in a sugar, it therefore follows that the configuration of the rest of the molecules in glucose and fructose must be the same. Hence the configuration of D(–)-fructose is (XV).

	D(–)-fructose (XV)		$\xrightarrow{C_6H_5NHNH_2}$		osazone	
	CH_2OH				$CH{=}NNHC_6H_5$	
	CO				$CO{=}NNHC_6H_5$	
HO	—	H		HO	—	H
H	—	OH		HO	—	H
H	—	OH		H	—	OH
	CH_2OH				CH_2OH	

D(–)-fructose
(XV)

osazone

$$\xleftarrow{C_6H_5NHNH_2}\quad \begin{array}{c} CHO \\ H-C-OH \\ HO-C-H \\ H-C-OH \\ H-C-OH \\ CH_2OH \end{array}$$

D(+)-glucose

The configurations of the other 2-ketohexoses are:

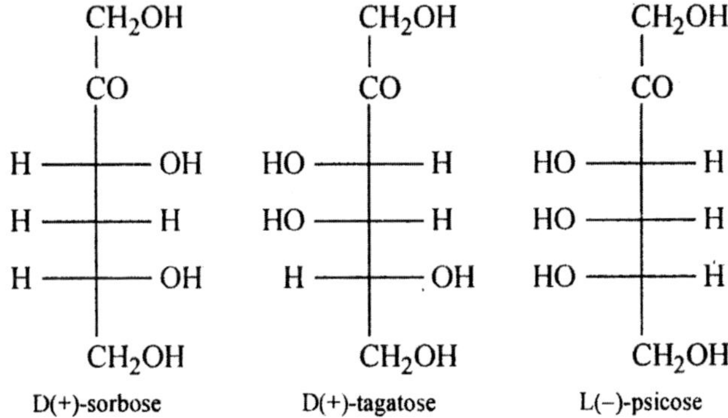

D(+)-sorbose D(+)-tagatose L(–)-psicose

Determination of the Size of Sugar-rings

E. Fischer (1893) refluxed glucose with excess of methanol in the presence of a small amount of hydrochloric acid and obtained a white crystalline solid containing *one* methyl group. This compound was no longer reducing, and also did not form an osazone. Thus Fischer had prepared methyl glucoside. Ekenstein (1894) isolated a second isomer from the same reaction, and Fischer explained the existence of these two isomers by suggesting a *ring* structure for them, and followed Tollens in believing that they were five-membered rings. There was no experimental evidence for the existence of a five-membered ring, and it was shown to be incorrect by Haworth, Hirst and their co-workers (1926 onwards).

They proved that the above glucosides were *six-membered* rings, *i.e.*, glucopyranosides. Their method was to fully methylate the glucoside (I), hydrolyse the product, methyl tetramethyl-α-D-ghicoside (II) to tetramethyl-α-D-glucose (III), oxidise this with bromine-water at 90°C to the lactone (IV), and then to oxidise this with nitric acid. The product obtained was shown to be xylotrimethoxy-glutaric acid (V) this can be obtained directly by tlie oxidation of methylated xylose). The most reasonable interpretation of these results is in accordance with the assumption that methyl glucoside is a six- and *not* a five-membered ring. Thus:

$$
\begin{array}{lcl}
\begin{array}{l}
CHOCH_3 \\
| \\
CHOH \\
| \\
CHOH \quad O \\
| \\
CHOH \\
| \\
CH \\
| \\
CH_2OH \\
\quad (I)
\end{array}
& \xrightarrow{(CH_3)_2SO_4} &
\begin{array}{l}
CHOCH_3 \\
| \\
CHOCH_3 \\
| \\
CHOCH_3 \quad O \\
| \\
CHOCH_3 \\
| \\
CH \\
| \\
CH_2OCH_3 \\
\quad (II)
\end{array}
\xrightarrow{HCl}
\end{array}
$$

$$
\begin{array}{lclcl}
\begin{array}{l}
CHOH \\
| \\
CHOCH_3 \\
| \\
CHOCH_3 \quad O \\
| \\
CHOCH_3 \\
| \\
CH \\
| \\
CH_2OCH_3 \\
\quad (III)
\end{array}
& \xrightarrow{Br_2/H_2O} &
\begin{array}{l}
CO \\
| \\
CHOCH_3 \\
| \\
CHOCH_3 \quad O \\
| \\
CHOCH_3 \\
| \\
CH \\
| \\
CH_2OCH_3 \\
\quad (IV)
\end{array}
& \xrightarrow{HNO_3} &
\begin{array}{l}
CO_2H \\
| \\
CHOCH_3 \\
| \\
CHOCH_3 \\
| \\
CHOCH_3 \\
| \\
CH_2H \\
\quad (V)
\end{array}
\end{array}
$$

Fischer (1914) also prepared methyl glucoside by dissolving glucose in methanol and letting it stand at room temperature in the presence of hydrochloric acid. He now obtained compounds which Haworth *et al.*, using the above method, showed to be iue-membered rings, *i.e.*, methyl D-glucofuranosides.

Hudson (1937, 1939) has also determined the size of sugar-rings, but his method was to oxidise methyl glycosides with periodic acid. As we have seen, periodic acid splits 1,2-glycols, and *one molecule of acid* is used for *each pair of adjacent alcoholic groups*, *e.g.*,

$$R^1CHOHCHOHR^2 \xrightarrow{1\,HIO_4} R^1CHO + R^2CHO$$

$$R^1CHOHCHOHCHOHR^2 \xrightarrow{2\,HIO_4} R^1CHO + HCO_2H + R^2CHO$$

Compounds containing an aldehyde or ketonic group adjacent to an alcoholic group are disc attacked by periodic acid in a similar manner to glycols:

$$RCHOHCHO \xrightarrow{1\,HIO_4} RCHO + HCO_2H$$

$$R^1COCOHR_2 \xrightarrow{1\,HIO_4} R^1CO_2H + R^2CHO$$

Aldoses and pentoses are *completely* broken down by periodic acid, *i.e.*, *free* sugars behave as open-chain compounds in this reaction, *e.g.*, an aldopentose consumes four molecules of periodic acid, and the products are four molecules of formic acid and one molecule of formaldehyde (from the terminal CH_2OH group).

$$OCHCHOHCHOHCHOHCH_2OH \xrightarrow{4\,HIO_4} 4HCO_2H + CH_2O$$

Thus, estimating the periodic acid used and the formic acid and/or formaldehyde produced will indicate the number of free adjacent oxidisable groups (CHOH, CHO or CO). On the other hand, the ring in *glycosides* is stable in the presence of periodic acid. Thus, the oxidation of methyl glucoside (produced under reflux conditions; see above) uses two molecules of periodic acid and produces *one* molecule of formic acid. Only one structure fits these facts, namely, a *six-membered* ring:

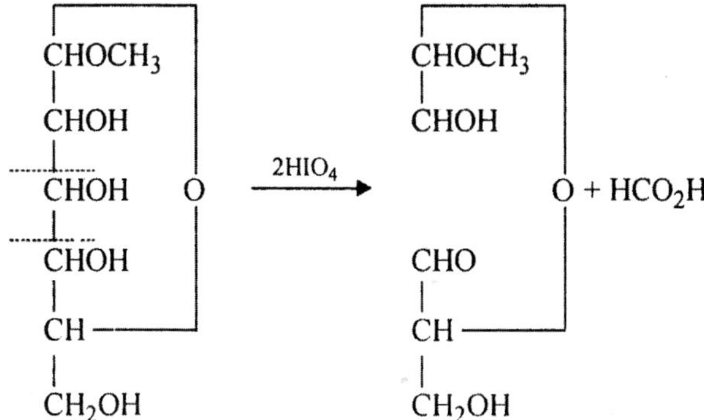

The other product of the oxidation has also been isolated and characterised by Hudson. This periodic acid oxidation method is now very widely used for structure investigation of carbohydrates.

Since methylation of sugars under different conditions produces different-sized rings, the question is: what is the size of the ring in the original sugar? Various sources of evidence, *e.g.*, X-ray analysis, show that the normal sugars are *pyranoses*.

One other point will be mentioned here, and that is the configuration of the *first* carbon atom. When the open-chain structure is closed to form the ring, two configurations of the new asymmetric carbon atom are possible, the one with the hydrogen to the left (the α-anomer), and the one with the hydrogen to the right (the β-). This is an arbitrary arrangement, but it has now been shown that the α- and β-anomers actually have the configurations originally arbitrarily assigned to them, *e.g.*. X-ray analysis of α-D-glucose has shown that the 1,2-hydroxyl groups are in the *cis*-position (McDonald *et al.*, 1950).

Since the L-series are the enantiomers of the D-series, in the former the a-anomer has the hydrogen on the first carbon atom to the right and the β-anomer has the hydrogen to the *left*. These are *opposite* to those of the D-series, *i.e.*, the configuration of C-l in an α-D-aldose is identical with that of the P-L-aldose.

Disaccharides

All the disaccharides are crystalline solids, soluble in water, and fall into two classes, the *reducing* sugars and the *non-reducing* sugars.

Just as methanol forms methyl glycosides with the monosaccharides, so can other hydroxy-compounds form similar unions with the monosaccharides. Since the latter are themselves hydroxy-compounds, it is possible that they can link up with themselves to form acetals, *i.e.*, glycosides in which the aglycon is another sugar molecule. A number of such compounds occur in nature, *e.g.*, sucrose, maltose and lactose.

Sucrose, cane-sugar, $C_{12}H_{22}O_{11}$, is one of the most important compounds commercially, and Is obtained from the sugar-cane and sugar-beet. In addition to crystalline sucrose, a syrup is always obtained which will not crystallise. This group, known as *molasses*, is also a commercial product.

Sucrose is a white crystalline solid, m.p. 180°C, soluble in water. When heated above its melting point, it forms a brown substance known as caramel. Concentrated sulphuric acid chars sucrose, the product being almost pure carbon. Sucrose is dextrorotatory, its specific rotation being +66-5°. On hydrolysis with dilute acids sucrose yields an equimolecular mixture of D(+)-glucose and D(–)-fructose:

$$\underset{}{C_{12}H_{22}O_{11}} + H_2O \xrightarrow{HCl} \underset{\text{glucose}}{C_6H_{12}O_6} + \underset{\text{fructose}}{C_6H_{12}O_6}$$

Since D(–)-fructose has a greater specific rotation than D(+)-glucose, the resulting mixture is laevorotatory. Because of this, the hydrolysis of cane-sugar is known as *the inversion of cane-sugar* (this is not to be confused with the *Walden inversion*), and the mixture is known as *invert sugar*. The inversion (*i.e.*, hydrolysis) of cane-sugar may also be effected by the enzyme *invertase* which is found in yeast.

Controlled oxidation of sucrose in alkaline solution with air gives D-arabonic acid. Oxidation of sucrose with nitric acid under different conditions gives either oxalic acid (80%), tartaric acid (40%), or glucaric acid (30%). Hydrogenation of sucrose under controlled conditions gives a mixture of mannitol and sorbitol (these may be separated by fractional crystallisation).

Sucrose is *not* a reducing sugar, *e.g.*, it will not reduce Fehling's solution; it does not form an oxime or an osazone, and does not undergo mutarotation. This indicates that neither the aldehyde group of glucose nor the ketonic group of fructose is free in sucrose. Thus a tentative structure of sucrose is one in which two molecules, glucose and fructose, are linked by the aldehyde group of the former and ketonic group of the latter. This has been amply confirmed by further work, and sucrose has been shown to be α-D-glucopyranosyl-β-D-fructofuranoside (*i.e.*, α-glucose is linked to β-fructose).

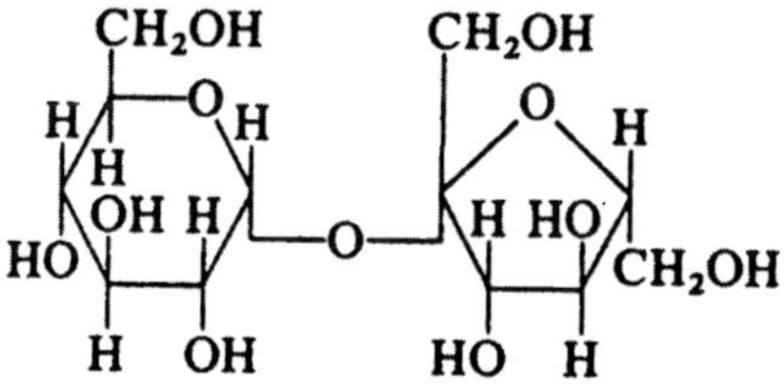

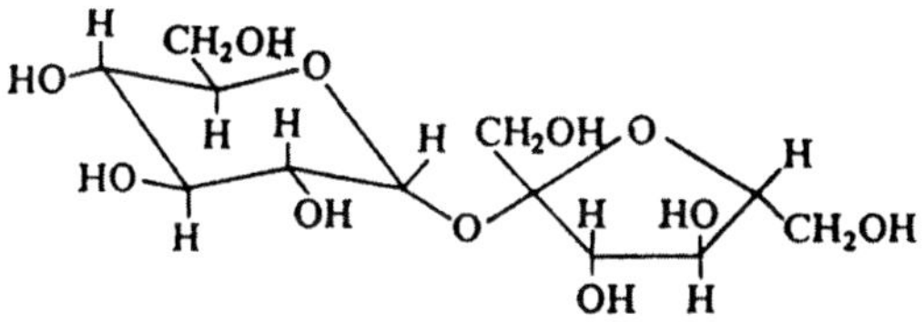

(+)-sucrose

Fig. 1.6

It should be noted that the fructose molecule in sucrose exists as the furanose form, and that when sucrose is hydrolysed, it is the pyrapose form of fructose which is isolated.

Maltose (*malt sugar*), $C_{12}, H_{22}O_{11}$, is produced by the action of malt (which contains the enzyme diastase) on starch:

$$(C_6H_{10}O_6)_8 \xrightarrow{\text{diastase}} \frac{n}{2} C_{12}H_{22}O_{11}$$

Maltose is a white crystalline solid, m.p. 160-165°C, soluble in water, and is dextrorotatory.

When it is hydrolysed with dilute acids or by the enzyme *maltose*, maltose yields two molecules of D(+)-glucose. Maltose is a reducing sugar, *e.g.*, it reduces Fehling's solution; it forms an oxime and an osazone, and undergoes mutarotation. This indicates that at least one aldehyde group (of the two glucose molecules) is free in maltose. Further work has shown that maltose is 4-O-α-D-glucopyranosyl-D-glucopyranose (*i.e.*, the reducing half is linked to the non-reducing half by an to-link). It should also be noted that the enzyme maltase splits only α-glycosidic links, and hence is used to ascertain the presence of this bond in disaccharides and polysaccharides.

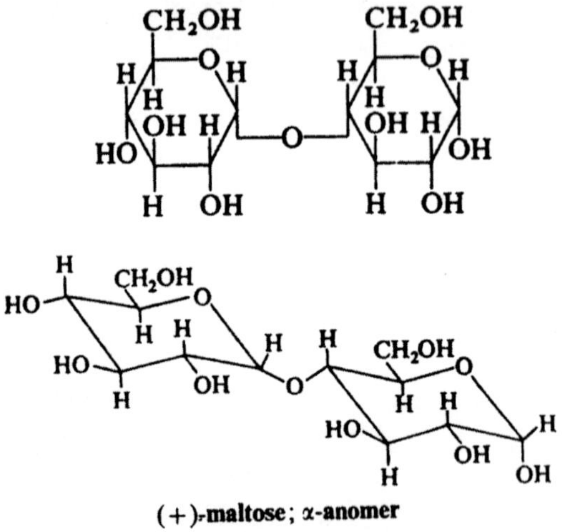

(+)-maltose; α-anomer

Fig. 1.7

Cellobiose, $C_{12}H_{22}O_{11}$, may be obtained from cellulose by acetylating good filter paper (which is almost pure cellulose) with acetic aphydride in the presence of concentrated sulphuric acid. The octa-acetate of cellobiose so obtained is hydrolysed with potassium hydroxide or with sodium ethoxide, whereupon cellobiose is produced.

Cellobiose is a white crystalline solid, m.p. 225°C, soluble in water, and dextrorotatory. When hydrolysed with dilute acids or by the enzyme *emulsin*, it yields two molecules of D(+)-glucose. It is a reducing sugar, forms an oxime and osazone, and undergoes mutarotation. Cellobiose is 4-O-β-D-glucopyranosyl-D-glucopyranose. Thus, the only difference between cellobiose and maltose is that in the former the glycosidic link is β, whereas in the latter it is α. Furthermore, the enzyme emulsin splits only β-glycosidic links (cf. maltase).

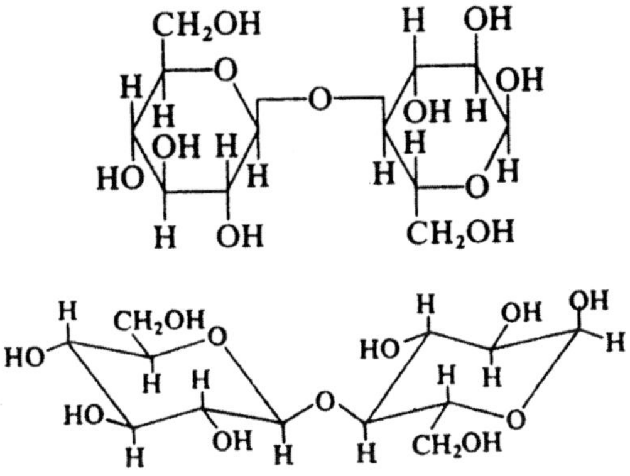

(+)-cellobiose; α-anomer

Fig. 1.8

Lactose (*milk-sugar*), $C_{12}H_{2}$,O, occurs in the milk of all animals, and is prepared commercially from whey by evaporation to crystallisation; whey is obtained as a by-product in the manufacture of cheese.

Lactose is a white crystalline solid, m.p. 203°C (with decomposition), soluble in water, and is dextrorotatory. It is hydrolysed by dilute acids or by the enzyme lactose, to an equimolecular mixture of D(+)-glucose and D(+)-galactose. Lactose is a reducing sugar, forms an oxime and osazone and undergoes mutarotation. It is 4-O-β-D-galactopyranosyl-D-glucopyranos It should also be noted that lactase is a P-glycosidase. *i.e.*. splits 8-glycosides (it has been shown to be identical with emulsin).

Sucrose maltose and lactose are three disaccharides which occur naturally. Cellobiose may be prepared from cellulose. Two other disaccharides which have been prepared are **melibiose** (from the trisaccharide *raffinose*) and **gentiobiose** (from the trisaccharide *gentianose*). These differ from the other disaccharides in that the two monosaccharide molecules are linked by the sixth carbon atom (the aldehyde carbon atom being number one) of the reducing monosaccharide:

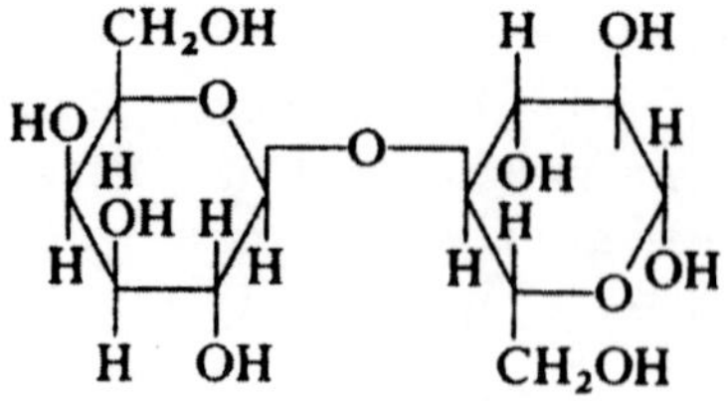

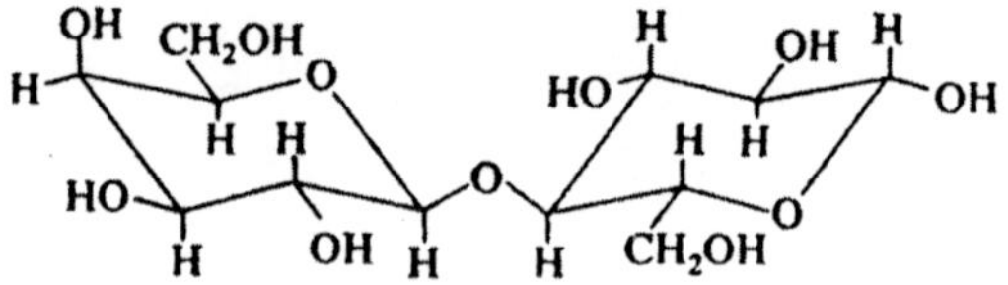

(+)-lactose; β-anomer

Fig. 1.9

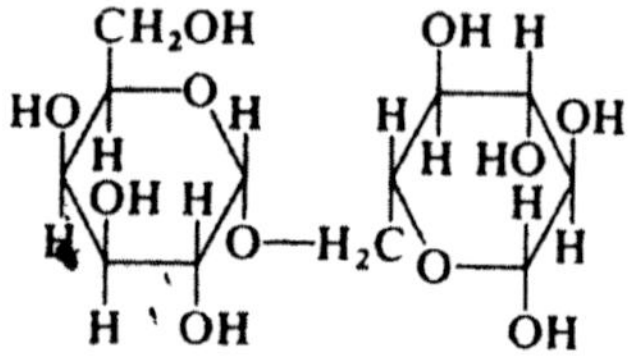

melibiose (β-anomer) **gentiobiose (β-anomer)**

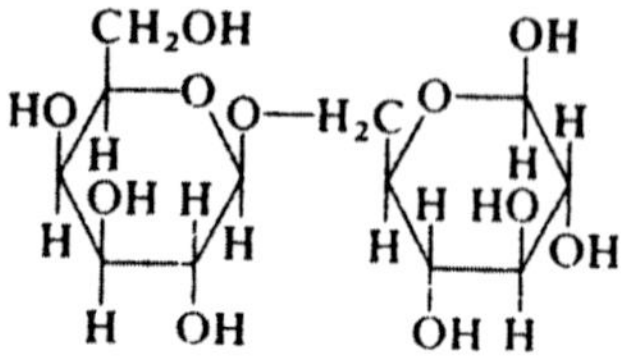

Fig. 1.10

Polysaccharides

The polysaccharides are high polymers of the monomeric sugars and have molecular weights that may range from a lew thousand to several millions.

Inulin occurs in many plants, *e.g.*. in the roots of the dandelion, in the tubers of the dahlia and in certain lichens. It is a white powder, insoluble in cold water, and in hot water forms a colloidal solution which does not form a gel on cooling. Inulin solutions do not give any colour with iodine. Inulin is hydrolysed by dilute acids to D(–)-fructose. Thus insulin is $(C_6H_{10},O_5\ )_n$. and; has a value of about 30 (obtained by chemical methods).

Starch, $(C_6H_{10}C_5)_8$, occurs in all green plants; the commercial sources of starch are maize wheat, barley, rice, potatoes and sorghum.

Starch consists of two fractions, one being known as α-**amylose** (the 'A' fraction), and the other as β-**amylose** or **amylopectin** (the 'B' fraction); the former comprises 10-20 per cent of starch, and the latter 80-90 per cent. a-Amylose is insoluble in water, and the solution gives a blue colour with iodine. This blue colour is believed to be due to the formation of an inclusion complex. An aqueous solution of α-amylose slowly forms a precipitate, since σ-amylose has a strong tendency to 'revert' to the insoluble state in solution. Amylopectin is soluble in water, is stable in contact with water, and gives a violet colour with iodine. α-Amylose and amylopectin are both hydrolysed to maltose by the enzyme diastase, and to D(+)-glucose by dilute acids (amylopectin gives about 50 per cent of maltose). Thus, both contain D-glucopyranose units joined by the α-glucosidic linkage.

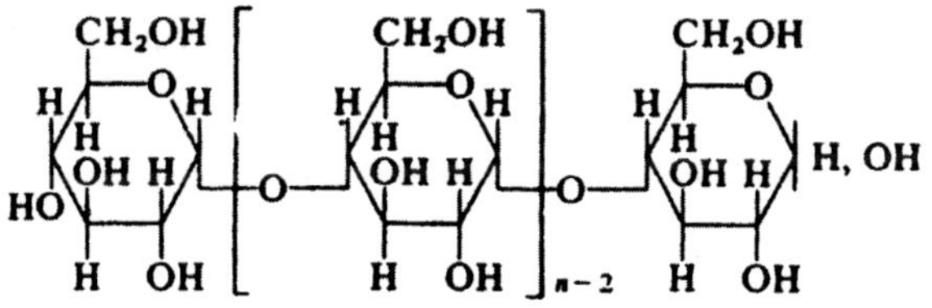

α-amylose

Fig. 1.11

α-Amylose consists of an unbranched chain, with a molecular weight varying between 10000 (n = ~60) and 1000000 (n = ~ 6000). The value of n depends-on the source and treatment of α-amylose. Amylopectin differs from α-amylose in that it contains *branched chains*, the branching occurring through 1,6-linkages (and other linkages) and the length of the unbranched sections being about 24-30 glucose units. The molecular weights recorded for amylopectin vary between 50000 and 10000000.

The dextrins, $(C_6H_{10}O_5)_n$, are produced by the partial hydrolysis of starch by boiling with water under pressure at about 250°. They are white powders, and are used for making adhesives and confectionery, for sizing paper, etc.

Glycogen, $(C_6H_{10}O_5)_n$, is found in nearly all animal cells, occurring mainly in the liver; it is the reserve carbohydrate of animals, and so is often known as '*animal starch*'. It has also been isolated from plant sources.

Glycogen is a white powder, soluble in water, the solution giving a purplish-red colour with iodine. On hydrolysis with dilute acid, glycogen gives D(+)-glucose. The molecular weight of glycogen has been given as 1000000 to 5000000, and glycogen contains highly branched chains.

Pectins are found in plant and fruit juices. Their characteristic property is the ability of their solutions to gelate, *i.e.*, form jellies. They have a high molecular weight, and are polygalacturonic acids (linked 1,4) with the carboxyl groups partially esterified with methanol.

Alginic acids occur in marine algae, and are polymannuronic acids (linked 1,4).

Cellulose, $(C_6H_{10}O_5)_n$, is the main constituent of the cell-wall of plants, and also occurs in certain animal tissues. It is the most widely distributed organic compound, and its main source is cotton (almost pure cellulose) and wood (which also contains *lignin*, which is not α-polysaccharide).

Cellulose is a white solid, insoluble in water but soluble in ammoniacal copper hydroxide solution (*Schweitzer's reagent*). Careful hydrolysis of cellulose gives cellobiose; it is also possible to isolate *cellotriose* (trisaccharide) and *cellotetrose* (tetrasaccharide). All of these saccharides, on further hydrolysis, yield only D(+)-glucose which exists in the β-form in cellulose (c. starch).

The molecular weight of cellulose varies between 20000 and 500000, and the compound consists of an unbranched chain.

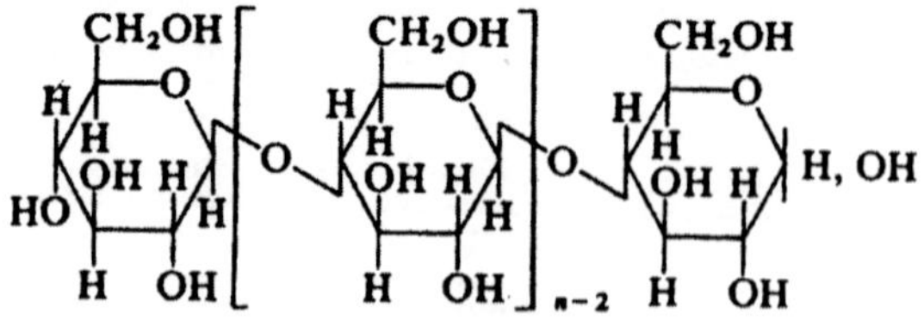

cellulose

Fig. 1.12

Artificial silk. The term rayon is used collectively to cover all synthetic or manufactured fibres from cellulose, but it is usually most often applied to viscose yarns.

Cellulose nitrates (*nitrocellulose*). 'The highest nitrate of cellulose theoretically possible is the trinitrate, $(C_6H_7N_3O_{11})_n$, which contains 14.4 percent nitrogen (each glucose unit contains three hydroxyl groups). The nitrates are prepared by the reaction of cellulose with a mixture of nitric and sulphuric acids, and the degree of 'nitration' depends on the concentrations of the acids and the time of the reaction. Cellulose trinitrate (12.2-13.2% N) is known as **gun-cotton** and is used in the manufacture of blasting explosives and smokeless powders. The lower nitrates, in the solid state, are known as **pyroxylin**, and their use depends on the nitrogen content, *e.g.*, (i) pyroxylin (10.8-11.0%N) mixed with camphor as plasticiser is converted into **celluloid**; (ii) *nitrocellulose lacquers* are made by dissolving pyroxylin (11.5-12.2%N), resins, pigments and plasticisers in a mixture of solvents.

Cellulose acetate (*celanese silk*). When acetylated with acetic anhydride in the presence of sulphuric acid, cellulose is converted into cellulose triacetate. When the reaction is complete, water is added to decompose the triacetate into the diacetate (approximately). The diacetate is washed, dried and dissolved in a mixture of organic solvents (of which acetone is generally the main constituent). The solution is then forced through a spinneret into a warm chamber, whereupon the solvent evaporates leaving behind fine threads of cellulose acetate.

Cellulose acetate is also used for making non-inflammable photographic and motion-picture films non-shatterable glass, lacquers and varnishes.

Esters other than the acetate are also used, *e.g.*, cellulose acetate butyrate (prepared by the esterification of cellulose with a mixture of acetic anhydride, acetic acid, and butyric anhydride) mixed with plasticisers forms a thermoplastic which is used for many purposes, *e.g.*, film and sheeting.

Viscose rayon. In the viscose process, cellulose is digested with sodium hydroxide solution, and there carbon disulphide is passed into the solution. A mixture of sodium cellulose xanthates, soluble if sodium hydroxide, is formed: R = cellulose:

$$R{-}OH + CS_2 + NaOH \longrightarrow S{=}C\begin{matrix} \diagup OR \\ \diagdown SNa \end{matrix} + H_2O$$

This alkaline solution has a high viscosity, and hence the silk obtained by this process was name viscose rayon. The viscose solution is forced through a spinneret into a sulphuric acid bath, whereupon cellulose is precipitated as fine threads.

Cellophane is made by extruding a viscose solution through a long narrow slit into an acid bath where upon cellulose is precipitated as very thin sheets. These sheets are made moisture-proof be coating with a transparent nitrocellulose lacquer. Some cellulose ethers, particularly the ethyl ether, are used in the manufacture of plastics.

When agitated with about 20 per cent aqueous sodium hydroxide, cellulose swells; apparent) cellulose combines with the sodium hydroxide to form a sodio-cellulose. This is unstable, and is readily decomposed into cellulose by the addition of water, but the cellulose has a number of it physical properties changed, *e.g.*, it absorbs dyes more readily than untreated cellulose. This 'reactive' form of cellulose is known as *mercerised cellulose.*

Cellulose may be oxidised by nitrogen dioxide to give a product which retains its original form a much of the original tensile strength. It appears that about half of the CH_2OH groups (alternately), oxidised to carboxyl, thereby giving a compound readily soluble in sodium hydroxide. Th s oxidic cellulose is useful in medicine *e.g.*, it possesses haemostatic properties, and is useful as sterile gauze.

EXERCISES

1. Outline the evidence to justify the structure $HOCH_2$-$(CHOH)_3CHO$ for arabinose. Is this a fair question? Comment.
2. Formulate the equations for the action of periodic acid on:

   (i) $CH_3CHOHCH_2OH$;

   (ii) $HOCH_2COCH_2OH$;

   (iii) $HOCH_2CHOHCHOHCH_2OH$;

   (iv) $HOCH_2CH_2CHOHCO_2H$;

   (v) $HO_2CCHOHCHOHCHO$;

   (vi) $OCHCH_2CHOMeCHOHCH_2OH$;

   (vii) $OCHCHOMeCHOHCH_2OEt$.

3. Convert:
   (i) D-erythrose to the next higher aldose;
   (ii) D-mannose to the next lower aldose;
   (iii) D-arabinose to the next higher ketose;
   (iv) D-fructose to the next lower aldose.
4. Designate, using R and S symbols, D (+) glucose.
5. Complete and comment:

   (i) D-glucose $\xrightarrow[\text{HCl}]{\text{EtSH}}$? $\xrightarrow[\text{NaOH}]{\text{Me}_2\text{SO}_4}$? $\xrightarrow[\text{CdCO}_3]{\text{HgCl}_2}$

   (ii) D-glyceraldehyde $\xrightarrow[\text{HCl}]{\text{Me}_2\text{CO}}$? $\xrightarrow{\text{CH}_2=\text{CHMgCl}}$? $\xrightarrow{\text{O}_3}$

   (iii) D-glucose $\xrightarrow[\text{EtONa}]{\text{MeNH}_2}$? $\xrightarrow{\text{H}_2\text{SO}_4}$? $\xrightarrow[\text{Ni}]{\text{H}_2}$
6. Draw the conformation of the α- and β-forms of:
   (i) D-arabinopyranose;
   (ii) D-glucopyranose. State in each case which is the more stable form.
7. α-Glucopyranose is oxidised by HIO_4 more rapidly than the β-anomer at the 1, 2-bond. Suggest a reason.
8. Complete the equations and comment.

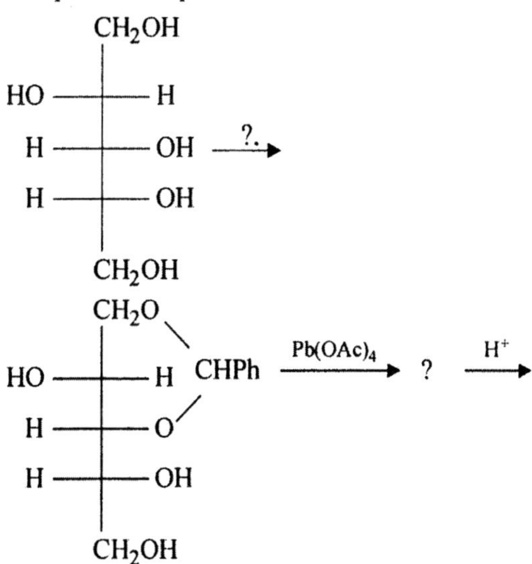

9. Ruthenium tetroxide is known is oxidise a single CHOH group to ketone. How could you convert $HOCH_2(CHOH)_4CHO$ into $HOCH_2(CHOH)_2COCHOHCHO$?

10. What sugars are obtained by epimerising, respectively. C-2, C-3, C-4 of D-glucose?

11. Glucose, mannose and fructose give identical osazones. Explain.

12. Compound A, $C_5H_{10}C_5$, gives a tetra-acetate with AC_2O, and oxidation of A with Br_2—H_2O gives an acid, $C_5H_{10}O_6$. Reduction of A with HI gives isopentane. What are possible structures of A and how could you distinguish between the possibilities?

13. Compound A, $C_5H_{10}O_4$, is oxidised by Br_2—H_2O to the acid, $C_5H_{10}O_5$, which readily forms a lactone. A forms a triacetate with Ac_2O and an osazone with $PhNHNH_2$. A is oxidised by HIO_4, only one molecule of which is consumed. What is the structure of A?

14. Compound A, $C_5H_{10}O_4$, on oxidation with Br_2—H_2O; gives acid B, $C_5H_{10}O_5$. A forms a triacetate (Ac_2O) and is reduced by HI to n-pentane. Oxidation of A with HIO_4 gives, among other products, 1 molecule of CH_2O and 1 molecule of HCO_2H. What are the possible structures of A and how could you distinguish between them?

2

UNSATURATED ALCOHOLS, ETHERS CARBONYL COMPOUNDS AND ACIDS

UNSATURATED ALCOHOLS

The simplest unsaturated alcohol is **vinyl alcohol**, CH_2=CHOH This is, however, unknown; all attempts to prepare it result in the formation of acetaldehyde (together with a small amount of ethylene oxide), *e.g.*, vinyl bromide on treatment with silver oxide in boiling water gives acetaldehyde (see the Erienmeyer rule):

$$CH_2=CHBr + \text{'AgOH'} \longrightarrow AgBr + [CH_2=CHOH] \longrightarrow CH_3CHO$$

Although vinyl alcohol itself is unknown, many of its derivatives have been prepared, and are quite stable. These derivatives may be prepared by the interaction between acetylene and the other reactant in the presence of a suitable catalyst; *e.g., vinyl chloride*:

$$C_2H_2 + HCl \xrightarrow{Hg^{2+}} CH_2 = CHCl$$

Vinyl chloride is manufactured by the thermal decomposition of ethylene dichloride at 600-650° C:

$$ClCH_2CH_2Cl \longrightarrow CH_2 = CHCl + HCl$$

Vinyl chloride and *bromide* are conveniently prepared in the laboratory by heating ethylene dichloride and dibromide, respectively, with ethanolic potassium hydroxide:

$$BrCH_2CH_2Br + KOH \xrightarrow{\text{ethanol}} CH_2 = CHBr + KBr + H_2O$$

Many of the vinyl compounds are used to make plastics. *Vinyon*, which is a thermoplastic, is a copolymer of vinyl chloride and vinyl acetate.

The halogen atom in vinyl halides is not reactive; vinyl halides do not undergo the usual double decomposition reactions, of the alkyl halides. Vinyl chlorides and bromides readily form Grignard reagents in *tetrahydrofuran* as solvent (Normant, 1954, 1957), and they react with lithium to form lithium compounds which undergo the usual reactions (Braude, 1950-1952), *e.g.*,

$$R_1^2C = CHBr \xrightarrow{Li} R_1^2C = CHLi \xrightarrow{R^2CHO} R_1^2C = CHCHOHR_2$$

The reason for the unreactivity of the halogen atom is not clear. Some believe it to be due to resonance through which the halogen atom acquires some double-bond character, and is thereby more strongly bound to the carbon atom due to the shortening of the C—Cl bond

$$CH_2 = CH—\ddot{\underset{..}{C}}l\!: \longleftrightarrow CH_2^-—CH = \overset{+}{\underset{..}{C}}l\!:$$

An alternative explanation is that due to Walsh (1947). As we have seen, an sp^2 hydridised carbon atom attracts electrons more than does an sp^3 hydridised carbon atom. Thus the Cl atom in the C—Cl group is more tightly bound in C=C—Cl than in C—C—Cl.

The non-reactivity of the chlorine atom in vinyl chloride may be explained from the M.O. point of view as follows. If the chlorine atom has sp^2 hybridisation, the C—Cl bond will be a σ-bond and the two lone pairs of electrons would occupy the other two sp^2 orbitals. This would leave a p orbital containing a lone pair, and this orbital copld now conjugate with the π-bond of the ethylenic link (Fig. 2.1). Thus two M.O.s will be required to accommodate these four π-electrons (cf. the carboxyl group). Furthermore, since chlorine is more electronegative than carbon, the electrons will tend to be found in the vicinity of the chlorine atom. Nevertheless, the chlorine atom has now lost 'full control' of the lone pair, and so is less negative than it would have been had there been no conjugation. Since two carbon atoms have acquired a share in the lone pair, each carbon atom acquires a small negative charge. Hence, owing to delocalisation of bonds (through conjugation), the vinyl chloride molecule has an increased stability.

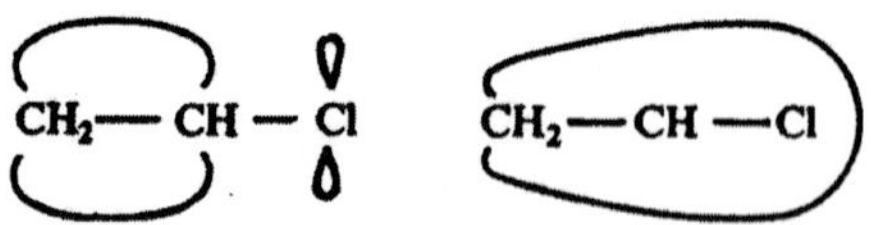

Fig. 2.1

Before the chlorine atom can be displaced by some other group, the lone pair must be localised again on the chlorine atom. This requires energy, and so the chlorine is more 'firmly bound' than had no conjugation occurred.

VinyMdene ddorides, $R_2C = CCl_2$. Dichloromethylene, formed by the action of potassium t-butoxide or an alkyl-lithium on chloroform, may be trapped by triphenylphosphine to form the ylid, dichloromethylene triphenylphosphorane (Speziale et at., 1960).

This readily undergoes the Wittig reaction to form a vinylidene chloride.

$$CHCl_3 \longrightarrow [CCl_2] \xrightarrow{Ph_3P}$$

$$= \quad CCl_2 \xrightarrow{R_2C=O} R_2C = CCl_2 + Ph_3P = O$$

Vinyl cyanide (*acrylonUrile*) is very useful for introducing thecyanoethyl group, —CH,CH.CN, by reaction with compounds containing an active methylene group. The best catalyst for **cyanoethylation** is benzyltrimethylammonium hydroxide, $C_6H_5CH_2\overset{+}{N}(CH_3)_3 \} OH^-$, but in many cases an aqueous or ethanolic solution of alkalis is effective (Bruson, 1942, 1943). Cyanoethylation is one example of the Michael condensation, *e.g.*,

$$MeCOOH_2Me + OH- \rightleftharpoons MeCO\bar{C}HMe + H_2O$$

$$MeCOCHMe + CH_2 = CH - C \equiv CH—C$$

$$\xrightleftharpoons{H_2O} \underset{\displaystyle CH_2CH=CN=H}{MeCOCHMe} \longrightarrow$$

$$\underset{\displaystyle CH_2CH_2=CN=H}{MeCOCHMe} \xrightarrow[\text{(ii) } CH_2=CHCN]{\text{(i) } OH^-} \underset{\displaystyle CH_2CH_2CN}{\overset{\displaystyle CH_2CH_2CN}{MeCOCHMe}} \quad (90\%)$$

This type of reaction offers a means of preparing a variety of compounds. Vinyl cyanide normally causes di- or tri-cyanoethylation of compounds which have an active methylene or methyl group respectively, but Campbell *et al.* (1956) have described conditions for mono-cyanoethylation. In addition to active methylene and methyl groups, compounds such as primary and secondary amines, alcohols, phenols, etc., can also undergo cyanoethylation.

Vinyl cyanide is manufactured by the direct combination of hydrogen cyanide with acetylene, and from propene as follows (acraldehyde is an intermediate):

$$2CH_2=CHCH_3+2NH_3+3O_2 \xrightarrow[450°C]{cat.} 2CH_2=CHCN+6H_2O$$

Allyl alcohol (*prop-2-en-1-ol*), b.p. 97°C, may be prepared as follows:

(i) By boiling allyl chloride with dilute sodium hydroxide solution under pressure (this is a commercial method).

$$CH_2=CHCH_2Cl + NaOH \xrightarrow{150°C} CH_2=CHCH_2OH + NaCl$$

Another commercial method is the isomerisation of propene oxide in the presence of lithium phosphate as catalyst.

(ii) By heating glycerol with formic acid or oxalic acid at 260° C:

$$\begin{array}{l} CH_2OH \\ | \\ CHOCO \\ | \\ CH_2OH \end{array} + \begin{array}{l} COOH \\ | \\ COOH \end{array} \longrightarrow \underset{\text{glyceryl monoxalate}}{\begin{array}{l} CH_2OCOCOOH \\ | \\ CHOH \\ | \\ CH_2OH \end{array}} \xrightarrow{-CO_2} \underset{\text{glyceryl monoformate}}{\begin{array}{l} CH_2OOCH \\ | \\ CHOH \\ | \\ CH_2OH \end{array}}$$

glyceryl monoxalate $\downarrow -H_2O_2$; glyceryl monoformate $\downarrow -CO_2 - H_2O$

$$\underset{\text{dioxalin}}{\begin{array}{l} CH_2OH \\ |\quad | \\ CHOCO \\ | \\ CH_2OH \end{array}} \xrightarrow{-2CO_2} \begin{array}{l} CH_2 \\ \| \\ CH \\ | \\ CH_2OH \end{array}$$

In practice it is better to use formic acid since this gives a higher yield (45-47 per cent).

A general synthesis of allylic alcohols is the selenium dioxide oxidation of alkenes of the type $R^1CH_2CH = CHR^2$.

Allyl alcohol has the properties of an unsaturated compound and a primary alcohol. It is oxidised to glycerol by dilute alkaline permanganate:

$$CH_2=CHCH_2OH + H_2O + [O] \longrightarrow HOCH_2CHOHCH_2OH$$

Allyl alcohol (and other α, β-unsaturated primary and secondary alcohols) may be con-veniently oxidised to the corresponding oxo compound with manganese dioxide or nickel peroxide, *e.g.*,

$$CH_2 = CHCH_2OH \xrightarrow{MnO_2} CH_2 = CHCHO$$

t-Butyl chromate, since it does not attack a double bond, is also a very useful reagent for oxidising unsaturated primary alcohols to the corresponding unsaturated aldehydes.

In general, oxidising agents attack the double bond in allyl alcohol as well as the alcoholic group. Oxidation, however, may be carried out by 'protecting' the double bond by bromination and then oxidising and debrominating with zinc dust in methanolic solution, *e.g.*,

$$CH_2 = CHCH_2OH \xrightarrow{Br_2} BrH_2CHBrCH_2OH \xrightarrow{HNO_3}$$

$$BrCH_2CHBrCO_2H \xrightarrow[MeOH]{Zn} CH_2 = CHCO_2H$$

Allyl alcohol adds on chlorine or bromine to form the corresponding 2, 3-dihalogeno-propan-l-ol. Allyl alcohol also adds on halogen acids and hypohalous acids, but the addition takes place *contrary* to Markownikoff's rule, *e.g.*, glycerol β-monochlorohydrin is formed with hypochlorous acid. This may be due to the presence of the oxygen atom which exerts a strong -I effect.

$$CH_2{=}CH{-}CH_2 \longrightarrow OH + HO{-}Cl \longrightarrow HOCH_2CHClCH_2OH$$

If we consider this addition from the point of view of the stabilities of the intermediate carbonium ions, $^{+}CH_2CHClCH_2OH$ (primary) and $ClCH_2\overset{+}{C}HCH_2OH$ (secondary), it might have been anticipated that the product would have arisen via the more stable secondary ion to give $ClCH_2CHOHCH_2OH$ (the Markownikoff product). There are, however, various complications, *e.g.*, the carbonium ion is probably not the classical one shown, but is a chlorinium (bridged) ion (it might also be the epoxide type?).

If a chlorinium ion, the hydroxide ion might be expected to attack at the CH_2 carbon atom, which is less hindered than the other carbon atom (C of the $CHCH_2OH$).

Allyl chloride (3-*chloropropene*), b.p. 45°C, is prepared industrially by the chlorination of propene at high temperature. It may be conveniently prepared in the laboratory by warming allyl alcohol with hydrochloric acid:

$$CH_2 = CHCH_2OH + HCl \longrightarrow CH_2 = CHCH_2Cl + H_2O$$

This reactivity of allyl alcohol is more characteristic of ar tertiary alcohol than a primary. This may be explained by an S_N1 mechanism operating because of the stability of the allyl carbonium ion (see below).

Allyl bromide (3-*bromopropene*), b.p. 70°C, is best prepared in the laboratory by distilling aqueous allyl alcohol with 48 per cent hydrobromic acid in the presence of sulphuric acid:

$$CH_2 = CHCH_2OH + HBr \longrightarrow CH_2 = CHCH_2Br + H_2O \text{ (92-96\%)}$$

Allyl iodide (3-*iodopropene*), b.p. 103.1°C, may be prepared by heating glycerol with a small amount ofhydriodic acid , or by the Finkelstein reaction:

$$CH_2 = CCH_2Cl + NaI \xrightarrow{\text{acetone}} CH_2 = CHCH_2I + NaCl$$

The halogen atom in the allyl halides is very reactive, and it has been found experimentall) that the position of the double bond with respect to the halogen atom determines the reactivit) of the halogen atom. In compounds of the type C = C—X, the halogen atom X is unreactive (see vinyl halides, above). In C = C—C—X, X is more reactive than in the alkyl halides. Where the halogen atom in an unsaturated compound is further removed from the double bond than it the allyl position, it behaves similarly to the halogen atom in alkyl halides; *e.g.*, 4-chlorobut-1-ene $CH_2 = CHCH_2CH_2Cl$, undergoes the usual reactions of the alkyl halides with reagents that do not affect the double bond.

The high reactivity of the allyl halides may be explained by the fact that when the halogen atom ionises, the allyl carbonium ion produced is a resonance hybrid, and consequently it is stabilised:

$$CH_2 = CH—CH_2^+ \longleftrightarrow \overset{+}{C}H_2—CH = CH_2$$

The allyl halides add on the halogen acids to form a mixture of the 1, 2- and 1, 3-dihalides. The addition of hydrogen bromide to allyl bromide has been studied in great detail (Kharasch, 1933):

$$CH_2 = CHCH_2Br + HBr \longrightarrow \underset{\text{1, 3-dibromide}}{BrCH_2CH_2Br} + \underset{\text{1, 2-dibromide}}{CH_3CHBrCH_2Br}$$

In the presence of peroxides, the 1,3-dibromide is obtained in 90 per cent yield. In the absence of peroxides, the l, 2-dibromide is obtained in 90 percent yield. When allyl bromide, which has not been carefully purified, is treated with hydrogen bromide, without special precautions being taken to exclude all traces of air, a mixture of the 1,2- and 1,3-dibromides is obtained.

The explanation for these results is as follows. The 1,3-dibromide is formed by a free-radical mechanism:

$$(RCO_2)_2 \longrightarrow 2RCOO\cdot \longrightarrow 2R\cdot + 2CO_2$$

$$R\cdot + HBr \longrightarrow RH + Br\cdot$$

$$Br\cdot + CH_2 = CHCH2Br \longrightarrow BrCH_2\dot{C}HCH_2Br$$

$$\xrightarrow{HBr} BrCH_2CH_2CH_2Br + Br\cdot \text{ ; etc.}$$

The alternative free radical, $\dot{C}H_2CHBrCH_2Br$, is a primary one, and since this is less stable than a secondary, the reaction proceeds via the latter.

The formation of the 1,2-dibromide proceeds by means of a polar mechanism, but now a proton is added in the first step to give the secondary carbonium ion, CH,CH$^+$CH$_2$Br, which is more stable than the alternative primary carbonium ion, $^+$CH$_2$CH$_2$CH,Br. Arguments based on the –I effect of the bromine atom would have anticipated the formation of the 1,3-dibromide (cf. with allyl alcohol, above).

A completely different approach to this problem of 1,2-addition is the use of the **Baker-Nathan effect** (1935). As we have seen, the general inductive effect of alkyl groups is $Me_3C > Me_2CH > MeCH_2 > Me$. This inductive order has been used satisfactorily to explain various physical data, etc. In some reactions, however, the inductive order is reversed, *e.g.*, Baker and Nathan (1935) examined kinetically the reaction between p-substituted benzyl bromides and pyridine. The reaction was carried out in acetone, and was shown to be entirely S_N2. Thus the reaction may be formulated:

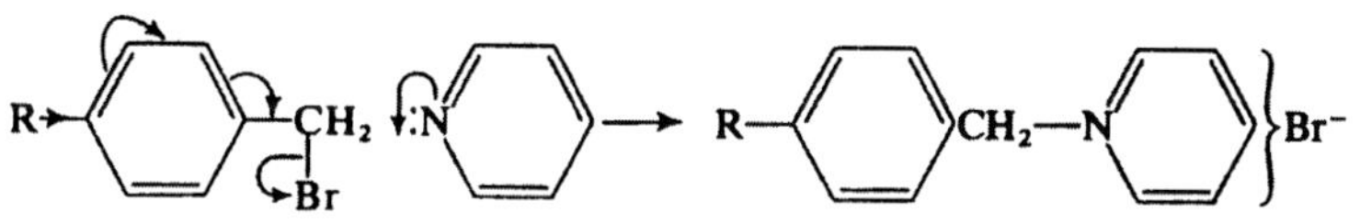

Fig. 2.2

The greater the +1 effect of the R group, the faster should be the reaction, but it was found that the rate order for the R groups was

$$Me > Et > isoPr \sim t\text{-}Bu$$

Thus the order of electron release is almost exactly the reverse of that given above. Therefore alkyl groups must possess some mechanism of electron-release in which the order is

$$\text{Me} > \text{Et} > \text{isoPr} > \text{t-Bu}$$

Further work showed that in all cases where the general inductive order of alkyl groups was reversed, the alkyl group was attached to an aromatic nucleus. This led Baker and Nathan (1935, 1939) to suggest that when the H—C bond is attached to an unsaturated carbon atom, the σ-electrons of the H—C bond become less localised by entering into partial conjugation with the attached unsaturated system, *i.e.*, σ, π-conjugation:

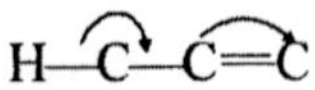

$$\text{H—C—C=C}$$

Thus there is conjugation between electrons of single and those of multiple bonds. This type of conjugation is known as **hyperconjugation**, and is a *permanent* effect (this name was given by Mulliken, 1941).

There are various ways of looking at the **hyperconjugative effect**. A widely used one is to regard the HùC bond as possessing partial ionic character due to resonance, *e.g.*, propene may be written as a resonance hybrid as follows:

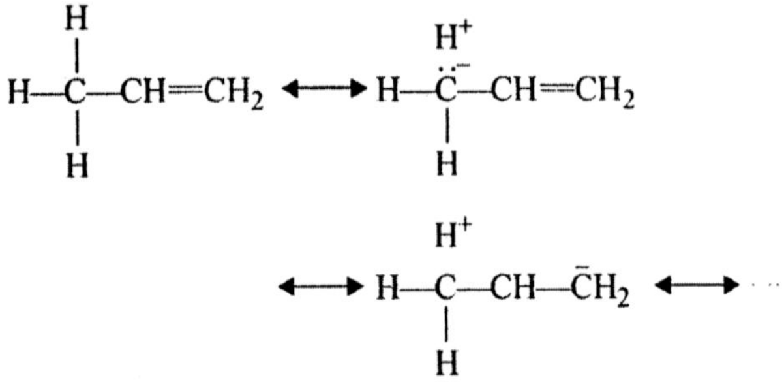

$$\begin{array}{c}\text{H}\\|\\\text{H—C—CH=CH}_2\\|\\\text{H}\end{array} \longleftrightarrow \begin{array}{c}\text{H}^+\\\\\text{H—}\overset{\cdot\cdot -}{\text{C}}\text{—CH=CH}_2\\|\\\text{H}\end{array}$$

$$\longleftrightarrow \begin{array}{c}\text{H}^+\\\\\text{H—C—CH—}\bar{\text{C}}\text{H}_2\\|\\\text{H}\end{array} \longleftrightarrow \cdots$$

From this point of view, hyperconjugation may be regarded as '*no bond resonance*'. The hydrogen atoms are not *free*; the effect is to increase the ionic character of the C—H bond, the electrons of which become partially delocalised through conjugation.

Various physical data have been explained by hyperconjugation, *e.g.*, bond lengths:

$$CH_3\text{—}CH_3 \qquad CH_3\text{—}C\equiv CH$$

Because of hyperconjugation, the CH_3—C bond has some partial double bond character and consequently is shortened. Also, since hyperconjugation stabilises a molecule (through resonance), propene, for example, should be more staJble than 'expected'.

This has been found to be so; the observed heat ofhydrogenation of propene is less than the calculated value.

Using this scheme of hyperconjugation, we are now in a position to explain the formation of the 1,2-dibromide from allyl bromide and hydrogen bromide by the polar mechanism.

$$\begin{array}{c}H\\|\\CH_2{=}CH{-}CHBr\end{array} \leftrightarrow \begin{array}{c}H^+\\ \\CH_2{=}CH{-}\ddot{\bar{C}}HBr\end{array} \leftrightarrow \begin{array}{c}H^+\\ \\\bar{C}H_2{-}CH{=}CHBr\end{array}$$

$$\xrightarrow{HBr} CH_3CHBrCH_2Br$$

The inductive effect of the bromine atom would tend to give the 1,3-dibromide (see above). If, however, the hyperconjugative effect is operating, and is *stronger* than the –I effect of the bromine atom, then the 1,2-dibromide will be formed.

An interesting point here is that with allyl alcohol, the –I effect of the hydroxyl group is stronger than the hyperconjugative effect.

From the M.O. point of view, hyperconjugation may be explained as follows: π-orbitals can overlap with other π-orbitals to produce conjugation. π-Orbitals, however, can also overlap to a certain extent with adjacent π-orbitals to form extended orbitals. When this occurs we have *hyperconjugation*, and this phenomenon is exhibited mainly by the hydrogen bonds in a *methyl* group when this group is attached to an unsaturated carbon atom.

The question is: how does this overlapping occur? Coulson (1942) treats the methyl group (Fig. 2.3a) as a 'compound' atom to form a 'group' orbital (b). An alternative 'group' orbital may also be formed as in (c). In this arrangement, the methyl group behaves as a group with a n-orbital, and this can conjugate with an adjacent re-bond in, *e.g.*, propene (Fig. 2.3b).

Hence, owing to partial delocalisation of bonds (σ and π) in this way, the propene molecule is more stable than 'expected'.

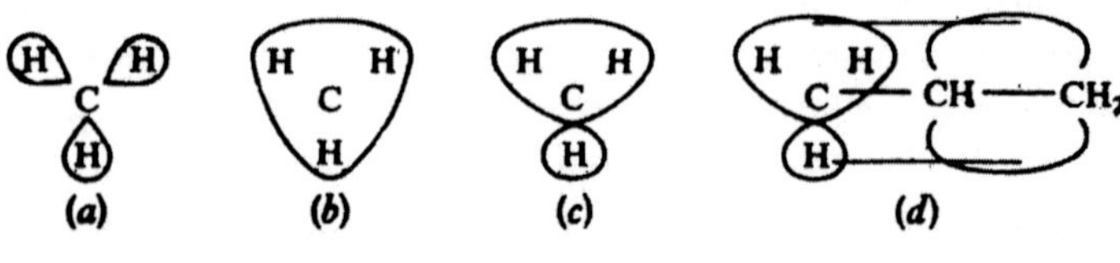

Fig. 2.3

The unusual reactivity of the halogen atom in allyl halides may be explained as follows. If the chlorine were to ionise, the carbon atom to which it was attached now has a positive charge and has only *six* electrons. Hence the π-bond covering the other two carbon atoms can extend to embrace this third carbon atom. The net result is an M.O. covering *three* carbon atoms, and so the delocalisation energy will be increased, *i.e.*, the new arrangement is stabilised, and behaves as if the chlorine atom is ionic Fig. 2.4.

$$CH_2{-}CH{-}CH_2{-}Cl \rightarrow \{CH_2{-}CH{-}CH_2\}^{+}Cl^{-} \rightarrow \{CH_2{-}CH{-}CH_2\}^{+}Cl^{-}$$

Fig. 2.4

There is, however, a certain amount of doubt that dipole moment data are evidence for hyperconjugation. Since the electronegativity of a carbon atom depends on its state of hybridisation, there will be a bond moment when two differently hybridised carbon atoms are linked together. Petro (1958) has calculated bond moments for several types of carbon carbon single bonds:

Bond	*Bond moment*
$C(sp^3)^+ — C(sp^2)^-$	0.681D
$C(sp^2)^+ — C(sp)^-$	1.151D
$C(sp^3)^+ — C(sp)^-$	1.48D

Ferreira (1960) has calculated the dipole moments of various molecules, *e.g.*, toluene, propene etc., on these principles, and has shown that the calculated values agree well with the experimental values. Thus it appears that dipole moment evidence for hyperconjugation is no valid.

It has also been found experimentally that all C—C single bonds are not the same length. This shortening has been attributed to hyperconjugation, but it can also be explained in term of different electronegatives of the two carbon atoms. Bloor and Gartside (1959) have calculated bond energies of C—C and C—H bonds for differently hybridised carbon atom, and since these vary with the nature of the hybrid state of the carbon atom this makes the evaluation of resonance energies from 'average' bond energies very uncertain.

It is quite possible, however, since bond moments, bond lengths and bond energies are ground-state properties, that hyperconjugation exists in excited states and in transition states of molecules, and therefore is important for the interpretation of reactivity of molecules.

Mulliken (1959) has concluded that the explanation offered above is partly right, but does not believe the π-delocalisation shortenings are negligible. He has made some calculations and has attributed 40 per cent of observed shortenings to π-electron resonance and 60 per cent to hybridisation differences. On the other hand, Ehrenson (1964), from his calculations, supports the theory that hyperconjugation may be the major driving force in some reactions where carbonium ion products are formed.

Crotyl alcohol (*crotonyl alcohol, but-2-en-l-ol*), b.p. 118°C, may be prepared by the reduction of crotonaldehyde with aluminium isopropoxide:

$$CH_3CH = CHCHO \xrightarrow{[(CH_3)_2CHO]_3Al} CH_3CH = CHCH_2OH \text{ (85–90\%)}$$

When treated with hydrogen bromide, crotyl alcohol produces a mixture ofcrotyl and methyl-vinylcarbinyl bromide:

$$CH_3CH = CHCH_2OH \xrightarrow{HBr} CH_3CH = CHCH_2Br + CH_3CHBrCH = CH_2$$

The formation of the rearranged proiduct is an example of anionotropy, and the most widely studied example is the three-carbon system, *i.e.*, the allylic system. Nucleophilic substitutior reactions of allylic halides may occur by the S_N2 mechanism, and when this is operating substitution proceeds normally.

When, however, nucleophilic substitution reactions are carried out under conditions which favour the unimolecular mechanism, then the product may be a mixture of two isomers,

the rearranged product being an example of the *allylic rearrangement*. The mechanism by which this rearrangement occurs is known as SN)', and is believed to occur via a resonance hybrid (ambident) cation:

$$MeC = CHCH_2OH \xrightleftharpoons{H^+} MeCH = CHCH_2OH_2^+ \xrightleftharpoons[+H_2O]{-H_2O}$$

$$MeCH{=}CHCH_2^+ \longleftrightarrow Me\overset{+}{C}HCH{=}CH_2$$

$$\downarrow Br^- \qquad\qquad \downarrow Br^-$$

$$\underset{S_N1}{MeCH{=}CHCH_2Br} \qquad \underset{S_N1'}{MeCHBrCH{=}CH_2}$$

There is much evidence to support this S_N1' mechanism, *e.g.*, Hughes *et al.* (1948) examine the conversion of α-methylallyl and crotyl chloride into α-methylallyl and crotyl ether by reaction with ethanolic sodium ethoxide. The concentration of the sodium ethoxide was made so small as to give first-order kinetics, or so large as to give second-order kinetics. The result were:

$MeCHClCH{=}CH_2$ —1st order→ (mixture) ←1st order— $MeCH{=}CHCH_2Cl$

$MeCHClCH{=}CH_2$ —2nd order→ $MeCH(OEt)CH{=}CH_2$ (100%)

$MeCH{=}CHCH_2Cl$ —2nd order→ $MeCH{=}CHCH_2OEt$ (100%)

1st order (both): $\begin{cases} MeCH(OEt)CH{=}CH_2 \ (13\%) \\ + \\ MeCH{=}CHCH_2OEt \ (87\%) \end{cases}$

With second-order substitution, each of chloride gives its own unrearranged ether (thus the S_N2 mechanism). With first-order substitution, however, each chloride gives the *same* mixture isomeric ethers. This latter result may be explained by assuming that *two* mechanisms a operating, S_N1 (without rearrangement) and S_N1' (with rearrangement), both proceeding through a common intermediate ambident cation.

In addition to the S_N1' mechanism, there is also the S_N2', *i.e.*, a bimolecular nucleophilic substitution with allylic rearrangement. This has been difficult to demonstrate, but a good example of it has been given by de la Mare *et al.* (1953). These authors have suppressed the S_N2 reaction by steric factors and reduced the S,1 reaction by means of a reagent of high nucleophilic activity and a solvent of low ionising power. They found that the reaction between α, α-dimethylallyl chloride and sodium thiophenoxide in ethanol gives 62 per cent of rearranged product by a second-order reaction (*i.e.*, S_N2'):

$$Me_2C(Cl)—CH=CH_2 \quad SPh \longrightarrow Me_2C=CHCH_2SPh + NaCl$$

One other problem that will be discussed here is the case of ion pairs in the allylic rearrangement (see also). Young *et al.* (1951) have shown that α, α-dimethylallyl chloride in acetic acid undergoes acetolysis accompanied by rearrangement to γ, g-dimethylallyl chloride:

$$Me_2CClCH = CH_2 \xrightarrow{MeCO_2H} \text{acetolysis products} + Me_2C = CHCH_2Cl$$

The rate of the isomerisation was shown to be proportional to the concentration of the starting chloride only. Thus a possible mechanism is the S_N1'. If this were so, then the rate of isomerisation would be affected by addition of chloride ions (since the first step is ionisation). Experiment showed that the rate was unaffected by added chloride ions, and it was also shown that when an excess of radioactive chloride ion was added, the rate of isomerisation was far greater than the incorporation of Cl^{*-}. It thus appears that the chloride ion is set free during the isomerisation and is mainly the 'first ion' to recombine, even though the solution contains a large excess of Cl^{*-} ions. Young *et al.*, explained this result by assuming that the starting chloride undergoes ionisation to form an ion-pair, and this is followed by internal return:

$$Me_2CClCH=CH_2 \rightleftharpoons \underset{\text{ion pair}}{[Me_2C\cdots CH\cdots CH_2]^+Cl^-}$$

$$\rightleftharpoons Me_2C=CHCH_2Cl$$

Nevertheless, some ion pairs escape from the solvent cage, and these Tree' carbonium ions are capable of undergoing acetolysis and combination with Cl^{*-}.

A number of unsaturated alcohols occur in essential oils. The essential oils are volatile oils with pleasant odours, and are obtained from vaiious plants to which they impart their characteristic odours. They are widely used in the perfume industry.

The simplest acetylenic alcohol is *propargyl alcohol* (prop-2-yn-l-ol), b.p. 114°C. This may be prepared from 1,2,3-tribromopropane:

$$BrCH_2CHBrCH_2Br \xrightarrow[KOH]{\text{ethanolic}} CH_2 = CBrCH_2Br$$

$$\xrightarrow[KOH]{\text{aqueous}} CH_2 = CBrCH_2OH \xrightarrow[KOH]{\text{ethanolic}} CH \equiv CCH_2OH$$

Propargyl alcohol (together with butynediol) is prepared by the interaction of acetylene and formaldehyde in the presence of silver or cuprous acetylide as catalyst:

$$C_2H_2 + HCHO \xrightarrow{Ag_2C_2} CH \equiv CCH_2OH$$

$$\xrightarrow[HCHO]{Ag_2C_2} HOCH_2C \equiv CCH_2OH$$

Propargyl alcohol behaves in many ways like acetylene, *e.g.*, it adds on two or four bromine atoms, and forms the silver or cuprous compounds when treated with an ammoniacal solution of silver nitrate or cuprous chloride. It may be catalytically reduced to allyl and n-propyl alcohols.

Unsaturated Ethers

The simplest unsaturated ether is methyl vinyl ether, which is prepared industrially by passing acetylene into methanoi at 160-200°C in the presence of 1-2 per cent potassium methoxide, and under pressure sufficient to prevent boiling:

$$C_2H_2 + CH_3OH \longrightarrow CH_3OCH = CH_2$$

The acetylene is diluted with nitrogen to prevent explosions.

Methyl vinyl ether is a very reactive gas, b.p. 5-6°C. It is hydrolysed rapidly by dilute acid at room temperature to give methanoi and acetaldehyde (c/. the ethers), but is stable in alkaline solution. It undergoes many addition reactions at the double bond, *e.g.*,

It readily polymerises, and is used for making polyvinyl ether plastics.

Divinyl ether (*vinyl ether*), b.p. 39°C, may be prepared by heating ethylene chlorohydrin with sulphuric acid, and then passing the product, 2,2'-dichlorodiethyl ether, over potassium hydroxide at 200-240°C:

It is used as an anaesthetic, being better for this purpose than diethyl ether.

Unsaturated Aldehydes

The carbonyl (C = 0 stretch) absorption band of C, B-unsaturated aldehydes occurs in the region 1705-1680 cm^{-1} (s). It is therefore readily distinguished from the carbonyl group in saturated (acyclic) aldehydes. The molecular formula will also help here.

The t-values of both α- and β-protons in α, β-unsaturated carbonyl compounds occur at lower field than those in alkenes (4.3-5.3), and the value for the β-proton is also dependent on whether it is cis or irons with respect to the carbonyl group. The usual range for α-protons is τ ~ 3.5-4.2 and for β-protons, τ ~ 2.4-3.5. This downfield shift for both α- and β-protons may be attributed to the –I effect of the CO group, and the greater shift for the β-proton may be explained by conjugation which results in a small positive charge on the β-carbon atom, thereby increasing deshielding of a β-proton:

$$—CH = CH—C = O \longleftrightarrow —\overset{+}{C}H—CH = C—O^-$$

Since shifts for protons, attached to an unsaturated carbon atom further along the chain are fairly close to those of protons in alkenes, this may be used to show whether a double bond is α, β with respect to the carbonyl group.

The simplest unsaturated aldehyde is **acraldchyde** (*acrolein, prop-2-en-l-al*), b.p. 52°C. This is conveniently prepared by heating glycerol with potassium hydrogen sulphate, which is the most satisfactory dehydrating agent for this reaction:

$$HOCH_2CHOHCH_2OH \xrightarrow{-2H_2O} CH_2 = CHCHO\ (33–48\%)$$

Acraldehyde may also be readily prepared by the oxidation of allyl alcohol with manganese dioxide or nickel peroxide.

Acraldehyde is prepared industrially in many ways, *e.g.*,

(i) by passing a mixture of acetaldehyde and formaldehyde vapour over lithium phosphate as catalyst:

$$CH_3CHO + CH_2O \longrightarrow CH_2 = CHCHO + H_2O$$

(ii) By the direct oxidation of propene over cuprous oxide on alumina:

$$CH_2 = CHCH_3 + O_2 \longrightarrow CH_2 = CHCHO + H_2O$$

Acraldehyde is unstable, readily polymerising to a white solid. It is an α, β-unsatuiated aldehyde, and undergoes many of the usual reactions of an alkene and an aldehyde, *e.g.*, it adds on two halogen atoms or a molecule of halogen acid (contrary to Markownikoff's rule; *cf.* allyl alcohol):

$$CH_2 = CHCHO + Br_2 \longrightarrow BrCH_2CHBrCHO$$

$$CH_2 = CHCHO + HCl \longrightarrow ClCH_2CH_2CHO$$

The addition of hydrogen bromide is unaffected by the presence of peroxides. Acraldehyde reduces ammoniacal silver nitrate, forms a cyanohydrin with hydrogen cyanide, and a phenyl-hydrazone with phenylhydrazine:

$$CH_2 = CHCHO + [O] \xrightarrow{AgNO_3} CH_2 = CHCO_2H$$

$$CH_2 = CHCHO + HCN \longrightarrow CH_2 = CHCH(OH)CN$$

$$CH_2 = CHCHO + C_6H_5NHNH_2 \longrightarrow CH_2 = CHCH = NNHC_6H_5 + H_2O$$

Many of the reactions of α, β-unsaturated aldehydes are 1,2-additions, but (α, β-unsaturated ketones tend to undergo 1,4-additions. Aldehydes are more reactive than ketones towards nucleophiles, and the kinetically controlled product with aldehydes is 1,2-addition and with ketones 1,4-addition. With the latter, rearrangement occurs if part of the addendum is a proton to give the thermodynamically controlled product.

Acraldehyde undergoes the Tischenko reaction to form allyl acrylate:

$$2CH_2 = CHCHO \xrightarrow{(C_2H_5O)_3Al} CH_2 = CHCO_2CH_2CH = CH_2$$

It does not, however, undergo the normal aldol condensation; instead, in the presence of alkali, cleavage of the molecule takes place:

$$CH_2{=}CHCHO + OH^- \rightleftharpoons HO{-}CH_2{-}\bar{C}HCHO$$

$$\overset{H_2O}{\rightleftharpoons} H{-}OCH_2CH_2CHO + OH^2$$

$$\rightleftharpoons O^-{-}CH_2{-}CH_2CHO + H_2O$$

$$\rightleftharpoons CH_2O + \bar{C}H_2CHO$$

$$\overset{H_2O}{\rightleftharpoons} CH_3CHO + OH^-$$

The first part of the reaction is the reversal of dehydration, and the second part is the reversal of the aldol condensation. This type of reaction is characteristic of α, β-unsaturated aldehydes.

When acraldehyde is reduced with metal and acid, there are obtained an unsaturated alcohol, a saturated aldehyde, a saturated alcohol and a compound formed by bimolecular reduction:

$$CH_2 = CHCHO \xrightarrow{e;H^+} CH_2 = CHCH_2OH + CH_3CH_2CHO$$
$$+ CH_3CH_2CH_2OH + CH_2 = CHCHOHCHOHCH = CH_2$$

A good yield of the bimolecular product is obtained by reducing acraldehyde with magnesium amalgam (*cf.* pinacol). Acraldehyde is reduced by sodium amalgam to n-propanol and by aluminium isopropoxide (Meerwein-Ponndorf-Verley reduction) to allyl alcohol. It should be noted that metal and acid do not usually reduce a double bond; it is only when the double bond is in the α, β-position with respect to a carbonyl group that it is reduced in this way (see also below).

The product obtained by the catalytic hydrogenation of α, β-unsaturated carbonyl compounds depends largely on the nature of the catalyst. Saturated carbonyl compounds are obtained with Pd—C at room temperature, unsaturated alcohols with PtO_2—FeCl—(CH_3CO_2), Zn and saturated alcohols with Ni, Pd—C, or PtO_2 at elevated temperatures and hydrogen under pressure.

α, β-Unsaturated carbonyl compounds may be reduced to unsaturated alcohols by lithium aluminium hydride, sodium borohydride, or best of all, by aluminium hydride.

Acraldehyde adds on to butadiene to form a cyclic compound (see the *Diels-Alder* reaction.

An interesting property of α, β-unsaturated aldehydes, ketones and acids is that they often undergo abnormal ozonolysis:

$$R_1^2C = CR^2CR^3 = O \longrightarrow R_1^2CO + R^2CO_2H + R^3CO_2H$$

$$R_1^2C = CR^2CO_2H \longrightarrow R_1^2CO + R^2CO_2H + CO_2$$

Crotonaldchyde (*but-2-en-l-al*), b.p. 104°C, may be prepared by heating aldol alone, better, with a dehydrating agent, *e.g.*, zinc chloride. The best yield is obtained by distilling aid with acetic acid as catalyst:

$$\begin{array}{cc} \begin{array}{c} H \quad CH_3 \\ \diagdown\diagup \\ C \\ \| \\ C \\ \diagup\diagdown \\ H \quad CHO \end{array} & \qquad \begin{array}{c} CH_3 \quad H \\ \diagdown\diagup \\ C \\ \| \\ C \\ \diagup\diagdown \\ H \quad CHO \end{array} \\ \textit{cis} & \qquad \textit{trans} \end{array}$$

Crotonaldehyde closely resembles acraldehyde in its chemical properties. It exists, however in two geometrical isomeric forms, cis and trans: A number of unsaturated aldehydes occur naturally in the essential oils. many accompanying corresponding unsaturated alcohol.

The simplest acetylenic aldehyde is **propargylaldehyde** (propiolaldehyde. propynal). b.p. 60°C. This may be prepared from acraldehyde, the aldehyde group of which is 'protected' by acetal formation.

$$CH_2=CHCHO+2C_2H_5OH \overset{HCl}{\rightleftharpoons} H_2O+CH_2=CHCH(OC_2H_5)_2$$

$$\xrightarrow{Br_2} BrCH_2CHBrCH(OC_2H_5)_2$$

$$\xrightarrow[KOH]{ethanolic} CH \equiv CCH\ (OC_2H_5)_2 \xrightarrow[HCl]{dilute} CH \equiv CCHO$$

It may also be obtained by the controlled oxidation of propargyl alcohol.

Propargylaldehyde undergoes cleavage when treated with sodium hydroxide (cf. acraldehyde above) :

$$CH \equiv CCHO + NaOH \longrightarrow CH \equiv CH + HCO_2Na$$

Unsaturated Ketones

The carbonyl infra-red absorption band (C = O stretch) in α, β-saturated ketones occurs in the region 1685-1665 cm^{-1} (s). This distinguishes these compounds from or, α, β-saturated aldfehydes (NMR spectra).

The simplest unsaturated ketone is **methyl vinyiketone** (but-3-en-2-one), b.p. 79°C. This may be prepared by means of an aldol condensation between formaldehyde and acetone, the product being dehydrated by heat.

$$HCHO + CH_3COCH_3 \xrightarrow{NaOH} HOCH_2CH_2COCH_3$$

$$\xrightarrow{-H_2O} CH_2 = CHCOCH_3$$

It may also be prepared by the action of acetic, anhydride on ethylene in the presence of zinc chloride as catalyst:

$$CH_2=CH_2+(CH_3CO)_2O \xrightarrow{ZnCl_2} CH_2=CHCOCH_3+CH_3CO_2H$$

Methyl vinyl ketone can be prepared by the action of acetic anhydride on vinylmagnesium bromide in tetrahydrofuran at –60° to –70°C (Martin, 1957); this is a general method for preparing vinyl ketones:

$$2CH_2 = CHMgBr + 2(CH_3CO)_2O \longrightarrow$$

$$2CH_2 = CHCOOCH_3 + MgBr_2 + (CH_3CO_2)_2Mg \quad (60\text{–}80\%)$$

It is manufactured by hydrating vinylacetylene in the presence of dilute sulphuric acid and mercuric sulphate (cf. acetaldehyde):

$$CH_2 = CHC \equiv CH + H_2O \longrightarrow CH_2= CHCOCH_3$$

Methyl vinyl ketone polymerises on standing and is"used commercially as the starting material for plastics.

Mesityl oxide (4-methylpent-3-en-2-one), b.p. 130°C, may be prepared by distilling diacetone cohol (*q.v.*) with a trace of iodine:

$$(CH_3)_2C(OH)C_2COCH_3 \xrightarrow{-H_2O} (CH_3)_2 = CHCOCH_3 \quad (90\%)$$

Stross *et al.* (1947) have shown that mesityl oxide prepared this way is a mixture of mesityl oxide and isomesityl oxide, $CH_2 = C(CH_3)CH_2$ $COCH_3$.

Mesityl oxide may also be prepared as follows:

$$(CH_3)_2C = CH_2 + CH_3COCl \xrightarrow{ZnCl_2} (CH_3)_2CClCH_2COCH_3$$

$$\xrightarrow{-HCl} (CH_3)_2C = CHCOCH_3$$

It is used as a solvent for oils, gums, etc.

Phorone (2,6-dimethylhepta-2,5-dien-4-one), m.p. 28°C, may be prepared by the action of hydrochloric acid on acetone:

$$3CH_3COCH_3 \xrightarrow{HCl} (CH_3)_2C = CHCOCH = C(CH_3)_2 + 2H_2O$$

Isophorone (3,5,5-tnmethylcyclohex-2enone), b.p. 215°C, is formed when acetone is distilled will aqueous sodium hydroxide under pressure.

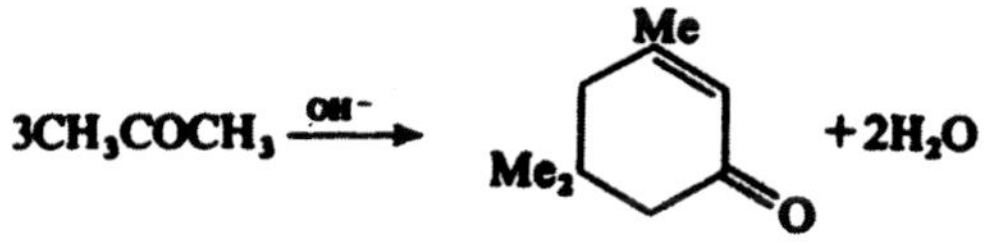

$$3CH_3COCH_3 \xrightarrow{OH^-} \quad \text{Me} \quad \text{Me}_2 \quad \text{O} \quad + 2H_2O$$

Fig. 2.5

It is used as a solvent.

Catalytic reduction and reduction wrth metal hydrides of α, β-unsaturatedketones have been described on. Reduction with metal and acid, or better, with magnesium amalgam, gives bimolecular products (cf. acraldehyde, above):

$$2(CH_3)_2C{=}CHCOCH_3 \xrightarrow{Mg/Hg} \begin{array}{c}(CH_3)_2CCH_2COCH_3\\ |\\ (CH_3)_2CCH_2COCH_3\end{array}$$

The addition of halogen to α, β-unsaturated ketones occurs by the normal electrophilic mechanism across the 3,4-positions. The addition of halogen acids, however, occurs by nucleophilic attack by the 1,4-mechanism.

The α,β-unsaturated ketones add on ammonia, primary amines and secondary amines to form β-amino-compounds with ammonia, mesityl oxide forms diacetonamine and phorone forms triacetonamine:

$$(CH_3)_2C{=}CHCOCH_3 + NH_3 \longrightarrow (CH_3)_2C(NH_2)CH_2COCH_3$$

$$(CH_3)_2C{=}CHCOCH{=}C(CH_3)_2 + 2NH_3 \longrightarrow$$

$$\left[(CH_3)_2C(NH_2)CH_2COCH_2C(NH_2)(CH_3)_2\right] \xrightarrow{-NH_3} OC\begin{matrix} CH_2{-}C(CH_3)_2 \\ \quad NH \\ CH_2{-}C(CH_3)_2 \end{matrix}$$

All the evidence is in favour of a 1,4-addition mechanism when the addendum is of the type H—Z, the product being an enol which then tautomerises to the stable keto form, the overall result being 3,4-addition, *e.g.*, with halogen acids:

$$R^1{-}\overset{4}{C}H{=}\overset{3}{C}H{-}\overset{2}{C}R^2{=}\overset{1}{O} \xrightarrow{HX} X^- + R^1{-}CH{=}CH{-}CR^2{=}\overset{+}{O}H$$

$$\longleftrightarrow R^1\overset{+}{C}H{-}CH{=}CR^2{-}OH$$

$$\xrightarrow{X^-} R^1{-}CHX{-}CH{=}CR^2{-}O{-}H$$

$$\longrightarrow R^1{-}CHX{-}CH_2{-}CR^2{=}O$$

The 1,4-addition intermediate is the kinetically controlled product and the 3,4-product is the thermodynamically controlled product.

Similarly, α, β-unsaturated ketones add on hydrogen cyanide to form β-cyano-compounds $R^1CH(CN)CH_2COR^2$; and sodium hydrogen sulphite to form β-sulphonic acids. $R^1CH(SO_3Na)CH_2COR^2$. The reaction with hydroxylamine usually results in a mixture of the oxime. $R^1CH = CHC(=NOH)R^2$ and the β-hydroxylaminoketone, $R^1CH(NHOH)$-CH_2COR^2 Trialkylboranes also undergo 1,4-addition, *e.g.*,

$$R_1^3B + CH_2 = CHCOR^2 \longrightarrow R_1CH_2CH = CR_2OBR_2^1$$

$$\xrightarrow{H_2O} R^1CH_2CH_2COR^2$$

A number of factors affect the direction of these addition reactions, an important one being steric hindrance. Steric hindrance at the carbonyl group favours 1,4-addition, whereas steric hindrance at the 3- and/or 4-positions favours 1,2-addition.

α,β-Unsaturated ketones (and aldehydes) are more difficult to epoxidise than alkenes, but are epoxidised by alkaline hydrogen peroxide.

One of the most important reactions that α,β-unsaturated ketones undergo is the Michae **condensation** (1887). This is the addition reaction between an α, β-unsaturated keto-compound and a compound with an active methylene group, *e.g.*, malonic ester, acetoacetic ester, etc. The condensation is carried out in the presence of a base, *e.g.*, sodium ethoxide, or secondary amines. The mechanism of the reaction is believed to be (using malonic ester and mesityl oxide as an example):

$$CH_2(CO_2Et)_2 + EtO^- \rightleftharpoons \bar{C}H(CO_2Et)_2 + EtOH$$

$$Me_2C{=}CH{-}\overset{\overset{\Large O}{\|}}{C}Me + \bar{C}H(CO_2Et)_2 \rightleftharpoons Me_2\underset{\underset{\large CH(CO_2Et)_2}{|}}{C}{-}CH{=}\overset{\overset{\large O^-}{|}}{C}Me \xrightleftharpoons{EtOH}$$

$$Me_2\underset{\underset{\large CH(CO_2Et)_2}{|}}{C}{-}CH{=}\overset{\overset{\large O-H}{|}}{C}Me + EtO^- \rightleftharpoons Me_2\underset{\underset{\large CH(CO_2Et)_2}{|}}{C}{-}CH_2{-}CH_2COMe + EtO^-$$

With this example, after the Michael condensation has occurred, it is followed by an internal Claisen condensation and leads to the formation of dimedone.

$$Me_2\underset{\underset{\large CH(CO_2Et)_2}{|}}{C}{-}CH_2COMe \xrightarrow{EtO^-} \begin{array}{c} CH_2 \\ / \quad \backslash \\ Me_2C \qquad CO \\ | \qquad\quad | \\ CH \qquad \bar{C}H_2 \\ / \quad \backslash \\ EtO_2C \qquad \underset{\underset{\large O}{\|}}{C}{-}OEt \end{array} \xrightarrow{-EtO^-}$$

$$Me_2C\begin{array}{l} \diagup CH_2{-}CO \diagdown \\ \diagdown \underset{\underset{\large CO_2{-}Et}{|}}{CH}{-}CO \diagup \end{array}CH_2 \xrightarrow[\text{(ii) } H^+]{\text{(i) KOH}} Me_2C\begin{array}{l} \diagup CH_2{-}CO \diagdown \\ \diagdown CH_2{-}CO \diagup \end{array}CH_2$$

The description of the Michael condensation given above is a limited one. The term now includes a large variety of addenda and acceptors, the a, unsaturated compounds may be ketones, esters, cyanides, etc., and the addenda may be malonic ester, E.A.A., cyanoacetic ester, nitroalkanes, etc.

The Michael condensation, as shown above, is reversible. Thus a Michael *retrogression, i.e.*, reversal, may occur to produce compounds different from the starting compounds, *e.g.* starting with benzylidenemalonic ester and E.A.A. in the presence of a large amount of sodiun ethoxide results in a mixture containing the reactants, the Michael condensation product, *and* benzylideneacetoacetate and malonic ester (produced by the retrogression).

If the reaction is carried out with the last two compounds as starting materials, the same mixture is obtained. Thus:

$$PhCH{=}C(CO_2Et)_2 + MeCOCH_2CO_2Et$$

$$\underset{}{\overset{EtO^-Na^+}{\rightleftharpoons}} \begin{array}{c} PhCHCH(CO_2Et)_2 \\ | \\ MeCOCHCO_2Et \end{array} \overset{EtO^-Na^+}{\rightleftharpoons}$$

$$\begin{array}{l} PhCH{-}\bar{C}H(CO_2Et)_2 \\ | \\ MeCO\,CCO_2Et \end{array} \rightleftharpoons \begin{array}{c} PhCH \\ \| \\ MeCOCCO^2Et \end{array} + \begin{array}{c} \bar{C}H(CO_2Et)_2 \\ \downarrow\uparrow EtOH \\ CH_2(CO_2Et)_2 + EtO^- \end{array}$$

Unsaturated Monocarboxylic Acids

Nomenclature

The longest carbon chain containing the carboxyl group and the double bond is chosen, and the position of the double bond with respect to the carboxyl group is indicated by number, *e.g.*,

$CH_3CH = CHCH_2CC_2H$ pent-3-enoic acid

Alternatively, the acid is named as a substitution product of the alkene, *e.g.*,

$CH_3CH = CHCC_2H$ prop-l-ene-l-carboxylic acid

In practice the unsaturated acids are usually known by their common names.

Many methods are available for preparing unsaturated monocarboxylic acids, but each one depends on the position of the double bond in the acid. A general method for preparing α, β-unsaturated acids is by the **Knoevenagel reaction** (1898). This is the reaction between aldehydes and compounds with active methylene groups in the presence of an organic base. The reaction may stop at stage:

(i) or proceed to stage

(ii) via a Michael condensation involving the initial product.

$$\text{(i) } RCHO + CH_2(CO_2C_2H_5)_2 \xrightarrow{\text{base}} RCH = C(CO_2C_2H5)_2 + H_2O$$

$$\text{(ii) } RCH{=}C(CO_2C_2H_5)_2 + CH_2(CO_2C_2H_5)$$

$$\xrightarrow{\text{base}} RCH\begin{matrix} \diagup CH(CO_2C_2H_5)_2 \\ \\ \diagdown CH(CO_2C_2H_5)_2 \end{matrix} + H_2O$$

Reaction (i) is favoured by using equivalent amounts of aldehyde and ethyl malonate in the presence of pyridine (Doebner, 1900). Reaction (ii) is t~voured by using excess of ethyl malonate in the presence of piperidine, and when the aldehyde is aliphatic. Furthermore, it appears that the term Knoevenagel reaction is taken to mean the condensation when unsaturated compounds are produced. Obviously then, to prepare α, β-unsaturated acids, (i) must be used, followed by hydrolysis and heating.

$$RCHO + CH_2(CO_2C_2H_5)_2 \xrightarrow{\text{pyridine}} RCH = C\ (CO_2C_2H_5)_2$$

$$\xrightarrow[\text{(ii) HCl}]{\text{(i) KOH}} RCH = C(CO_2H)_2 \xrightarrow{150\text{–}200°C}$$

$$RCH = CHCO_2H + CO_2$$

In practice, it is usually sufficient to treat the aldehyde with malonic acid in the presence of pyridine; *e.g.*, acetaldehyde gives crotonic acid:

$$CH_3CHO + CH_2(CO_2H)_2 \xrightarrow{\text{pyridine}} CH_3CH = CHCO_2H + CO_2 + H_2O \quad (60\%)$$

The mechanism of this reaction has been the subject of much discussion. When a *tertiary* base, *e.g.*, pyridine, is used as catalyst, then the mechanism is believed to be similar to that of the aldol condensation:

$$CH_2(CO_2Et)_2 + B \rightleftharpoons BH^+ + :\bar{C}H(CO_2Et)_2$$

$$\xrightarrow{RCH=O} RCH(O^-)CH(CO_2Et)_2 \overset{BH^-}{\rightleftharpoons}$$

$$B + RCH(OH)CH(CO_2Et)_2 \rightleftharpoons BH^+ + RCH(OH)\text{—}\bar{C}(CO_2Et)_2$$

$$\longrightarrow RCH{=}C(CO_2Et_2) + OH^-$$

When a primary or secondary base is used as catalyst, the mechanism may still possibly be as given above, but a complication arises from the fact that carbonyl compounds can form addition products with such bases and these addition products may therefore be intermediates in the reaction.

Ethyl malonate condenses with aldehydes only. On the other hand, ethyl cyanoacetate condenses with ketones in the presence of an amino-acid (as catalyst) such as 3-aminopropionic acid; *e.g.*, acetone forms isopropylidenecyanoacetic ester:

$$(CH_3)_2CO + CH_2(CN)CO_2C_2H_5 \xrightarrow{-H_2O} (CH_3)_2C{=}C(CN)CO_2C_2H_5$$

Since cyano-compounds are readily hydrolysed to the corresponding acid, the above condensation may be used to prepare α, β-unsaturated acids of the type $R_2C{=}CHCC_2H$

α, β-Unsaturated acids may also be prepared by heating α-bromo-acids with ethanolic potassium hydroxide, or better still, with potassium t-butoxide (Cason *et al.*, 1953); or by heating β-hydroxyacids with aqueous sodium hydroxide:

$$RCH_2CHBrCO_2H + KOH \xrightarrow{\text{ethanol}} RCH = CHCO_2H + KBr + H_2O$$

$$RCHOHCH_2CO_2H \xrightarrow{\text{NaOH}} RCH = CHCO_2H + H_2O$$

β, γ-Unsaturated acids may be prepared by heating an aldehyde with sodium succinate and acetic anhydride at 100°C. The product is a γ-alkylparaconic acid, and when this is heated at 150°C, it eliminates carbon dioxide to form the β, γ-unsaturated acid (Fittig *et al.*, 1885). This is a variation of the Perkin reaction, and succinic anhydride (formed by interaction of sodium succinate and acetic anhydride) is probably the reacting species.

$$RCHO + \begin{matrix} CH_2CO \\ | \quad\quad \searrow \\ | \quad\quad\quad O \\ | \quad\quad \nearrow \\ CH_2CO \end{matrix} \longrightarrow \left[\begin{matrix} RCH—CHCO \\ | \quad\quad | \quad \searrow \\ | \quad\quad | \quad\quad O \\ | \quad\quad | \quad \nearrow \\ OH \quad CH_2CO \end{matrix} \right]$$

$$\xrightarrow{-H_2O} \left[\begin{matrix} RCH{=}C—CO \\ | \quad\quad \searrow \\ | \quad\quad\quad O \\ | \quad\quad \nearrow \\ CH_2—CO \end{matrix} \right]$$

$$\left[\begin{matrix} R—CH{=}C—CO_2H \\ H \quad | \\ O \quad\quad CH_2 \\ \searrow \quad \nearrow \\ C \\ \| \\ O \end{matrix} \right] \xrightarrow{\text{lactonises}} \begin{matrix} RCH—CHCO_2H \\ | \quad\quad\quad | \\ O \quad\quad CH_2 \\ \searrow \quad \nearrow \\ C \\ \| \\ O \end{matrix} \xrightarrow[150^\circ C]{2 \text{ mol.}}$$

$$RCH{=}CHCH_2CO_2H + \begin{matrix} RCH—CH_2 \\ | \quad\quad\quad | \\ O \quad\quad CH_2 \\ \searrow \quad \nearrow \\ C \\ \| \\ O \end{matrix} + CO_2$$

Unsaturated alcohols may be oxidised to the corresponding unsaturated acid, provided the double bond is protected, *e.g.*, acrylic acid from allyl alcohol.

Oxidation of unsaturated aldehydes produces unsaturated acids. The oxidation, however, cannot be carried out with the usual oxidising agents, such as acid or alkaline permanganate, acid dichromate, etc., since these will attack the double bond. A useful oxidising agent for unsaturated aldehydes is ammoniacal silver nitrate; *e.g.*, crotonaldehyde is oxidised to crotonic acid:

$$CH_3CH = CHCHO + [O] \xrightarrow[AgNO_3]{\text{ammoniacal}} CH_3CH = CHCO_2H$$

Unsaturated methyl ketones may be oxidised to unsaturated acids by sodium hypochlorite (cf. the haloform reaction), *e.g.*,

$$RCH = CHCOCH_3 \xrightarrow{NaOCl} CH_3CH = CHCO_2H$$

The carbonyl (C = O stretch) infra-red absorption band in α, β-unsaturated acids occurs in the region 1715-1690 cm^{-1} (s). This readily, distinguishes these compounds from α, β- unsaturated ketones, but is not so clear cut with α, β-unsaturated aldehydes, and may also lead to confusion with the saturated monocarboxylic acids.

However, the absorption region of the double bond (C = C stretch), provided it is not conjugated, is 1650-1630 cm^{-1} (v). Thus, although this does not indicate its position, it does show that a double bond is present. The molecular formula should distinguish unsaturated acids from the saturated acids and from α, β-unsaturated aldehydes (and ketones) [N.M.R. spectra].

General Reactions of the Unsaturated Acids

The esters of α, β-unsaturated acids undergo the same addition reactions (including the Michael condensation) at the double bond as the α, β-unsaturated ketones. When reduced by the Bouveault-Blanc method or with a large excess of sodium borohydride in methanoi, α, β-unsaturated esters are converter into the corresponding saturated alcohol:

$$CH_3CH = CHCO_2C_2H_5 \xrightarrow{e;H^+} CH_3CH_2CH_2OH$$

Lithium aluminium hydride, however, generally reduces these esters to the corresponding *unsaturated* alcohols. Esters in which the double bond is further removed from the ester group are reduced to the corresponding unsaturated alcohol; *e.g.*, butyl oleate is reduced to oleyl alcohol:

$$CH_3(CH_2)_7CH = CH(CH_2)_7CO_2CO_4H_9$$

$$\xrightarrow{Na/C_4H_9OH} CH_3(CH_2)_7CH = CH(CH_2)_7CH_2OH \ (82–84\%)$$

All the unsaturated acids may be reduced catalytically to saturated acids.

α, β- and α, γ- unsaturated acids add on halogen acid, the halogen atom becoming attached to the unsaturated carbon atom which is further from the carboxyl group:

$$CH_2 = CHCO_2H + HCl \longrightarrow ClCH_2CH_2CO_2H$$

$$CH_3CH = CHCH_2CO_2H + HBr \longrightarrow CH_3CHBrCH_2CH_2CO_2H$$

This mode of addition (which may be contrary to MarkownikofTs rule, as in the case of acrylic acid) may be ascribed to the inductive effect of the carboxyl group (cf. allyl alcohol, etc.).

On the other hand, addition of halogen acid to γ, δ-unsaturated acids of the type $CH_2 = CHCH_2CH_2CO_2H$ takes place in accordance with Markownikoff's rule. This must be due to the fact that the inductive effect of the carboxyl group ceases to be felt beyond the β-carbon atom.

When unsaturated acids are boiled with alkali, the double bond tends to move so as to form the α, β-unsaturated acid:

$$CH_3CH = CHCH_2CO_2H \xrightarrow{NaOH} CH_3CH_2CH = CHCO_2H$$

An interesting example of this migration of the double bond is the hydrolysis of allyl cyanide with boiling alkali to produce crotonic acid. But-3-enoic acid is probably formed first, and this then rearranges to crotonic acid:

$$CH_2 = CHCH_2Br \xrightarrow{CuCN} CH_2 = CHCH_2CN \xrightarrow{NaOH}$$
$$[CH_2 = CHCH_2CO_2H] \xrightarrow{NaOH} CH_3CH = CHCO_2H$$

The movement of a double bond to the α, β-position shows that the latter acid is thermodynamically more stable, *e.g.*, Linstead *et al.* (1943) showed that the migration is reversible and that th(equilibrium between the sodium salts of vinylacetic acid and crotonic acid is 98 per cent on the side of the latter

$$CH_2 = CHCH_2CO_2^- \rightleftharpoons CH_3CH = CHCO_2^-$$

The mechanism of this shift is believed to be:

$$CH_2 = CHCH_2CO_2^- \underset{}{\overset{OH^-;-H^+}{\rightleftharpoons}} CH_2 = CH\bar{C}HCO_2^-$$
$$\longleftrightarrow \bar{C}H_2CH = CO_2^- \overset{H^+}{\rightleftharpoons} CH_3CH = CHCO_2^-$$

When unsaturated acids are fused with alkali, cleavage of the chain takes place with the formation of two acids/one of which-is always acetic acid. This again indicates the migration of the double bond to the α, β-position *e.g.*, oleic acid gives acetic and palmitic acids:

$$CH_3(CH_2)_7CH = CH(CH_2)_7CO_2H \xrightarrow{KOH} [CH_3(CH_2)_{14}CH = CHCO_2H]$$

$$\xrightarrow{KOH} CH_3(CH_2)_{14}CO_2H + CH_3CO_2H$$

α, β-Unsaturated esters add on aliphatic diazo-compounds to form pyrazoline derivatives; *e.g.*. acrylic ester reacts with diazomethane to give pyrazoline-3-carboxylic ester:

$$CH_2{=}CHCO_2C_2H_5 + CH_2N_2 \longrightarrow \begin{array}{l} CH_2\text{——}CHCO_2C_2H_2 \\ \;| \qquad\quad | \\ HC \qquad NH \\ \;\; \diagdown\!\!\diagdown \; N \; \diagup \end{array}$$

α, β-Unsaturated acids may be degraded by the Hofmann method using a modified procedure. The α, β-unsaturated acid amide in methanol is treated with an alkaline solution of sodium hypochlorite, the urethan produced being hydrolysed in acid solution to give the aldehyde in good yield:

$$RCH = CHCONH_2 \xrightarrow[MeOH]{NaOCl} RCH = CHNHCO_2Me$$

$$\xrightarrow{H_3O^+} RCH = CHNH_2 \rightleftharpoons RCH_2CH = NH \xrightarrow{H_3O^+} RCH_2CO$$

On the other hand, β, γ- and γ, δ-unsaturated acid amides produce the corresponding unsaturated primary amine, but in poor yield:

$$RCH = CHCH_2CONH_2 \longrightarrow RCH = CHCH_2NH_2$$

Baizer *et al.* (1964) have shown that α, β-unsaturated acids may be *hydrodimerised* electrochemically in faintly aqueous alkaline solution ($X = CO_2R$, CN, etc.)

$$2\;{>}C{=}\overset{|}{C}{-}X \xrightarrow[2H_2O]{2e} X{-}\underset{H}{\overset{|}{\underset{|}{C}}}{-}\overset{|}{\underset{|}{C}}{-}\overset{|}{\underset{|}{C}}{-}\underset{H}{\overset{|}{\underset{|}{C}}}{-}X$$

This reaction can be carried out with two different substrates and so offers a method of preparing variety of 1,4-bifunctional compounds.

γ, δ- and δ, ε-Unsaturated acids form lactones in the presence of concentrated sulphuric acid.

Acrylic acid (*prop-2-enoic acid*), b.p. 141°C, may be prepared:

(i) By the oxidation of allyl alcohol or acraldehyde.

(ii) By heating β-hydroxypropionic acid with aqueous sodium hydroxide:

$$HOCH_2CH_2CO_2H \xrightarrow{NaOH} CH_2 = CHCO_2H + H_2O$$

Alternatively, ethylene cyanohydrin may be heated with sulphuric acid:

$$HOCH_2CH_2CN \xrightarrow{H_2SO_4} HOCH_2C_2CO_2H \xrightarrow{-H_2O} CH_2 = CHCO_2H$$

Vinyl cyanide also gives acrylic acid on hydrolysis

$$CH_2 = CHCN \xrightarrow{H^+} CH_2 = CHCO_2H$$

Acrylic acid is oreoared industrially in manv wavs. and so are its esters, *e.g.*

(i) $CH_3CO_2 \xrightarrow{\text{pyrolysis}} CH_2{=}C{=}O \xrightarrow{CH_2O}$

$$\begin{matrix} CH_2-C{=}O \\ | \quad\;\; | \\ CH_2-O \end{matrix} \longrightarrow \begin{cases} \xrightarrow{H_3PO_4} CH_2{=}CHCO_2H \\ \xrightarrow[H_2PO_4]{ROH} CH_2{=}CHCO_2R \end{cases}$$

(ii)

$$\overset{\;O\;}{CH_2-CH_2} \xrightarrow[Ni(CO)_4]{CO} CH_2{=}CHCO_2H \xrightarrow{ROH} CH_2{=}CHCO_2R$$

$$\text{(epoxide)} \longrightarrow HOCH_2CH_2CN \longrightarrow CH_2{=}CHCO_2R$$

Acrylic acid, on standing, slowly polymenses to polyacrylic acid, which is a solid.

Methacrylic acid (2-methylprop-2-enoic acid), m.p. 15°C, may be prepared by removing a molecule of hydrogen bromide from α-bromoisobutyric acid:

$$CH_3\overset{\overset{CH_3}{|}}{C}BrCO_2H \xrightarrow[-HBr]{\text{ethanolic KOH}} CH_2{=}\overset{\overset{CH_3}{|}}{C}CO_2H$$

Methyl methacrylate is very important commercially, since it polymerises to polymethyl methacrylate under the influence of heat. This polymer is tough, transparent, and can be moulded; one of its trade names its *perspex*. One industrial method of preparing methyl methacrviate is as follows:

$$(CH_3)_2CO \xrightarrow{HCN} (CH_3)(OH)C(CN)$$

$$\xrightarrow[H_2SO_4]{CH_3OH} CH_2{=}\overset{\overset{CH_3}{|}}{C}CO_2CH_3 + NH_4HSO_4$$

Methyl acrviate also Holvmerises. but this Dolvmer is softer than ihat from the methacrviate.

Crotonic acid and **isocrotonic acid** are geometrical isomers, crotonic acid, the trans isomer, and isocrotonic the cis:

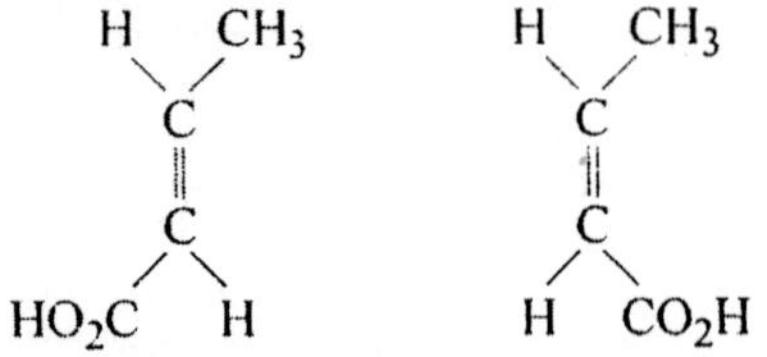

$$\begin{array}{cc} \text{H} \quad \text{CH}_3 & \text{H} \quad \text{CH}_3 \\ \diagdown\,\diagup & \diagdown\,\diagup \\ \text{C} & \text{C} \\ \| & \| \\ \text{C} & \text{C} \\ \diagup\,\diagdown & \diagup\,\diagdown \\ \text{HO}_2\text{C} \quad \text{H} & \text{H} \quad \text{CO}_2\text{H} \end{array}$$

crotonic acid   isocrotonic acid

The more stable form is crotonic acid, m.p. 72°C. Isocrotonic acid, m.p. 15°C, slowly changes into crotonic acid when heated at 100°C.

Crotonic acid (*trans*-but-2-enoic acid) may be prepared by oxidising crotonaldehyde with ammoniacal silver nitrate; by heating β-hydroxybutyric acid with sodium hydroxide; or by the Knoevenagel reaction:

$$CH_3CHO+CH_2(CO_2H)_2 \xrightarrow[20°C]{\text{pyridine}} CH_3CH{=}CHCO_2H+CO_2+2O \ (60\%)$$

Isocrotonic acid (*cis*-but-2-enoic acid) may be prepared by the action of sodium amalgam on β-chloroisocrotonic ester which is obtained from acetoacetic ester by reaction with phosphorus pentachloride:

$$CH_3COCH_2CO_2C_2H_5 \xrightarrow{PCl_5} CH_3CCl_2CH_2CO_2C_2H_5 \xrightarrow{-HCl}$$

$$CH_3CCl = CHCO_2C_2H \xrightarrow{Na/Hg} CH_3CH = CHCO_2C_2H_5$$

Angelic and **tiglic acids** are geometrical isomers with the structure $CH_3CH = C\ (CH_3)\ CO_2H$ (2-*methylbut*-2-*enoic acid*):

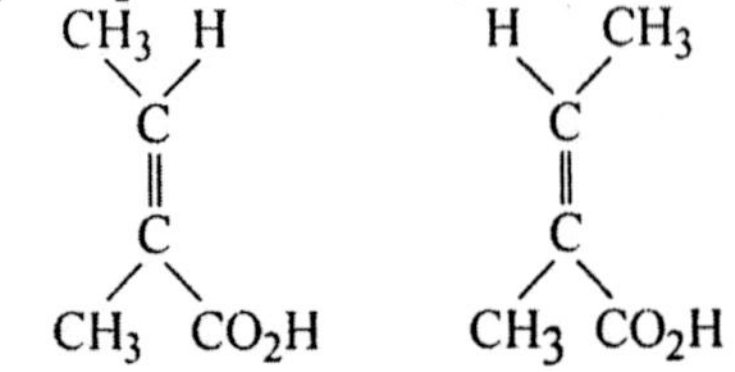

$$\begin{array}{cc} \text{CH}_3 \quad \text{H} & \text{H} \quad \text{CH}_3 \\ \diagdown\,\diagup & \diagdown\,\diagup \\ \text{C} & \text{C} \\ \| & \| \\ \text{C} & \text{C} \\ \diagup\,\diagdown & \diagup\,\diagdown \\ \text{CH}_3 \quad \text{CO}_2\text{H} & \text{CH}_3 \quad \text{CO}_2\text{H} \end{array}$$

tiglic acid, m.p. 64° C   isocrotonic acid

Undecylenic acid (*dec*-9-*ene*-1-*carboxylic acid*), m.p. 24-5°C, may be obtained by the destructive distillation of ricinoleic acid (11-*hydroxyheptadec*-8-*ene*-1-*carboxy tic acid*), which occurs as the glyceride ester in castor oil. Heptanal is the other product, 10 per cent yield of each being obtained:

$$CH_3(CH_2)_5CHOHCH_2CH = CH(CH_2)_7CO_2H$$

$$\longrightarrow CH_3(CH_2)_5CHO + CH_2 = CH(CH_2)_8CO_2H$$

When ricinoleic acid is heated with sodium hydroxide in air, octan-2-ol and sebacic acid are obtained:

$$CH_3(CH_2)_5CHOHCH_2CH = CH(CH_2)_7CO_2H$$

$$\xrightarrow[\text{air}]{\text{NaOH}} CH_3(CH_2)_5CHOHCH_3 + HO_2C(CH_2)_8CO_2H$$

When. ricinoleic acid (castor oil) is treated with concentrated sulphuric acid, it gives a complex mixture consisting of the hydrogen sulphate of ricinoleic acid in which the hydroxyl group is esterified, and a compound in which the sulphuric acid has added to the double bond: esterification and addition do not occur together in the same molecule of ricinoleic acid. The product, which is known as *Turkey-red oil*, is used as a wetting agent.

Oleic acid: *Theptadec-8-ene-1-caroxylic acid*), m.p. 16° C, occurs as the glyceryl ester in oils and fats. CO_2H reduction coverts it into stearic acid. Cold dilute alkaline permanganate, or hydrogen proxide in acetic acid, converts oleic acid into 9, 10-dihydroxystearic acid, which, on oxidation of periodic acid or lead tetra-acetate, gives nonanal (perlargonic aldehyde) and azelaic half aldehyde. These products show that the position of the double bond in oleic acid is 9, 10. The same products are also obtained on ozonolysis of oleic acid.

$$Me(CH_2)_7CH = CH(CH_2)_7CO_2H \xrightarrow[\text{AcOH}]{H_2O_2}$$

$$Me(CH_2)_7CHOHCHOH(CH_2)_7CO_2H$$

$$\xrightarrow{HIO_4} Me(CH_2)_7CHO + OHC\ (CH_2)_7CO_2H$$

Oleic acid is the *cis* form. The *trans* form is **elaicdic acid**, m.p. 51°C, which may be obtained by the action of nitrous acid on oleic acid:

$$\begin{array}{ccc} H\quad (CH_2)_7CH_3 & & CH_3(CH_2)_7\quad H \\ \diagdown\ \diagup & & \diagdown\ \diagup \\ C & & C \\ \| & \xrightarrow{HNO_2} & \| \\ C & & C \\ \diagup\ \diagdown & & \diagup\ \diagdown \\ H\quad (CH_2)_7CO_2H & & H\quad (CH_2)_7CO_2H \end{array}$$

Linoloic acid (*heptadeca*-8, 11-*diene*-1-*carboxylic acid*), $CH_3\,(CH_2)_4$ $CH = CHCH_2CH = CH(CH_2)_7$COaH, b.p. 228°C/14 mm, occurs as the glyceryl ester in linseed oil, hemp oil, etc.; it has been synthesised by Raphael *et al*. (1950).

Linolemc acid (heptadeca-8, 11, 14-triene-1-carboxylic acid), CH, $CH_2CH = CHCH_2CH = CHCH_2CH =CH(CH_2)_7CO_2H$, is a liquid which occurs as the glyceryl ester in, *e.g.*, poppy-seed oil. Oils containing linoleic and linolenic acids are drying oils. Linolenic acid is the most abundant and widespread triene acid in Nature; it has been synthqised by Weedon *et al.* (1956).

Eruic acid, m.p. 34°C, is the cis form of *heneicos*-12-ene-1-carboxylic acid, $CH_3(CH_2)_7CH = CH(CH_2)_{11}CO_2H$. It occurs as the glyceryl ester in various oils, *e.g.*, rape oil, cod-liver oil, etc. It is converted into the trans isomer brassidic acid, m.p. 61-5°C, by the action of nitrous acid.

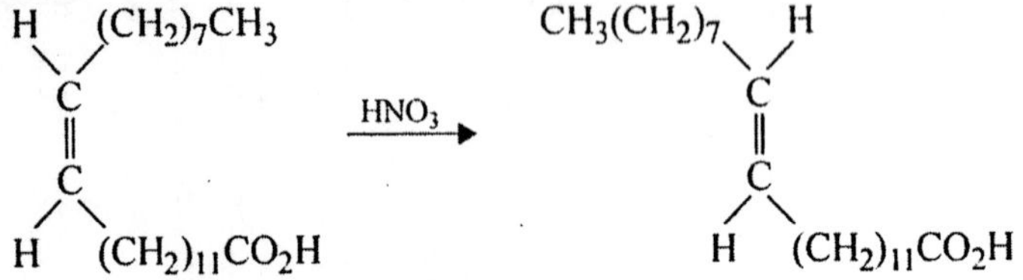

$$\begin{array}{c} H \quad (CH_2)_7CH_3 \\ \diagdown \diagup \\ C \\ \| \\ C \\ \diagup \diagdown \\ H \quad (CH_2)_{11}CO_2H \end{array} \xrightarrow{HNO_3} \begin{array}{c} CH_3(CH_2)_7 \quad H \\ \diagdown \diagup \\ C \\ \| \\ C \\ \diagup \diagdown \\ H \quad (CH_2)_{11}CO_2H \end{array}$$

Nervonic acid (*tricos-14-ene-1-carboxylic acid*), $CH_3(CH_2)_7CH = CH(CH_2)_{13}\ CO_2H$, occurs in human brain-tissue and in fish oils. Both the cis and trans forms are known, but it appears to be the cis acid, m.p. 43°C, which occurs naturally.

ArachMonic acid (*nonadeca*-4, 7, 10; 13 *retraene-1-carboxylic acid*), $CH_3(CH_2)_3$, $(CH_2CH = CH_2)_4CH_2)CO_2H$, has been isolated from phosphatides.

Propiolic acid (*proparylic acid, prop-2-yonic acid*), m.p. 9°C, is the simplest acetylenic carboxylic acid. It is conveniently prepared by the action of dry carbon dioxide on sodium acetylide:

$$CH \equiv CNa \xrightarrow{CO_2} CH \equiv CCO_2Na \xrightarrow[H_2SO_4]{\text{dilute}} CH \equiv CCO_2H$$

It is reduced by sodium amalgam to propionic acid and it adds on halogen acid to give the β-halogeno-acrylic acid. It forms the silver or cuprous salt when treated with an ammoniacal solution of silver nitrate or cuprous chloride, respectively. When exposed to sunlight, it polymerises to *trimesic acid* (benzenel,3,5-tricarboxylic acid; *cf.* acetylene):

oxylic acid; *cf.* acetylene):

$$3CH{\equiv}CCO_2H \longrightarrow C_6H_3(CO_2H)_3 \text{ (benzene ring with } CO_2H, HO_2C, CO_2H)$$

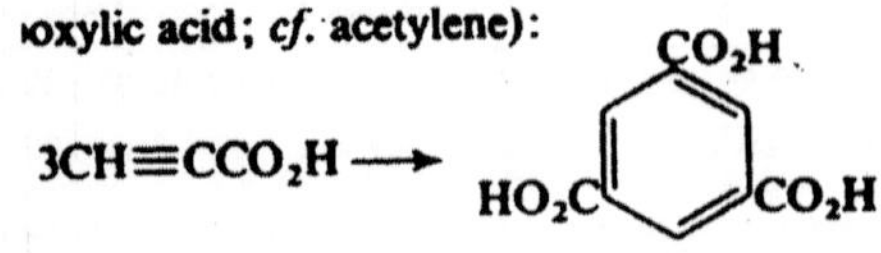

Fig. 2.6

α, β-Acetylenic acids are stronger acids than their ethylenic and saturated analogues due to the acetylene bond exerting a strong electron-attracting effect on the *sp* hybridised carbon atom.

The Principle of Vinylogy

According to Fuson (1935), the principle of vinylogy may be described as follows: If E_1 and E_2 represent non-metallic elements, then in the compound of the type $A—E_1 = E_2$ or $A—E_1 \equiv E_2$, if a structural unit of the type $\left(—\overset{|}{\underset{|}{C}}=\overset{|}{\underset{|}{C}}\right)_n$ is interposed between A and E_1, the function of E_1 remains qualitatively unchanged, but that of E_2 may be usurped by the carbon atom attached to A. Thus

$$A—\left(\underset{|}{C}=\underset{|}{C}\right)_n—E_1=E_2 \quad \text{or} \quad A—\left(\underset{|}{C}=\underset{|}{C}\right)_n—E_1\equiv E_2$$

form vinylogous series. This may be illustrated by the compounds ethyl acetate and ethyl crotonate. These are vinyloeues (*n* being equal to 1):

$$CH_3—\underset{\underset{OC_2H_5}{|}}{C}=O \qquad CH_3—(CH=CH)—\underset{\underset{OC_2H_5}{|}}{C}=O$$

A is equivalent to CH_3, and $E_1 = E_2$ to $—\underset{\underset{OC_2H_5}{|}}{C}=O$

Ethyl acetate condenses with ethyl oxalate to form *oxatacetic ester* (this, of course, is an example of the Claisen condensation):

$$C_2H_5OOCCOOC_2H_5 + CH_3COOC_2H_5 \xrightarrow[\text{ether}]{\text{Na}} C_2H_5OOCCOCH_2COOC_2H_5 + C_2H_5OH$$

In the same way, ethyl crotonate also condenses with ethyl oxalate:

$$C_2H_5OOCCOOC_2H_5 + CH_3(CH=CH)COOC_2H_5 \xrightarrow[\text{ether}]{\text{Na}}$$

$$C_2H_5OOCCOOCH_2(CH=CH)COOC_2H_5 + C_2H_5OH$$

The principle of vinylogy is readily understood in terms of the inductive effect, *e.g.*,

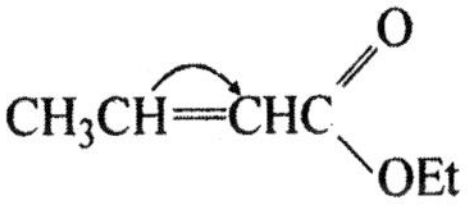

$$CH_3CH=CHC\begin{matrix} \nearrow O \\ \searrow OEt \end{matrix}$$

Thus the methyl group behaves like an active methylene group due to the ù1 effect of the carbethoxyl group, which is readily transmitted through the —CH = CH— group. At the same time, the –I effect is also assisted by the increased electronegativity of the sp^2 hybridised carbon atom adjacent to the methyl group.

Ketens (Ketenes)

Ketens are compounds which are characterised by the presence of the grouping $>C{=}C{=}O$. If the compound is of the type RCH = C = O, it is known as an *aldoketen*, and if $R_2C = C = O$, then a *ketoketen.*

Ketoketens are generally prepared by debrominating an α-bromoacyl bromide with zinc, *e.g.*, dimethyiketen from α-bromoisobutyryl bromide:

$$(CH_3)_2CBrCOBr + Zn \longrightarrow (CH_3)_2C = C = O + ZnBr_2$$

Aldoketens are usually prepared by refluxing an acid chloride in pyridine solution, *e.g.*,

$$CH_3CH_2COCl + C_5H_5N \longrightarrow CH_3CH = C = O + C_5H_5NH+ Cl^-$$

The simplest member of the keten series is **keten** (kefene), $CH_2 = C = O$, and may be prepared by debrominating bromoacetyl bromide with zinc:

$$CH_2BrCOBr + Zn \longrightarrow CH_2 = C = O + ZnBr_2$$

It is, however, usually prepared by the thermal decomposition of acetone (ethyl acetate or acetic anhydride may also be used as the starting material):

$$CH_3COCH_3 \longrightarrow CH_2 = C = CO + CH_4$$

The thermal decomposition of acetone has been shown to be a free-radical chain reaction.

Keten is a colourless, poisonous, pungent gas, b.p. –41°C, which oxidises in air to the unstable peroxide, the structure of which may be as shown. It rapidly polymerises to diketeh, and undergoes photochemical decomposition to give methylene.

```
CH2—C=O
|    |
|    |
O————O
```

Fig. 2.7(a)

The reactions of keten are generally those of an acid anhydride; it acetylates most compounds with an active hydrogen atom.

The mechanism of acetylations by 'keten can be generalised in terms of the addition of a nucleophile as the rate-determining step:

$$CH_2{=}C{=}O \quad H{-}Z \underset{}{\overset{\text{fast}}{\rightleftharpoons}} Z^- + CH_2{=}C{-}OH \xrightarrow{\text{slow}}$$

$$CH_2{=}C(OH)(Z) \xrightarrow{\text{fast}} CH_3{-}C({=}O){-}Z$$

(i) When keten is passed into water, acetic acid is formed slowly:

$$CH_2 = CO + H_2O \longrightarrow CH_3CO_2H$$

(ii) When keten is passed into glacial acetic acid, acetic anhydride is formed:

$$CH_2 = CO + CH_3CO_2H \longrightarrow (CH_3CO)_2O$$

By means of this reaction a mixed anhydride may be obtained, one group of which, of course must be acetyl:

$$RCO_2H + CH_2 = CO \longrightarrow RCOOCOCH_3$$

(iii) Keten reacts with aliphatic or aromatic hydroxy-compounds:

$$CH_2O = CO + ROH \longrightarrow CH_3CO_2R$$

The reaction, however, is best carried out in the presence of a catalyst, *e.g.*, sulphuric acid.

It also acetylates enolisable compounds, *e.g.*,

$$CH_3COCH_2CO_2C_2H_5 \xrightarrow[H_2SO_4]{CH_2=CO}$$

$$CH_3C(OCOCH_3){=}CHCO_2C_2H_5 + CH_2{=}C(OCOCH_3)CH_2CO_2C_2H_5$$

(iv) Keten reacts with ammonia to form acetamide, and with primary or secondary amines to form N-alkyl-acetamides

$$CH_3\ CONH_2 \xleftarrow{NH_3} CH_2 = CO \xrightarrow{RNH_2} CH_3CONHR$$

Primary amino-groups are acetylated extremely readily, and hence it is possible to acetylate this group selectively in compounds containing both an amino and hydroxyl group, *e.g.*, *p*-aminophenol

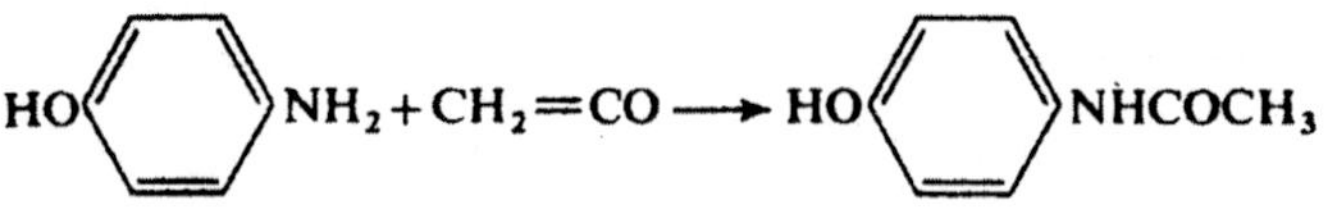

$$HOC_6H_4NH_2 + CH_2{=}CO \longrightarrow HOC_6H_4NHCOCH_3$$

Fig. 2.7(b)

A particularly useful reaction of this type is the acetylation of amino-acids:

$$RCH(NH_2)CO_2H + CH_2 = CO \longrightarrow RCH(NCHOCH_3)CO_2H$$

Amides are also acetylated by keten in the presence of sulphuric acid, but at elevated temperatures (in the absence of sulphuric acid) the cyanide is formed.

$$RCONH_2 + CH_2 = CO \longrightarrow RCN + CH_3CO_2H$$

(v) Thiols are acetylated (preferably in the presence of sulphuric acid as catalyst):

$$RSH + CH_2 = CO \longrightarrow RCOCH_3$$

(vi) Keten reacts with Grignard reagents to form methyl ketones:

$$RMgX + CH_2 = CO \longrightarrow RCOCH_3$$

In addition to behaving as an acetylating reagent, keten behaves as an unsaturated compound, *e.g.*,

(i) Keten adds on bromine to form bromoacetyl bromide:

$$CH_2 = CO + Br_2 \longrightarrow BrCH_2COBr$$

The halogeno-acetyl halide is also formed by the interaction of keten and phosphorus pentahalides:

$$CH_2 = CO + PCl_5 \longrightarrow ClCH_2COCl + PCl_3$$

(ii) Keten adds on halogen acid to form the acetyl halide:

$$CH_2 = CO + HX \longrightarrow CH_3COX$$

Alkyl halides also react with keten in the presence of charcoal at 100°C to form acid chlorides:

$$RX + CH_2 = CO \longrightarrow CH_2COX$$

This offers a means of stepping up a seriSs.

(iii) Keten reacts with aldehydes to form β-lactones; *e.g.*, with formaldehyde, β-*propiolactone* is formed:

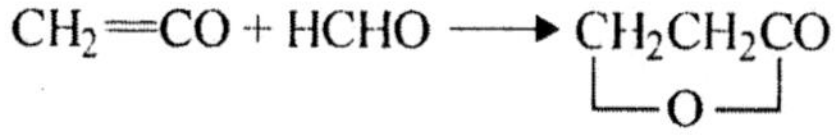

$$CH_2{=}CO + HCHO \longrightarrow \underset{\llcorner\!-O-\!\lrcorner}{CH_2CH_2CO}$$

Ketones also form β-lactones, but in this case a catalyst is usually required, *e.g.*, zinc chloride.

Diketen is readily formed by passing keten into acetone cooled in solid carbon dioxid (yield: 50-55 per cent). Diketen is a pungent-smelling liquid, b.p. 127°C, which, when strongly heated (550-600°C), is depolymerised to keten.

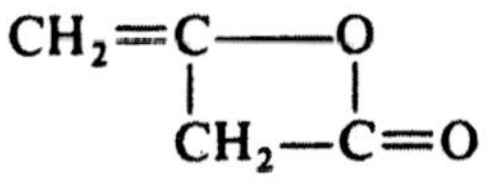

Various structures have been suggested for diketen, but based mainly on N.M.R. spectroscopy (which showed the presence of two methylene groups) and X-ray analysis, the structure accepted is that shown (vinylaceto-β-lactone). Diketen may be further polymerised to dehydroacetic acid.

$$\begin{array}{ccc} CH_2{=}C & - & O \\ | & & | \\ CH_2 & - & C{=}O \end{array}$$

vinylaceto-β-lactone

Diketen reacts with alcohols to form esters of acetoacetic acid, and this reaction is now being used to prepare acetoacetic ester industrially. It also reacts with primary amines to form N-substituted acetoacetamides, *e.g.*, with aniline:

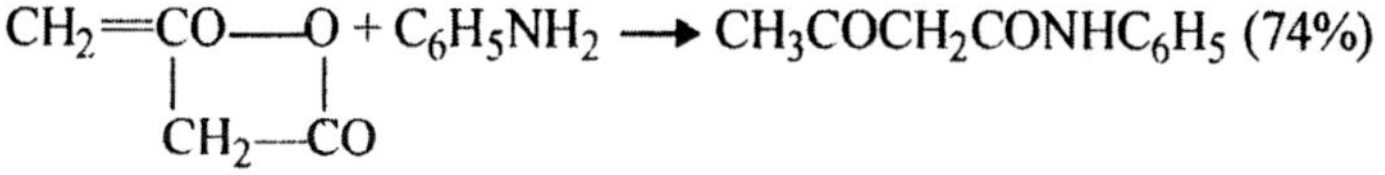

$$\begin{array}{ccc} CH_2{=}CO & - & O \\ | & & | \\ CH_2 & - & CO \end{array} + C_6H_5NH_2 \longrightarrow CH_3COCH_2CONHC_6H_5\ (74\%)$$

Diketen also undergoes the Friedel-Crafts reaction with benzene to form benzoylacetone

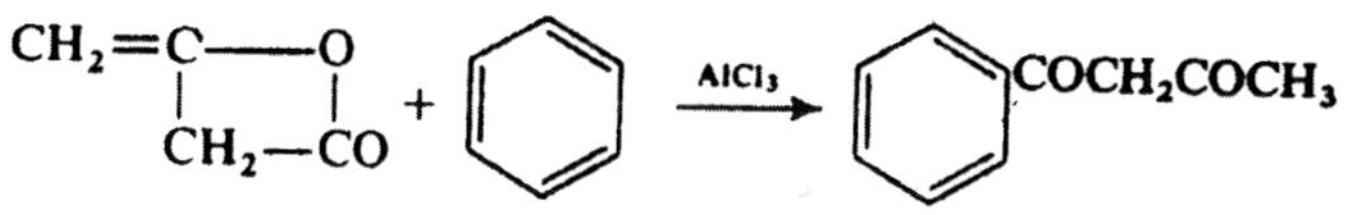

$$\begin{array}{ccc} CH_2{=}C & - & O \\ | & & | \\ CH_2 & - & CO \end{array} + C_6H_6 \xrightarrow{AlCl_3} C_6H_5COCH_2COCH_3$$

Fig. 2.8

MONOHYDRIC ALCOHOLS

An alcohol is a compound that contains one or more *hydroxyl* groups, *i.e.*, alcohols are hydroxy-derviatives of the alkanes. They are classified according to the number of hydroxyl groups present. Monohydric alcohols contain one hydroxyl group; dihydric, two; trihydric, three; etc. When the alcohols contain four or more hydroxyl groups, they are usually called polyhydric alcohols.

The monohydric alcohols from an homologous series with the general formula $C_nH_{2n+2}O$, but, since their functional group is the hydroxyl group, their general formula is more satisfactorily written as $C_nH_{2n+1}OH$ or ROH.

NOMENCLATURE

The simpler alcohols are commonly known by their trivial names, which are obtained by naming the alcohol as a derivative of the alkyl group attached to the hydroxyl group, *e.g.*, CH_3OH, methyl alcohol; $CH_3CH_2CH_2OH$. n-propyl alcohol; $CH_3CH(OH)CH_3$, isopropyl alcohol; $(CH_3)_3COH$, t-butyl alcohol.

Another system of nomenclature considers the alcohols as derivatives of methyl alcohol, which is named *carbinol* or *methanol*, *e.g.*, $CH_3CH_2CHOHCH_3$, ethylmethylcarbinol or ethylmethylmethanol.

In the I.U.P.A.C. system of nomenclature, the longest carbon chain containing the hydroxyl group is chosen as the parent hydrocarbon. The class suffix is-*ol*, and the positions of side-chains and the hydroxyl group are indicated by numbers, the lowest possible number being given to the hydroxyl group, *e.g.*, CH_3OH, methanol; C_2H_5OH, ethanol; $CH_3CH_2CH_2OH$, Propan-1-ol; $(CH_3)_2CHCHOHCH_3$, 3-methylbuttan-2-ol. According to this system the alcohols are referred to as the *alkanols*.

Monohydric alcohols are subdivided into primary, secondary and tertiary alcohols according as the alkyl group attached to the hydroxyl group is a primary, secondary or tertiary group, respectively. Primary alcohols contain the *primary alcoholic group* ---CH_2OH, *e.g.*, ethanol, CH_3CH_2OH; secondary alcohols, the *secondary alcoholic group* CH(OH), *e.g.*, isoprepanol, $(CH_3)_2CHOH$; and tertiary alcohols the *tertiary alcoholic group* $\equiv C(OH)$, *e.g.*, t-butanol, $(CH_3)_3COH$.

GENERAL METHODS OF PREPARATION

1. By the hydrolysis of an alkyl halide with aqueous alkali or silver oxide suspended in water:

$$RX + \text{‘AgOH’} \longrightarrow ROH + AgX$$

2. By the hydrolysis of esters with alkali:

$$R^1CO_2R^2 + KOH \longrightarrow R^1CO_2K + R^2OH$$

This method is important industrially for preparing certain alcohols that occur naturally as esters. Since tertiary alkyl halides give large amounts of alkene on hydrolysis, they are best converted into alcohols as follows:

$$R_3CX + CH_3CO_2Ag \xrightarrow[\text{heat}]{\text{EtOH}} AgX + CH_3CO_2\dot{C}R_3$$

$$\xrightarrow{\text{NaOH}} CH_3CO_2Na + R_3COH$$

3. By heating ethers with dilute sulphuric acid under pressure, *e.g.*, diethyl ether forms ethanol:

$$(C_2H_5)_2O + H_2O \xrightarrow{H_2SO_4} 2C_2H_5OH$$

This method is important industrially, since ethers are formed as by-products in the preparation of certain alcohols (see ethanol and propanols).

4. By the reduction of aldehydes, ketones or esters by means of excess of sodium and ethanol or n-butanol as the reducing agent (*Bouveault-Blanc reduction*, 1903), *e.g.*,

(*i*) *Aldehydes:* $RCHO \xrightarrow{e;\ H^+} RCH_2OH$ (g.-v.g.)

(*ii*) *Esters:* $R^1CO_2R^2 \xrightarrow{e;\ H^+} R^1CH_2OH + R^2OH$ (g.)

(*iii*) *Ketones:* $R_2CO \xrightarrow{e;\ H^+} R_2CHOH$ (g.)

The Bouvealult-Blanc reduction is believed to occur in steps involving the transfer of one electron at a time; *e.g.*,

$$RC(=O\cdot)—OEt \xrightarrow{Na} R\ddot{C}^{-}(—O\cdot)—OEt \xrightarrow{EtOH} RCH(—O\cdot)—OEt \xrightarrow{Na} RCH(—O^-)—OEt \longrightarrow$$

$$OEt^- + RCH{=}O \xrightarrow{Na} R\ddot{C}H—O\cdot \xrightarrow{EtOH} RCH_2—O\cdot \xrightarrow{Na}$$

$$RCH_2—O^- \xrightarrow{EtOH} RCH_2OH$$

Reduction with Metallic Hydrides

Many complex metallic hydrides reduce various functional groups, and their umber is increasing rapidly. The most versatile reagent is **lithium aluminium hydride** (LAH). This reduces most functional groups, Table 6.1, but does not normally reduce the olefinic bond. An unusual feature of this regent is its reduction of the *carboxyl group* to primary alcohol.

Reductions with lithium aluminium hydride are usually carried out in ethereal solutions, the compound in ether being added to the lithium aluminium hydride solution. In certain cases the reverse addition is necessary, *i.e.*, the hydride solution is added to the solution of the compound to be reduced.

Sodium borohydride, which is insoluble in ether, is used in *ethanolic* solution to reduce *carbonyl* compounds, the only important exception being the carboxyl group. It also does not normally reduce esters, but reduction to primary alcohol can often be effected by use of a large excess of reagent in methanol.

The reduction of the carbonyl group by LAH or sodium borohydride occurs in a stepwise manner, each step involving hydride ion transfer, *e.g.*,

$$R_2C{=}O \;\; H{-}AlH_3 \longrightarrow \left[R_2\overset{O}{\underset{|}{C}}H + AlH_3 \right] \longrightarrow R_2CH{-}O{-}AlH_3 \xrightarrow{R_2CO}$$

$$(R_2CHO)_2AlH_2 \xrightarrow{R_2CO} (R_2CHO)_3AlH \xrightarrow{R_2CO} (R_2CHO)_4Al^- \xrightarrow{H^+} 4R_2CHOH$$

Experimental work has shown that the rate of reduction is decreased as the size of R increases. One contributing factor is steric hindrance, and this is made use of by employing reducing complexes containing large groups, *e.g.*, when the substrate contain is two carbonyl groups ad one is sterically hindered, lithium tri-t-butoxyaluminium hydride, reduce only the nonhindered carbonyl group.

This reagent is generally used in tetrahydrofuran (THF) solution. In addition to steric hindrance, there is also the inductive effect in the anion, *e.g.*,

$$(R{-}O{\leftarrow})_3Al{\leftarrow}H \quad \rangle C{=}O$$

The electronegative oxygen atom exerts a strong – I effect and consequently makes hydride ion transfer more difficult. Hence, the reactivity of lithium tri-t-butoxyaluminium hydride is lower than that of LAH.

Many other metallic hydrides are also reducing reagents, and because of their different reactivities, they offer a means of *selective* reduction.

Lithium borohydride (in ether or THF) is more reactive than sodium borohydride but less than LAH.

Lithium tri-t-butoxyaluminium hydride (in THF) useful for reducing acid chlorides to aldehydes at --78 C.

Sodium aluminium hydride (in THF) is similar to LAH, but it reduces esters to aldehydes.

Diborane (in THF) reduces many functional groups.

Aluminium hydride (in ether or THF) is similar to LAH, but is better for reducing α, β-unsaturated carbonyl compounds to unsaturated alcohols.

The reducing power of a *given* metallic hydride is affected by the nature of the solvent and the presence of certain other compounds, *e.g.*, **LAH and aluminium chloride** in ether form AlH_3, AlH_2Cl, or $AlHCl_2$ according to the proportion of reagents used:

$$3LiAlH_4 + AlCl_3 \longrightarrow 4AlH_3 + 3LiCl$$

$$LiAlH_4 + AlCl_3 \longrightarrow 2AlH_2Cl + LiCl$$

$$LiAlH_4 + 3AlCl_3 \longrightarrow 4AlHCl_2 + LiCl$$

Hence, the reducing power of the reagent will depend on which one is actually present; all are milder reducing reagents than LAH, *e.g.*, LAH--$AlCl_3$ does not reduce alkyl halides. It has also been shown that a solution of LAH in pyridine which has been allowed to stand will reduce a carbonyl group but not a carboxyl or carbalkoxyl group. Thus, LAH has been converted into a milder reducing reagent, the structure of which appears to be uncertain.

Sodium borohydride-aluminium chloride (in diethylene glycol dimethyl ether, *i.e.*, diglyme is more reactive than sodium borohydride itself.

Catalytic hydrogenation. Many functional groups are reduced catalytically the most common catalysts being nickel, platinum, palladium, rhodium, and ruthenium; other catalysts used are copper chromite and copper-barium-chromium oxide. The catalytic activity of a given metal is dependent on its method of preparation, *e.g.*, nickel. It also depends on other factors; (*a*) the presence of other compounds which may either increase (*promoters*) or decrease or inhibit (*poisons*) catalytic activity; (*b*) the nature of the solvent: neutral, basic, or acidic.

In addition to having different reactivities, catalysts have different selectivities, *e.g.*, carboxylic acids can be hydrogenated to alcohols with a Ru---C catalyst but not with a Pt catalyst. In general, catalysts are deposited on the surface of a support such as charcoal, alumina, calcium carbonate, etc., and the activity of the catalyst may be changed by changing the support.

5. Primary, secondary and tertiary alcohols may be prepared by means of a Grignard reagent and the appropriate carbonyl compound.

6. A number of alcohols are obtained by fermentation processes.

7. In recent years synthetic methods have become very important for preparing various alcohols commercially:

(i) By the hydration of alkenes .

(ii) By heating a mixture of carbon monoxide and hydrogen under pressure in the presence of a catalyst, *e.g.*, zinc chromite plus small amounts of alkali metal or iron salts. A mixture of alcohols containing methyl, ethyl, n-propyl, isobutyl and higher-branched alcohols is obtained, the individuals being separated by factional distillation.

(iii) The **Oxo process** (*Oxo synthesis, carbonylation* or *hydroformylation reaction*). A mixture of alkene, carbon monoxide and hydrogen, under pressure and elevated temperature, in the presence of a catalyst, forms aldehydes. A common catalyst is cobalt carbonyl hydride, $[CoH(CO)_4]$, and the product is a mixture of isomeric straight-chain and branched-chain aldehydes (the former predominating). These are reduced catalytically to the corresponding alcohols, *e.g.*, propene gives a mixture of n- and isobutanols:

$$2MeCH = CH_2 + 2CO + 2H_2 \longrightarrow Me(CH_2)_2CHO + Me_2CHCHO$$

$$\xrightarrow[Cu-Zn]{H_2} Me(CH_2)_2\,CH_2OH + Me_2CHCH_2OH$$

The aldehydes are first separated by fractional distillation.

(*iv*) Methanol, ethanol, propanols and butanols are prepared industrially by the oxidation of natural gas.

Most of the methods given above can be used for the preparation of any particular class of alcohol: it is only a question of starting with the appropriate compound, *e.g.*, all three types may be prepared via Grignard reagents; *primary* alcohols from formaldehyde ($R^1 = R^2 = H$), *secondary* from aldehydes ($R^1 = R$; $R^2 = H$), and *tertiary* from ketones (R^1 and R^2 = alkyl groups).

$$R^1C(=O)R^2 + R^3MgX \longrightarrow R^1C(OMgX)(R^2)-R^3 \xrightarrow{H^+} R^1R^2R^3COH$$

General Properties of the Alcohols

The alcohols are neutral substances: the lower members are liquids, and have a distinctive smell and a burning taste; the higher members are solids and are almost odourless.

In a group of isomeric alcohols, the primary alcohol has the highest boiling point and the tertiary the lowest, with the secondary having an intermediate value. The lower members are far less volatile than is to be expected from their molecular weight, and this is due to association through hydrogen bonding *extending over a chain of molecules,* thus, giving rise to a large molecule' the volatility of which would be expected to below:

$$\begin{array}{ccccccc} R & & R & & R & & R \\ | & & | & & | & & | \\ O & -H- & O & -H- & O & -H- & O-H \end{array}$$

The lower alcohols are very soluble in water, and the solubility diminishes as the molecular weight increases. Their solubility in water is to be expected, since the oxygen atom of the hydroxyl group in alcohols can form hydrogen bonds with the water molecules. In the lower alcohols the hydroxyl group constitutes a large part of the molecule, whereas as the molecular weight of the alcohol increases the hydrocarbon character of the molecule increases, and hence the solubility in water decreases. This, however; is not the complete story; the structure of the carbon chain also plays a part, *e.g.*, n-butanol is fairly soluble in water, but t-butanol is miscible with water in all proportions.

General Reactions of the Alcohols

1. Alcohols react with organic and inorganic acids to form *esters*:

$$R^1CO_2H + R_2OH \longrightarrow R^1CO_2R^2 + H_2O$$

Esters of the halogen acids are, as we have seen, the alkyl halides.

The order of reactivity of an alcohol with a given acid is primary alcohol > secondary > tertiary, provided the mechanism is bimolecular for all cases. With a given alcohol, the order of reactivity of the halogen acids in the bimolecular mechanism is HI > HBr > HCl.

Let us consider the following example:

$$EtOH + H^+ + X \underset{}{\overset{fast}{\rightleftharpoons}} X^- + EtOH_2^+ \rightleftharpoons \overset{\delta-}{X} \cdots Et \cdots \overset{\delta+}{O}H_2 \xrightarrow{fast}$$

$$X—Et + H_2O$$

The order of reactivity of the halogen acids is attributed to the variation in polarisability of the anion; the greater the size of the anion, the greater is its polarisability. What this amounts to is that the larger the anion, the more readily it can donate a lone pair to form a covalent bond. Also, the nature of the alkyl group will influence the mechanism of halide formation (in aqueous solution) in the same way as it does their hydrolysis. Protonated t-alcohols readily eliminate a molecule of water to form a carbonium ion and so react by the S_N1 mechanism. Straight-chain primary alcohols will favour the S_N2 mechanism, but if branched-chain, then the S_N1 mechanism may operate. Secondary alcohols would be expected to react by both methanisms.

2. Alcohols react with phosphorus halides to form alkyl halides.

3. Alcohols combine with phenyl isocyanate to form phenyl-substituted urethans:

$$C_6H_5NCO + ROH \longrightarrow C_6H_5NHCO_2R$$

4. Strongly electropositive metals (K, Na, Mg, Al, Zn) liberate hydrogen from alcohols to form *alkoxides*, *e.g.*, sodium reacts with ethanol to form sodium ethoxide:

$$2C_2H_5OH + 2Na \longrightarrow 2C_2H_5O^-Na^+ + H_2$$

Alkoxides are white deliquescent solids, readily soluble in water with hydrolysis:

$$RO^-Na^+ + H_2O \rightleftharpoons ROH + NaOH$$

Alkoxides react with carbon disulphide to form xanthates:

$$RONa + S{=}C{=}S \longrightarrow RO{-}C({=}S){-}\bar{S}Na^+$$

The fact that alcohols liberate hydrogen by the action of metals shows that alcohols can behave as acids, but since they do not affect the pH of water, they are weaker acids than water.

This accounts for alkoxides being hydrolysed by water and for the fact that the ethoxide ion (the cB of EtOH) is a stronger base than the hydroxide ion (the cB of H_2O).

Since alkyl groups have a +I effect, there will be an increased electron displacement towards the oxygen atom in going from primary to secondary to tertiary alcohol. This may be represented (qualitatively) as follows:

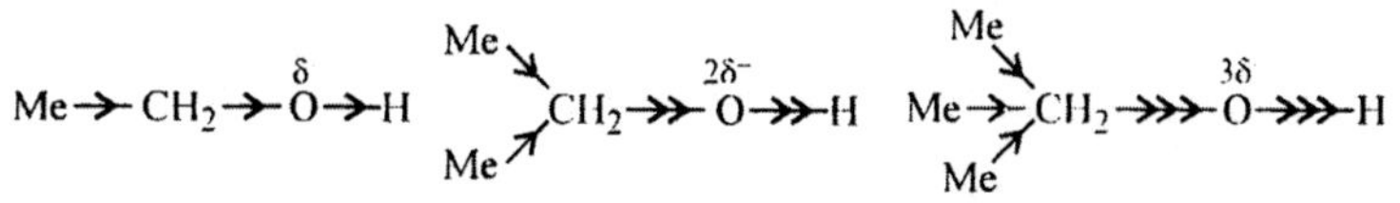

$$Me \rightarrow CH_2 \rightarrow \overset{\delta}{O} \rightarrow H \qquad \begin{matrix} Me \searrow \\ Me \nearrow \end{matrix} CH_2 \twoheadrightarrow \overset{2\delta^-}{O} \twoheadrightarrow H \qquad Me \rightarrow \overset{Me}{\underset{Me}{CH_2}} \ggg \overset{3\delta}{O} \ggg H$$

The greater the negative charge on the oxygen atom, the closer is the covalent pair in the O—H bond driven to the hydrogen atom and consequently separation of a proton becomes increasingly difficult. Thus the acid strengths of alcohols will be in the order: prim > s > t. This is the order of reactivity of alcohols towards metals, and with t-alcohols the reaction with sodium is so slow that it is better to use the more electropositive potassium for these alcohols, *e.g.*, potassium t-butoxide is the usual salt prepared from t-butanol.

From what has been said above, it can be seen that the tendency for the C—O bond to break will be the reverse of that for the O—H bond, *i.e.*, reactions involving the breaking of the C—O bond will follow the order of reactivity: t > s > prim. This order would be expected from the consideration of the stabilities of the carbonium ions produced .

If we now consider β-substituted alcohols containing strong –I groups, the acidity of the alcohol is increased, *e.g.*, pK_a of ethanol of ethanol is ~ 18 whereas that of trifluoroethanol is 12.4.

$$Me \rightarrow CH_2 \rightarrow O—H \qquad CF_3 \lll CH_2 \lll O—H$$

The Me group is weakly + I, whereas the CF_3 group is strongly – I.

Thus the covalent pair is pulled away from the oxygen atom in the C—O bond in the latter compound, and consequently the covalent pair i the O—H bond is drawn towards the O atom, thereby facilitating release of the hydrogen as a proton.

A number of alkoxides are important as synthetic reagents; *e.g.*, sodium ethoxide, C_2H_5ONa; aluminium ethoxide, (C_2H_5O) Al; aluminium t-butoxide $[(CH_3)_3 CO]_3Al$. The aluminium alkoxides may be conveniently prepared by the action of aluminium amalgam or aluminium shavings o the alcohol.

5. Primary and secondary alcohols may be acetylated with acetyl chloride, *e.g.*, ethanol gives ethyl acetate:

$$CH_3COCl + C_2H_5OH \longrightarrow CH_3CO_2C_2H_5 + HCl$$

With tertiary alcohols the reaction is often accompanied by dehydration of the alcohol to alkene or by the formation of a tertiary alkyl chloride; *e.g.*, t-butanol gives a good yield of t-butyl chloride:

$$(CH_3)_3\ COH + CH_3COCl \longrightarrow (CH_3)_3\ CCl + CH_3CO_2H$$

However, in the presence of a base such as dimethylaniline, the ester is produced in good yield.

$$CH_3COCl + (CH_3)_3COH \xrightarrow{PhNMe_2} CH_3CO_2C(CH_3)_3 + PhNMe_2\ HCl \quad (63\text{–}68\%)$$

A possible explanation for this is that t-butyl esters are very readily decomposed by hydrochloric acid as follows:

$$MeCOCl + Me_3COH \longrightarrow MeCOOCMe_3 + HCl \longrightarrow MeCO_2H + Me_3CCl$$

In the presence of a base, the HCl is removed as base· HCl, and consequently the second step cannot occur.

Acetylation can also be carried out with acetic anhydride in the presence of a catalyst, *e.g.*, pyridine. This method is successful for prim and s-alcohols, but fails with t-alcohols, *e.g.*, t-butanol is not acetylated. However, t-alcohols can be acetylated if *p*-toluenesulphonic acid is used as acetalyst.

6. Alcohols may be oxidised, and the products of oxidation depend on the class of the alcohol and on the nature of the oxidising agent.

Halogens, in *aqueous* solution, oxidise alcohols to carbonyl compounds. According to Swain *et al.* (1961), the mechanism involves hydride ion removal:

$$Br\text{—}Br\ \ H\text{—}CHMe\text{—}\ddot{O}\text{—}H \longrightarrow Br^- + HBr + MeCH{=}\overset{+}{O}H \xrightarrow{-H^+} MeCHO$$

In *alkaline* solution, the mechanism proposed is:

$$MeCH_2O\text{—}H\ \ \bar{O}H \rightleftharpoons H_2O + MeCH_2\text{—}O^-$$

$$MeCH_2\text{—}\bar{O} + Br_2 \longrightarrow MeCH{=}O + HBr + Br^-$$

In both cases, the carbonyl compound is then halogenated under the catalytic influence of acid or alkali.

On the other hand, alcohols are oxidised by hydrogen peroxide and ferrous sulphate to diols which, according to Coffman *et al.* (1960), are produced by dimerisation of free-radical intermediates, *e.g.*, t-butanol forms, 2,5-dimethylhexane-2,5-diol.

$$Fe^{2+} + H2O2 \longrightarrow Fe^{3+} + HO^- + HO\cdot$$

$$(CH_3)_2C(OH)CH_3 \xrightarrow[-H_2O]{\cdot OH} (CH_3)_2C(OH)CH_2\cdot \xrightarrow{2} (CH_3)_2C(OH)CH_2CH_2C(OH)(CH_3)_2$$

t-Butanol is particularly useful since all the alkyl hydrogens are equivalent and so only one free radical is produced.

7. Alcohols may be dehyrated to alkenes, the ease of dehydration being t > s > prim. Dehydration may be effected by heat alone (400–800°C), but in the presence of a catalyst, lower temperatures may be used, *e.g.*, when passed over heated alumina, t-alcohols are dehydrated at about 150°C, s-alcohols at 250°C and prim.-alcohols at 350°C. Also, primary alcohols are dehydrated by concentrated sulphuric acid at about 170°C, and secondary and tertiary alcohols by boiling dilute sulphuric acid (this is used to avoid polymerisation of the alkene). Isomerisation usually occurs with dehydration by acid or alumina, but in the latter case, isomerisation is suppressed by the addition of a small amount of pyridine.

The mechanism of dehydration with acid is described on the basis that a carbonium ion is formed as an intermediate, it can be seen why the ease of dehydration is t > s > prim.; this is the order of stabilities of the carbonium ions.

With secondary ad tertiary alcohols, dehydration may occur in two ways, *e.g.*,

$$CH_3CH_2CH(OH)CH_3 \xrightarrow[(1:1H_2SO_4)]{-H_2O} \begin{cases} CH_3CH_2CH{=}CH_2 \\ CH_3CH{=}CHCH_3 \end{cases}$$

Experiment shows that hydrogen attached to the adjacent carbon atom joined to the least number of hydrogen atoms is eliminated most easily. Thus, in the above reaction, the main product is but-2-ene (65–80 per cent). This elimination therefore occurs i accordance with Saytzeff's rule for the dehydrohalogenation of alkyl halides, and the reason is the same. However, when thorium oxide or lanthanide metal oxides are used as the dehydrating agents, the yield of 1-alkene from 2-alcohols is generally above 98 per cent (Lundeen *et al.*, 1963).

As mentioned above, rearrangement often occurs with acid-catalysed dehydration. All three types of alcohol may behave in this way via a carbonium ion that may undergo methyl and/or hydride ion 1,2-shift. In the following examples, only methyl shifts are shown (all the possible hydride ion shifts lead finally to the same products resulting from some of the competing methyl shifts).

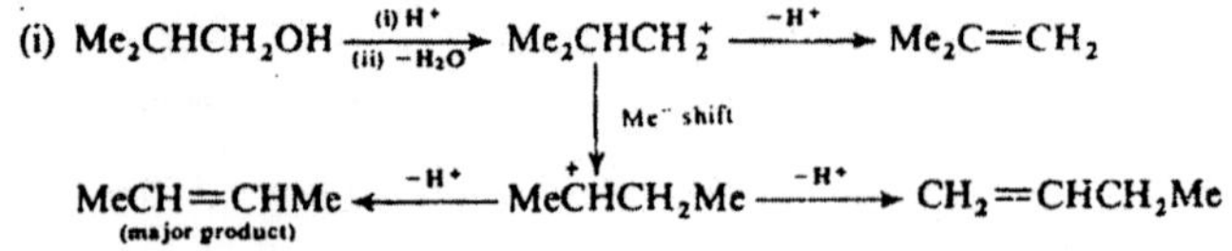

$$\text{(i) } Me_2CHCH_2OH \xrightarrow[\text{(ii) } -H_2O]{\text{(i) } H^+} Me_2CHCH_2^+ \xrightarrow{-H^+} Me_2C{=}CH_2$$

$$\downarrow \text{Me}^- \text{ shift}$$

$$\underset{\text{(major product)}}{MeCH{=}CHMe} \xleftarrow{-H^+} Me\overset{+}{C}HCH_2Me \xrightarrow{-H^+} CH_2{=}CHCH_2Me$$

N.B. An illustration of hydride ion shift is:

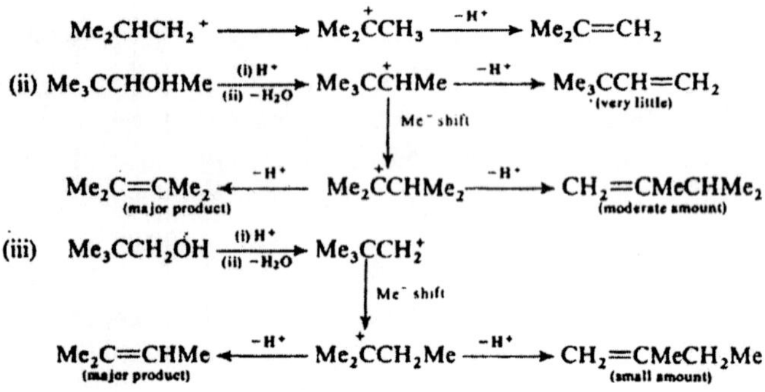

$$Me_2CHCH_2^+ \longrightarrow Me_2\overset{+}{C}CH_3 \xrightarrow{-H^+} Me_2C{=}CH_2$$

$$\text{(ii) } Me_3CCHOHMe \xrightarrow[\text{(ii) } -H_2O]{\text{(i) } H^+} Me_3C\overset{+}{C}HMe \xrightarrow{-H^+} \underset{\text{(very little)}}{Me_3CCH{=}CH_2}$$

$$\downarrow \text{Me}^- \text{ shift}$$

$$\underset{\text{(major product)}}{Me_2C{=}CMe_2} \xleftarrow{-H^+} Me_2\overset{+}{C}CHMe_2 \xrightarrow{-H^+} \underset{\text{(moderate amount)}}{CH_2{=}CMeCHMe_2}$$

$$\text{(iii) } Me_3CCH_2OH \xrightarrow[\text{(ii) } -H_2O]{\text{(i) } H^+} Me_3CCH_2^+$$

$$\downarrow \text{Me}^- \text{ shift}$$

$$\underset{\text{(major product)}}{Me_2C{=}CHMe} \xleftarrow{-H^+} Me_2\overset{+}{C}CH_2Me \xrightarrow{-H^+} \underset{\text{(small amount)}}{CH_2{=}CMeCH_2Me}$$

Fig. 2.9

In each case, the major product is in accordance with Saytzeff's rule. Another point to note is that the rearrangements occur extremely rapidly. In fact, in the examples given, rearrangement is faster than proton elimination. Furthermore, Phelan *et al.* (1967) have shown that ease of anion migration is Ph >> Me > H.

When a double bond is produced in the product and is accompanied by a 1,2-shift, the reaction is said to be a *retropinacol rearrangement*. This type of rearrangement, when occurring in *open chain* compounds, is also sometimes called the *Wagner rearrangement*.

Alcohols may also be converted into alkenes via their esters or xanthates, but in these cases no rearrangements occur, *e.g.*, $Me_3CCH(OAc)Me \longrightarrow Me_3CCH{=}CH_2$ (*cf.* the action of acid on the alcohol, above).

Dehydration may be used as the first step in the conversion of a primary alcohol of suitable structure into a secondary or tertiary alcohol, or a s-alcohol of suitable structure into a t-alcohol, *e.g.* (*N. B.* HI additions are in accordance with Markownikoff's rule):

$$\text{(i) } MeCH_2CH_2OH \xrightarrow[350°C]{Al_2O_3} MeCH{=}CH_2 \xrightarrow[\text{(ii) AgOH}]{\text{(i) HI}} MeCHOHMe$$

$$\text{(ii) } Me_2CHCH_2OH \xrightarrow[350°C]{Al_2O_3} Me_2C{=}CH_2 \xrightarrow[\text{(ii) AgOH}]{\text{(i) HI}} Me_3COH$$

(*iii*) $Me_2CHCHOHMe \xrightarrow[350°C]{Al_2O_3} Me_2C{=}CHMe \xrightarrow[(ii)\ AgOH]{(i)\ HI}$

$Me_3C(OH)CH_2Me$

The 'reverse' procedure, *i.e.*, the conversion t → s → prim. alcohol can be carried out via hydroboration, *e.g.*,

$$Me_2C(OH)CH_2Me \xrightarrow{-H_2O} Me_2C{=}CHMe \xrightarrow{B_2H_6}$$

$$(Me_2CHCHMe{-})_2BH \xrightarrow{heat} (Me_2CHCH_2CH_2{-})_2BH$$

$$\downarrow H_2O_2;\ OH^- \qquad\qquad \downarrow H_2O_2;\ OH^-$$

$$Me_2CHCHOHMe \qquad\qquad Me_2CHCH_2CH_2OH$$

It is also possible to step down the alcohol series by first dehydrating to the alkene, which is then subjected to ozonolysis, *e.g.*,

(*a*) $RCH_2CH_2OH \xrightarrow[350°C]{Al_2O_3} RCH{=}CH_2 \xrightarrow[(ii)\ Zn/H^+]{(i)\ O_3} RCHO \xrightarrow[cat.]{H_2}$

RCH_2OH

(*b*) $RCHOHCH_3 \xrightarrow[300°C]{ThO_2} RCH{=}CH_2 \cdots\cdots\rightarrow RCH_2OH$

8. Alcohols combine with acetylene in the presence of mercury compounds as catalyst to form acetals:

$$2ROH + CH{\equiv}CH \xrightarrow{Hg^{2+}} CH_3CH(OR)_2$$

If, however, the reaction is carried out in the presence of potassium alkoxides at high temperature and under pressure, vinyl ethers are obtained.

METHODS OF DISTINGUISHING BETWEEN THE THREE CLASSES OF ALCOHOLS

1. By means of *oxidation*. The nature of the oxidation products of an alcohol depends on whether the alcohol is primary, secondary of tertiary.

(*i*) A *primary* alcohol on oxidation first gives an *aldehyde*, and this, on further oxidation, gives an acid. *Both aldehyde and acid contain the same number of carbon atoms as the original alcohol, e.g.*,

$$CH_3CH_2OH \xrightarrow{[O]} CH_3CHO \xrightarrow{[O]} CH_3CO_2H$$

(*ii*) A *secondary* alcohol, on oxidation, first gives a *ketone with the same number of carbon atoms as the original alcohol.* Ketones are fairly difficult to oxidise, but prolonged atoms than the original alcohol, *e.g.*, methyl-n-propylmethanol gives first pentan-2-one, and then a mixture of acetic and propionic acids:

$$CH_3CHOHCH_2CH_2CH_3 \xrightarrow{[O]} CH_3COCH_2CH_2CH_2CH_3 \xrightarrow{[O]}$$
$$CH_3CO_2H + CH_3CH_2CO_2H$$

(*iii*) *Tertiary* alcohols are resistant to oxidation in neutral or alkaline solution, but are readily oxidised by acid oxidising agents to a mixture of *ketone and acid, each containing fewer carbon atoms than the original alcohol.*

$$(CH_3)_2C(OH)CH_2CH_3 \xrightarrow{[O]} (CH_3)_2CO + CH_3CO_2H$$

The oxidising agents usually used for oxidising alcohols are: acid dichromate, acid or alkaline potassium permanganate, and dilute nitric acid.

The accepted mechanism for the oxidation of primary and secondary alcohols by dichromate and pemanganate involves an ester intermediate, *e.g.*, with acid dichromate:

$$Cr_2O_7^{2-} + H_2O \rightleftharpoons 2HCrO_4^-$$

$$Me_2CHOH + HCrO_4^- + 2H^+ \rightleftharpoons Me_2CH—O—CrO_3H_2^+ + H_2O$$

$$\underset{\underset{H_2O\ \ H}{|}}{Me_2C}—O—CrO_3H_2^+ \xrightarrow{slow} Me_2CO + H_3O+ + H_2CrO_3$$

The rate-determining step is the removal of hydrogen from the intermediate ester: This has been demonstrated by Westheimer *et. al.* (1949), who showed that Me_2CDOH is oxidised at about 1/7 the rate of either Me_2CHOH or $(CD_3)_2CHOH$. Thus, the r/d step involves the fission of the C—H bond (this accounts for the kinetic isotope effect).

This mechanism cannot operate with t-alcohols, since there is no H—COH group present. According to Sager (1956), the mechanism involves dehydration, *e.g.*,

$$Me_3COH \xrightarrow[(ii)\ -H_2O]{(i)\ H^-} Me_3C+ \xrightarrow{-H^+} Me_2C{=}CH_2 \xrightarrow{[O]} Me_2\overset{O}{C—CH_2} \xrightarrow{[O]}$$
$$Me_2CO + HCO_2H$$

2. The three classes of alcohols differ in their behaviour when the vapour is passed over copper at 300°C:

(*i*) A primary alcohols is dehydrogenated to an aldehyde, *e.g.*,

$$CH_3CH_2OH \xrightarrow[300°C]{Cu} CH_3CHO + H_2$$

(*ii*) A secondary alcohol is dehydrogenated to a ketone, *e.g.*,

$$CH_3CH(OH)CH_3 \xrightarrow[300°C]{Cu} CH_3COCH_3 + H_2$$

(*iii*) A tertiary alcohol is dehydrated to a alkene, *e.g.*,

$$(CH_3)_2C(OH)CH_2CH_3 \xrightarrow[300°C]{Cu} (CH_3)_2C{=}CHCH_3 + H_2O$$

Methyl alcohol, methanol (*carbinol*) is prepared industrially by several methods. The earliest method was the destructive distillation of wood, whereby tar and an aqueous fraction known as *pyroligneous acid* are obtained.

Pyroligneous acid contains methanol, acetone and acetic acid, and all three compounds may be obtained by suitable treatment. It was this method which gave rise to the name 'wood spirit' for methanol. The modern methods are synthetic.

(*i*) Synthesis gas is passed at a pressure of 200 atmospheres over a catalyst containing the oxides of copper, zinc and chromium at 300°C:

$$CO + 2H_2 \longrightarrow CH_3OH$$

If the proper precautions are taken, the yield of methanol is almost 100 per cent, an its purity is above 99 per cent. By changing the catalyst and the ratio of carbon monoxide to hydrogen, methanol and a variety of higher alcohols are produced.

(*ii*) By the catalytic oxidation of methane. A mixture of methane and oxygen (ratio by volume of 9 : 1) at a pressure of 100 atmospheres is passed through a copper tube at 200°C:

$$CH_4 + \tfrac{1}{2}O_2 \longrightarrow CH_3OH$$

Methanol is a colourless, inflammable liquid, b.p. 64°C, and is poisonous. It is miscible with water in all proportions, and is also miscible with most organic solvents.

It burns with a faintly luminous flame, and its vapour forms explosive mixtures with air or oxygen when ignited. It combines with calcium chloride to form $CaCl_2{\cdot}4CH_3OH$, and hence cannot be dried this way (*cf.* ethanol).

Methanol is used as a solvent for paints, varnishes, shellac celluloid cements, etc.; in the manufacture of dyes, perfumes, formaldehyde, etc. It is also used for making methylated spirit and automobile antifreeze mixtures.

Structure of methanol. Analysis and molecular-weight determinations show that the molecular formula of methanol is CH_4O. Assuming that carbon is quadrivalent, oxygen bivalent and hydrogen univalent, only one structure is possible:

$$\begin{array}{c} \quad H \\ \quad | \\ H-C-O-H \\ \quad | \\ \quad H \end{array} \quad \text{or} \quad CH_3OH$$

This is supported by all the chemical reactions of methanol, *e.g.*, (*i*) only one hydrogen atom is methanol is replaceable by sodium; this suggests that one hydrogen atom is in a different state of combination from the other three.

(*ii*) Methanol is formed from methyl chloride by hydrolysis with sodium hydroxide. Methyl chloride can have only the structure CH_3Cl.

It is reasonable to suppose that the methyl group in methyl chloride is unchanged by the action of dilute alkali, and that the reaction takes place by the replacement of the chlorine atom by a hydroxyl group.

(*iii*) The presence of the hydroxyl group is confirmed, for example, by the reaction between methanol and phosphorus pentachloride, when methyl chloride, hydrogen chloride and phosphoryl chloride are formed. Thus one oxygen atom (bivalent) and one hydrogen atom (univalent) have been replaced by one chlorine atom (univalent).

This implies that the oxygen and hydrogen atoms exist as a univalent group in methanol: the only possibility is as a hydroxyl group, OH. It is the hydrogen of the hydroxyl group which is displaced by sodium.

Infra-red spectroscopic studies of hyroxy-compounds show that they absorb (O—H stretch) in the region of 3650 - 3580 cm^{-1}(v). This is true only if there is no hydrogen bonding.

Intermolecular hydrogen bonding produces absorption)broad band) in the region 3350 – 3230cm^{-1}(v), whereas intramolecular hydrogen bonding absorbs in the region 3590 – 3420 cm^{-1} (v).

Moreover, the former shows changes in intensity and frequency shifts on dilution, whereas the latter is unaffected by dilution; thus the two may be distinguished. In addition to O—H stretching vibrations, there are also C—O stretching vibrations which are characteristic of the type of the type of alcohol: primary, close to 1050 (s); secondary, close to 1100 (s); tertiary, close to 1150 cm^{-1}(s). Thus the classes of alcohols may be identified in these regions.

Two infra-red spectra of ethanol are shown in Fig. 2.10 (as a film) and in Fig. 2.11 (as a 1 per cent solution in carbon tetrachloride). IN the former, all the OH groups are associated, but in the latter, some OH groups are free.

However, a further point that may be mentioned here is the τ-value for methylene protons in the —CH_2—O—group is between 6.5–6.65 p.p.m., and when the alcohol is acetylated, the τ-value changes to ~ 5.9. This change may be used to identify a primary alcohol.

Secondary alcohols can also be identified; > CH—O—(τ = ~ 6.1) and the acetylated alcohol (τ = ~ 5.0). These down-field shifts are due to the increased deshiedling effect of the acetoxy! group ($OOCCH_3$) compared to hydroxyl oxygen.

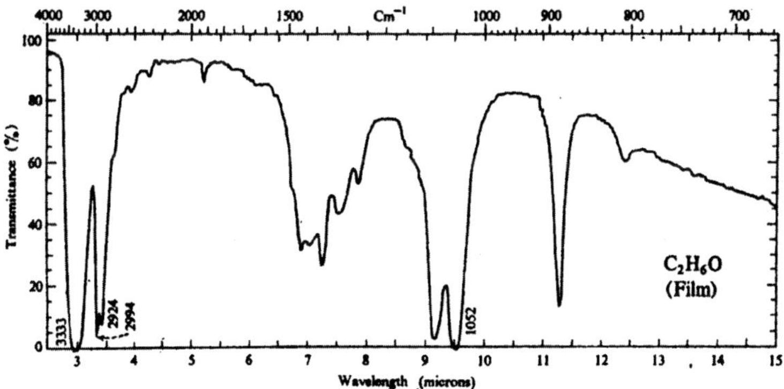

Fig. 2.10

3333 cm^{-1} : O—H (str.) association (through hydrogen bonding). 2994 and 2994 and 2994 and 2 924cm^{-1} : C—H (str.) in Me and/or CH_2. 1052cm^{-1} : C—O (str.); characteristic of primary alcohols.

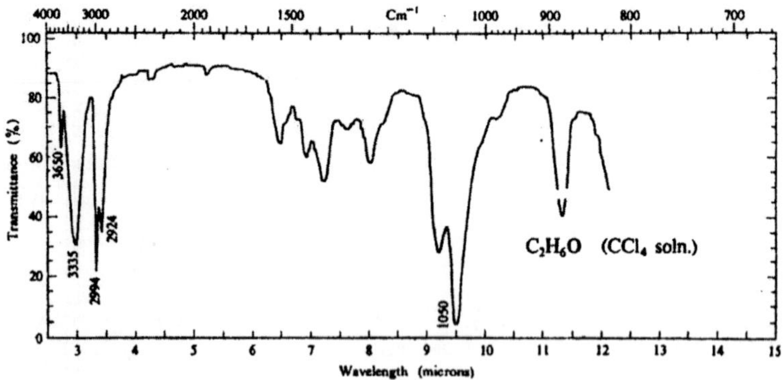

Fig. 2.11

3650cm^{-1} : O—H (str.); free OH. 3333, 2994, 2924, and 1050 cm^{-1}; As for Fig. 2.10

Mass spectra. The mass spectrum of ethanol but have we shall describe the mass spectra of alcohols from a general point of view. Usually, strong peaks are shown by fragmentation involving β-cleavage (*i.e.*, the C—COH bond), and the fragment containing the hydroxyl group is stabilised by resonance:

$$\left[R^1-\underset{R^3}{\overset{R^2}{\overset{|}{\underset{|}{C}}}}-\overset{\cdot+}{O}H\right] \longrightarrow R^{1\cdot} + \left[\begin{matrix} R^2 \\ \end{matrix}\!\!\!\!\!\!\!\!\!\!\!\!\!\!\!\; \underset{R^3}{\overset{R^2\diagdown}{}}\!\!\!\!C-\overset{\cdot+}{O}H \longleftrightarrow \underset{R^3}{\overset{R^2\diagdown}{}}\!\!\!\!C=\overset{+}{O}H\right]$$

The alkyl group with the heaviest mass is eliminated preferentially, but ions also appear for alternative eliminations. Further fragmentation may also occur as follows (R^2 = H, R_3= C_2H_5):

$$\left[CH_3-CH_2-CH=\overset{+}{O}H \longleftrightarrow CH_2\begin{matrix} \diagup CH_2-H \\ \diagdown \overset{+}{C}H-OH \end{matrix}\right] \longrightarrow C_2H_4 + [CH_2-OH]^+ \underset{m/e\ 31}{}$$

Hence, both *s*- and *t*-alcohols can also show peaks at *m/e* 31 (the most characteristic ion of alcohols).

Primary and secondary alcohols usually give weak molecular ions, and for tertiary alcohols the molecular ion is very weak or absent. Long-chain alcohols show peaks at M-18 due to the loss of a molecule of water and formation of a cyclic ion (cf. alkyl halides):

$$\left[C_n \begin{matrix} C{-}H \\ C{-}OH \end{matrix} \right]^{+\cdot} \longrightarrow \left[C_n \begin{matrix} C \\ | \\ C \end{matrix} \right]^{+\cdot} + H_2O$$

On the other hand, a McLafferty rearrangement can also occur with loss of water:

$$\left[\begin{matrix} & H & \\ & O & \\ CH_2 & & H \\ | & & | \\ CH_2 & & CHR \\ & CH_2 & \end{matrix} \right]^{+\cdot}_{M^{+}} \longrightarrow H_2O + CH_2{=}CH_2 + \underset{M-46}{[RCH{=}CH_2]}$$

In general, long-chain alcohols also show the fragmentation patterns of alkanes and alkenes.

Ethyl alcohol, ethanol, is prepared industrially by several methods, *e.g.*, ethylene (from cracked petroleum) is absorbed in concentrated sulphuric acid (98 per cent) at 75-80°C, under pressure (17.5–35kg/cm^2). Ethyl hydrogen sulphate and ethyl sulphate are formed:

$$C_2H_4 + (HO)_2SO_2 \longrightarrow C_2H_5OSO_2OH$$

$$C_2H_5OSO_2OH + C_2H_4 \longrightarrow (C_2H_5O)_2SO_2$$

The reaction mixture is then diluted with about an equal volume of water, and warmed. Hydrolysis takes place and ethanol together with some diethyl ether is formed:

$$C_2H_5OSO_2OH + H_2O \longrightarrow C_2H_5OH + H_2SO_2$$

$$(C_2H_5O)_2SO_2 + 2H_2O \longrightarrow 2C_2H_5OH + H_2SO_4$$

$$(C_2H_5O)_2SO_2 + C_2H_5OH \longrightarrow (C_2H_5)_2O + C_2H_5OSO_2OH$$

The ether is kept to a minimum by separating the ethyl sulphate from the reaction products, and hydrolysing it separately.

The hydrolysed liquids are distilled, and the aqueous ethanol distillate is concentrated by fractional distillation.

Ethanol is also manufactured by the *direct* hydration of ethylene with steam under pressure in the presence of a suitable catalyst, *e.g.*, phosphoric acid on a support.

$$C_2H_4 + H_2O \longrightarrow C_2H_5OH$$

Some ether (~ 5 per cent) is formed as a by-product.

The earliest method of preparing ethanol is by fermentation and this is still used for the manufacture of beer, wine, brandy, etc., and also a source of ethanol. The starting material is starch, which is obtained from sources depending on the particular country: common sources of starch are wheat, barely, potato, etc. Molasses is also used as the starting material for ethanol. The grain, *e.g.*, wheat or barley, is mashed with hot water, and then heated with malt (germinated barley) at 50°C for 1 hour. Malt contains the enzyme *diastase* which, by hydrolysis, converts starch into the sugar, maltose *(q.v.)*:

$$(C_6H_{10}O_5)_n + \frac{n}{2} H_2O \xrightarrow{\text{maltase}} \frac{n}{2} C_{12}H_{22}O_{11}$$

If molasses is used, then this step is unnecessary, since it contains carbohydrates already present as sugars which can be fermented.

The liquid is cooled to 30°C and fermented with yeast for 1–3 days. Yeast contains various enzymes, among which are *maltase*, which converts the maltose into glucose, and *zymase*, which converts the glucose into ethanol:

$$C_{12}H_{22}O_{11} + H_2O \xrightarrow{\text{maltase}} 2C_6H_{12}O_6$$

$$C_6H_{12}O_6 \xrightarrow{\text{zymase}} 2C_2H_5OH + 2CO_2$$

The carbon dioxide is recovered and sold as a by-product. The fermented liquor or 'wort', which contains 6-10 per cent ethanol and some other compounds, is fractioned into three fractions:

(*i*) *First runnings*, which consists mainly of acetaldehyde.

(*ii*) *Rectified spirit*, which is 93–95 per cent w/w ethanol.

(*iii*) *Final runnings* or *fusel oil*, which contains n-propyl, n-butyl, isobutyl, n-amyl, isoamyl and 'active] amyl alcohol.

Industrial alcohol is ordinary rectified spirit. *Methylated spirit* is of two kinds: (*a*) *Mineralised methylated spirit* is 90 per cent rectified spirit, 9 per cent methanol and 1 per cent petroleum oil, and a purple dye. (*b*) *Industrial methylated spirit* is 95 per cent rectified spirit and 5 per cent methanol, whose purpose is to '*denature*' the rectified spirit, *i.e.*, make it unfit for drinking purposes.

Absolute alcohol is 99.5 per cent ethanol, and is obtained from rectified spirit. When an aqueous solution of ethanol is fractionated, it forms a constant-boiling mixture containing 96 per cent ethanol from which 100 per cent ethanol may be obtained by adding a small amount of benzene, and then distilling. The first fraction is the ternary azeotrope, *i.e.*, a constant-boiling mixture containing three constituents, b.p. 64.8°C (water, 7.4 per cent; ethanol, 18.5 per cent; benzene, 74.1 per cent). After all the water has been removed, the second fraction that distils over is the binary azeotrope. b.p. 68.2°C (ethanol, 32.4 per cent; benzene, 67.6 per cent). After all the benzene has been removed, pure ethanol, b.p. 78.1°C, distils over.

Ethanol cannot be dried by means of calcium chloride, since a compound (an *alcoholate*) is formed, *e.g.*, $CaCl_2{\cdot}3C_2H_5OH$ (*cf.* methanol). Distillation of rectified spirit over calcium oxide, and then over calcium, gives absolute alcohol. This method is often used in the laboratory, and was formerly used industrially.

Ethanol is a colourless, inflammable liquid, b.p. 78·1°C. It is miscible with water in all proportions, and is also miscible with most organic solvents. Ethanol ad methanol resemble each other very closely, but they may be distinguished (*i*) by the fact that ethanol gives the haloform reaction (*q.v.*), whereas methanol does not; and (*ii*) ethanol gives acetic acid on oxidation; methanol gives formic acid. These two acids are readily distinguished from each other.

Ethanol is used for the preparation of esters, ether, chloral, chloroform, etc. It is also used as a solvent for gums, resins, paints, varnishes, etc., and as a fuel.

Structure of ethanol. Analysis and molecular-weight determinations show that the molecular formula of ethanol is C_2H_6O. Assuming that carbon is quadrivalent, oxygen bivalent, and hydrogen univalent, two structures are possible.

$$\underset{\text{(I)}}{CH_3-CH_2-OH} \qquad \underset{\text{(II)}}{CH_3-O-CH_3}$$

(*i*) Only one hydrogen atom in ethanol is replaceable by sodium or potassium. This indicates that one hydrogen atom is in a different state of combination from the other five. In (I), one hydrogen atom differs from the other five, but in (II) all the hydrogen atoms are equivalent.

(*ii*) When ethanol is treated with hydrochloric acid or phosphorus pentachloride, one oxygen atom (bivalent) and one hydrogen atom (univalent) are replaced by one chlorine atom (univalent) to give ethyl chloride, C_2H_5Cl. This implies the presence of a hydroxyl group (*cf.* methanol).

(*iii*) When ethyl chloride is hydrolysed with dilute alkali, ethanol is obtained. This reaction also indicates the presence of a hydroxyl group in ethanol.

(*iv*) Ethanol may be prepared as follows:

$$C_2H_6 \xrightarrow{Cl_2} C_2H_5Cl \xrightarrow{NaOH} C_2H_5OH$$

The arrangement of the six hydrogen atoms in ethane is known, and it is reasonable to suppose that five retain their original arrangement in ethyl chloride and ethanol, since these five hydrogen atoms do not enter (presumably) into the above reactions. Thus there is an ethyl group C_2H_5—in ethanol. This is so in (I), but not in (II).

(*v*) Structure (II) is definitely eliminated, since it can be shown that it is the structure of dimethyl ether (*q.v.*), a compound that has very little resemblance, physically or chemically, to ethanol.

Hence (I) is accepted as the structure of ethanol, and it accounts for all the known properties of ethanol.

It can be seen from the various examples given on structure determination, *e.g.*, methane, butanes, ethylene, methanol and ethanol, that the method of approach follows certain definite lines. First the molecular formula is obtained. Then, if the compound is simple–in the sense that it contains a small number of unlike atoms, and that the total number of atoms is also small–the valencies of the atoms present are assumed, and various possible structures are written down.

If the compound is 'simple' the number of possible structures will not be large (four or five at the most). Then by considering the chemical properties of the compound in question, the structure which best fits the observed facts is accepted as the correct one. If the compound is not 'simple', the procedure is to detect the presence of as many functional groups as possible; to degrade the compound into simpler substances whose structures are already known or which may be determined by further degradation; to build up structural formulae based on the facts obtained; and then to choose the structure which best fits the facts.

Finally a synthesis is attempted, and if successful, will usually give proof of the correctness of the structure suggested.

It should be noted that the arguments used for structure determination by purely chemical methods are based on the assumption that no rearrangement have occurred in the series of reactions performed. Thus, the possibility of rearrangement must always be borne in mind when carrying out chemical reactions.

As was pointed out physical methods are being used nowadays for structural determination. Once the molecular formula is obtained, functional groups may be identified, etc. In this way, the structures of many compounds have now been ascertained with a minimum of chemical evidence.

Propyl alcohols. Two isomeric propyl alcohols are possible, and both are known.

n-Propyl alcohol, propan-1-ol, n-propanol, was originally obtained from fusel oil (see above), but it is now produced by the hydogenation of carbon monoxide. A more recent method is by the catalytic reduction of propargyl alcohol (from acetylene and formaldehyde).

$$CH{\equiv}CCH_2OH + 2H_2 \xrightarrow{Ni} 2CH_3CH_2CH_2OH$$

n-Propanol, b.p. 97.4°C, is miscible with water, ethanol and ether. It is used in the preparation of propionic acid, toilet preparations such as lotions, etc.

Isopropyl alcohol, propan-2-ol, isoprpanol, is prepared industrially by passing propene (from cracked petroleum) into concentrated sulphuric acid, then diluting with water, and distilling off the isopropanol. Isopropyl ether is obtained as a by-product (*cf.* ethanol):

$$CH_3CH{=}CH_2 + (HO)_2SO_2 \longrightarrow (CH_3)_2CHOSO_2OH$$

$$(CH_3)_2CHOSO_2OH + CH_3CH{=}CH_2 \longrightarrow [(CH_3)_2CHO]_2SO_2$$

$$(CH_3)_2CHOSO_2OH + H_2O \longrightarrow (CH_3)_2CHOH + H_2SO_4$$

$$[(CH_3)_2CHO]_2SO_2 + 2H_2O \longrightarrow 2(CH_3)_2CHOH + H_2SO_4$$

$$[(CH_3)_2CHO]_2SO_2 + (CH_3)_2CHOH \longrightarrow [(CH_3)_2CH]_2O + (CH_3)_2CHOSO_2SO_4$$

Isopropanol is also prepared by *direct* hydration by passing a mixture of propene and steam at 220-250°C, and under pressure (220 atm), over a catalyst of tungsten oxide plus zinc oxide, on a silica carrier.

Isopropanol, b.p. 82.4°C, is soluble in water, ethanol and ether. It is used for preparing esters, acetone, keten, as a solvent, and for high-octane fuel.

Butyl alcohols. Four isomers are possible, and all are known.

*n-Butyl alcohol, butan-*1*-ol, n-butanol,* b.p. 117.4°C, is prepared industrially from propene by the Oxo process. It is widely used as a solvent.

Isobutyl alcohol, 2*-methylpropan-*1*-ol, n-butanol,* b.p. 108°C, is also obtained by the Oxo process. It behaves as a primary alcohol, but it readily rearranges due to the presence of a branched chain near the COH group, *e.g.*, when it is treated with hydrochloric acid, isobutyl chloride and t-butyl chloride are obtained:

$$(CH_3)_2CHCH_2OH \xrightarrow{HCl} (CH_3)_2CHCH_2Cl + (CH_3)_3CCl$$

s-Butyl alcohol, butan--2-ol, s-butano, b.p. 100°C, is prepared industrially by the hydration of 1- or 2-butene (from cracked petroleum) by means of concentrated sulphuric acid (*cf.* ethanol and isopropanol). It is used for the preparation of butanone, esters, and as a lacquer solvent.

t-Butyl alcohol, t-butanol (*trimethylmethanol*), m.p. 25.5°C, is prepared synthetically by the hydration of isobutene (from cracked petroleum). It is mainly used as an alkylating agent.

Amyl alcohols. Eight isomers are possible, and all are known:

1. $CH_3CH_2CH_2CH_2CH_2OH$, n-amyl alcohol, pentan 1-ol, n-pentanol, p.b. 138°C.
2. $(CH_3)_2CHCH_2CH_2OH$, isoamyl alcohol, isopentanol, p.b. 130°C.
3. $CH_3CH_2CH(CH_3)CH_2OH$, 'active' amyl alcohol, 2-methylbutan-1-ol, b.p. 128°C.
4. $(CH_3)_3CCH_2OH$, neopentyl alcohol, t-butylcarbinol, b.p. 113°C.
5. $CH_3CH_2CH_2CH(OH)CH_3$, pentan-2-ol, methyl-n-propylcarbinol, p.b. 120°C.
6. $CH_3CH_2CH(OH)CH_2CH_3$, pentan-3-ol, diethylcarbinol, b.p. 117°C.
7. $(CH_3)_2CHCH(OH)CH_3$, 3-methylbutan-2-ol, isopropylmethyl-carbinol, b.p. 114°C.

8. $(CH_3)_2C(OH)CH_2CH_3$, 2-methylbutan-2-oil, t-amyl alcohol, t-pentanol, ethyldimethylcarbinol, b.p. 102°C.

There amyl alcohols, *viz.*, n-pentanol, isopentanol and 'active' amyl alcohol, have been isolated from fusel oil (*q.v.*). The last two are the chief constituents of fusel oil, and all three are produced by the fermentation of protein matter associated with the carbohydrates in starch. This mixture of amyl alcohols (from fusel oil) is used for the preparation of esters (for artificial essences), scents, and as a laboratory reducing agent with sodium it is better than ethanol owing to its higher boiling point (this mixture of amyl alcohols will be referred to in future as isopentanol).

A mixture of amyl alcohols, known as *pentasol*, is prepared industrially by chlorinating at 200°C, and in the dark, a mixture of n- and isopentanes (from petroleum) to the amyl chlorides which are hydrolysed with dilute sodium hydroxide solution plus a little sodium oleate for emulsification, to the amyl alcohols. Seven isomeric amyl chlorides are theoretically possible, but in practices six are obtained. 2-Chloro-2-methylbutane (the only t-chloride produced) readily eliminates hydrogen chloride to form trimethylethylene. Pentasol (the mixture of six amyl alcohols) finds great use as a solvent in the lacquer industry.

Commercial 's-amyl alcohol' (80 per cent 2- and 20 per cent 3-pentanol) is made by hydrating 1-and 2-pentenes (from cracked petroleum).

A number of higher alcohols occur as esters in waxes, *e.g.*, cetyl (palmityl) alcohol (hexadecan-1-ol), $C_{16}H_{33}OH$, m.p. 49°C, occurs as the palmitate in spermaceti (obtained from the oil of the sperm whale); carnaubyl alcohol (tetracosan-1-ol), $C_{24}H_{49}OH$, m.p. 69°C, as esters in wool-grease; ceryl alcohol (hexacosan-1-ol), $C_{26}H_{53}OH$, m.p. 79°C, as the cerotate in chinese wax; myricyl (melissyl) alcohol (triacontan-1-ol), $C_{30}H_{61}OH$, m.p. 88°C, as esters in bees-wax.

A number of higher alcohols are now prepared industrially by the catalytic reduction of the ethyl (or glyceryl) esters of the higher monocarboxylic acids, particularly the alcohols lauryl (dodecan-1-ol), $C_{12}H_{25}OH$, m.p. 24°C; myristyl (tetradecan-1-ol), C14H29OH, m.p. 38°C; palmityl (hexadecan-1-ol), $C_{16}H_{33}OH$, m.p. 49°C; and stearyl (octaecan-1-ol), $C_{18}H_{37}OH$, m.p. 59°C. These alcohols are used in the form of sodium alkyl sulphates, $ROSO_2ONa$, as detergents, emulsifying agents, wetting agents, insecticides, fungicides, etc. These sodium salts lather well, and are not affected by hard water, and hence can be used as a soap substitute.

POLYCARBONYL COMPOUNDS

Tautomerism : *Acetoacetic ester* or *ethyl acetoacetate* (E.A.A.) is the ethyl ester of acetoacetic acid, CH_3COCH_2H, a b-ketonic acid. Acetoacetic ester was fist discovered by Geuther (1863), who prepared it by the action of sodium on ethyl acetate, and suggested the formula $CH_3C(OH){=}CHCO_2C_2H_5$ (b-hydroxycrotonic ester). In 1865, Fankland and Duppa, who, independently of Geuther, also prepared acetoacetic ester by the action of sodium on ethyl acetate, proposed the formula $CH_3COCH_2CO_2C_2H_5$ (b-ketobutyric ester). These two formulae immediately gave rise to two schools of thought, one upholding the Geuther formula, and the other the Frankland-Duppa formula, each school bringing forward evidence to prove its own claim, *e.g.*,

1. **Evidence in favour of the Geuther Formula** (*reactions of an unsaturated alcohol*). (*i*) When acetoacetic ester is treated with sodium, hydrogen is evolved and the sodium derivative is formed. This indicates the presence of a hydroxyl group.

(*ii*) When acetoacetic ester is treated with an ethanolic solution of bromine, the colour of the latter is immediately discharged. This indicates the presence of an olefinic double bond.

(*iii*) When acetoacetic ester is treated with ferric chloride, a reddish-violet colour is produced.

This is characteristic or compounds containing the group $-C(OH){=}C\langle$ (*cf.* phenols).

2. **Evidence in favour of the Frankland-Duppa formula** (*reactions of a ketone*).

(i) Acetoacetic ester forms a bisulphite compound with sodium hydrogen sulphite.

(*ii*) Acetoacetic ester forms a cyanohydrin with hydrogen cyanide.

Thus the remarkable position arose where it was possible to show that a given compound had two different formulae, each of which was based on a number of *particular* reactions. The controversy continued until about 1910, when chemists were coming to the conclusion that both formulae were correct, and that the two compounds existed together in equilibrium in solution (or in the liquid state):

$$CH_3COCH_2CO_2C_2H_5 \rightleftharpoons CH_3\overset{\overset{\displaystyle OH}{|}}{C}{=}CHCO_2C_2H_5$$

When a reagent which reacts with ketones is added to acetoacetic ester, the ketone form is removed. This upsets the equilibrium, and in order to restore the equilibrium mixture, the hydroxy-form of acetoacetic ester changes into the ketone form. Thus, provided sufficient reagent is added, acetoacetic ester reacts completely as the ketone from. Similarly, when a reagent which reacts with alkenes or with hydroxy-compounds is added in sufficient quantity, acetoacetic ester reacts completely as the hydroxy-form.

The problem was finally settled by Knorr (1911), who succeeded in isolating both forms. He cooled a solution of acetoacetic ester in light petrol to – 78°C, and obtained crystals which melted at – 39°C. This substance gave no coloration with ferric chloride and did not combine with bromine, and was therefore the pure ketone form corresponding to the Frankland-Duppa formula. Knorr then suspended the sodium derivative of acetoacetic ester in light petrol cooled to – 78°C, and treated his suspension with just enough hydrogen chloride to decompose the sodium salt. He now obtained a product which did not crystallise, but set to a glassy solid when cooled. This substance gave an intense coloration with ferric chloride, and was therefore the pure hydroxy-form corresponding to the Geuther formula.

Thus acetoacetic ester is a substance that does the duty of two structural isomers, each isomer being capable of changing rapidly into the other when the equilibrium is disturbed, *e.g.*, by the addition of certain reagents. This is a case of *dynamic isomerism*, and the name *tautomerism* (Greek: *same parts*) was given to this phenomenon by Laar (1885). The two forms are known as *tautomers* or *tautomerides*, the ketone isomer being called the *keto* form, and the hydroxy isomer the *enol* form. Hence this type of tautomerism is known as *keto-enol* tautomerism. (In the I.U.P.A.C. system of nomenclature the suffix *-en* indicates the presence of a double bond, and the suffix *-ol*, a hydroxyl group. The word *enol* is a combination of these suffixes and indicates the structure of this form.)

When one tautomer is more stable than the other under ordinary conditions, the former is known as the *stable* form, and the latter as the *labile* form. In practice, it is generally difficult to say which is the labile form, since very often a slight change in the conditions, *e.g.*, temperature or solvent, shifts the equilibrium from keto to enol or *vice-versa* . Tautomerism in the solid state is rare, and hence, in the solid state, one or other tautomer is normally stable, but in the liquid or gaseous state, or in solution, the two forms usually exist as an equilibrium mixture.

It has been found that the enol form is more volatile than the keto, ad that he change from enol to keto is extremely sensitive to catalysts. Meyer *et al.* (1920, 1921) found that traces of basic compounds were very effective catalysts. Thus the found that soft glass vessels were unsuitable for the separation of the keto and enol forms, since, when fractionated, the more volatile enol form rapidly changed into the original keto-enol mixture under he catalytic influence of the walls of the containing vessel. Meyer, however, succeeded in separating the enol form from the keto by carrying out the fractional distillation under reduced pressure in silica apparatus which had been thoroughly cleaned (free from dust, moisture, etc.). Distillation under these conditions is known as *aseptic distillation.*

In tautomerism in general thee may be an equilibrium between two or more forms. One may predominate, or all may be present to about the same extent, the concentration of each form depending on the temperature and the solvent (if in solution). Although it may not be possible to separate tautomers owing to the ease and rapidity of their interconversion or, as in many cases, due to one form being almost completely absent, the presence of more than one compound may be shown by special properties of each isomer. Experiments using deuterium exchange reactions have also the presence of keto-enol mixtures, *e.g.*, Klar (1934) showed that hydrogen is exchanged shown the presence of keto-enol mixtures, *e.g.*, Klar (1934) showed that hydrogen is exchanged slowly by deuterium when acetaldehyde is dissolved in D_2O. Since the C—H bond in alkanes is stable under these conditions, the inference is that a hydroxyl group is present in acetaldehyde, *e.g.*, some enol form is present. Acetone was found to undergo this exchanged more rapidly.

The methods used for the quantitative estimation of each form in any tautomeric equilibrium mixture, of which the keto-enol system is only one example, fall into two distinct groups, physical and chemical.

PHYSICAL METHODS

Physical methods of not disturb the equilibrium. for they do not depend on the removal of one form, and they should therefore be used wherever possible.

(*i*) The refractive index of the equilibrium mixture is determined experimentally. The refractive indices of both the keto and the enol forms are calculated (from a table of atomic refractions), and from these figures it is then possible to calculate the amount of each form present in the equilibrium mixture. In some cases the refractive index of each form may be obtained directly by isolating it, *e.g.*, acetoacetic ester.

(*ii*) If *one* from is an electrolyte, the electrical conductivity of the mixture is determined experimentally, and the amount of this form present may be calculated from the results, *e.g.*, nitromethane.

(*iii*) The composition of the equilibrium mixture may be determined by means of optical rotation measurements (*cf.* mutarotation).

(*iv*) The position of equilibrium in b-diketones and β-keto-esters has been determined by proton magnetic resonance.

CHEMICAL METHODS

Since chemical methods cause the removal of one form, it is necessary to use a reagent that reacts with this form faster than the rate of interconversion of the tautomers. Meyer (1911, 1912) found that in the case of keto-enol tautomerism, bromine reacts instantaneously with the enol form, and so slowly with the keto form in comparison with the enol, that the keto reaction may be ignored. Meyer introduced two procedures, the *direct* and the *indirect* method. In the *direct* method a weighed sample of the keto-enol mixture dissolved in ethanol is *rapidly* titrated with a dilute ethanolic solution of bromine at 0°C (to slow down the interconversion of the tautomers). The first appearance of excess of bromine indicates the end point.

The titration must be carried out rapidly; otherwise the keto form changes into the enol during the time taken for the titration. In any case, it has been found impossible to carry out the titration sufficiently rapidly to avoid the conversion of some keto into enol, and so this method always results in too high a value for the conversion of some keto into enol, and so this method always results in too high a value for the enol form.

More reliable results may be obtained by the *indirect* method. An *excess* of dilute ethanolic solution of bromine is added rapidly to the weighed sample dissolved in ethanol, and then an excess of 2-nanphthol dissolved in ethanol is added *immediately.*

By this means, the excess of bromine is removed almost instantaneously, and so the keto-enol equilibrium is not given time to be disturbed. Potassium iodide solution and hydrochloric acid are now added, and the liberated iodine is titrated with standard thiosulphate. The overall equation is:

$$CH_3COCHBrCO_2Et + 2I^- + H^+ \longrightarrow CH_3COCH_2CO_2Ft + I_2 + Br^-$$

The mechanism of bromination of ethyl acetoacetate is probably (*cf.* acetone):

$$Br{-}Br \;\; CH{=}C(OH)CH_3 \text{ (with } CO_2Et \text{ on CH)} \longrightarrow Br^- + CHBrC(={}^+OH)CH_3 \text{ (with } CO_2Et \text{ on CHBr)} \xrightarrow{-H^+} CH_3COCHBrCO_2Et$$

Gero (1960, 1961) has determined the enol content of monoketones by adding an excess of iodine monochloride in methanol to a mixture of the dry ketone and sodium hydrogen carbonate, then adding an excess of aqueous sodium iodide and titrating with standard thiosulphate.

Schwarzenbach and Wittwer (1947) have introduced the *flow-method for* the estimation of the enol mixture and an acidified bromide-bromate solution are simultaneously mixed and diluted in a mixing chamber, and the mixture made to flow past a platinum electrode.

The relative amounts of he two solutions are adjusted so that the potential measured at the platinum electrode shows a sharp rise–this corresponds to the end-point in the titration of the enol form by the bromine. The method gives good results of enol contents as low as 10^{-5} per cent. *e.g.*, the enol content of acetone in aqueous solution was found to be 2.5×10^{-4} per cent.

The enol content in keto-enol mixtures (in dipropyl ether) has been determined by means of lithium aluminium hydride; the hydrogen liberated is estimated (Hofling *et al.*, 1952). The values obtained are about 10 per cent higher than those obtained by the bromine titration or by physical methods.

Table 2.1

Compound	*Per cent enol (in ethanol)*
$CH_3COCH_2CO_2CH_3$	4.8
$CH_3COCH_2CO_2C_2H_5$	7.5
$CH_3COCH_2COCH_3$	76
$CH_3COCH(CH_3)COCH_3$	31
$C_6H_5COCH_2COC_6H_5$	96
$CH_2(CO_2C_2H_5)_2$	trace
Aldehydes of type RCH_2CHO	trace
Ketones of type RCH_2COCH_2R	race

Enolisation

The phenomenon of enolisation is exhibited by compounds containing either a methylene group, —CH_2— , or a methyne group, $>$CH— adjacent to a carbonyl group. The actual amount of enol from present depends on a number of factors; these are discussed later.

When the methylene or methine group is attached to two or three carbonyl groups, the hydrogen atom might migrate equally well to one or other carbonyl group.

This is not found to be so in practice for unsymmetrical compounds, one enol form being present exclusively, or largely predominating; *e.g.*, in acetoacetic ester the hydrogen atom migrates exclusively, or largely predominating *e.g.*, in acetoacetic ester the hydrogen atom migrates exclusively to the acetyl carbonyl group. When two or more enol forms are theoretically possible, ozonolysis may be used to ascertain the structure of the form present; *e.g.*, in hexane-2,4-dione, $CH_3COCH_2COCH_2CH_3$, the two possible enols are:

$$CH_3C(OH){=}CHCOCH_2CH_3 \quad (I) \qquad CH_3COCH{=}C(OH)CH_2CH_3 \quad (II)$$

Ozonolysis of (I) will give CH_3CO_2H and CH_3CH_2COCHO; (II) will give CH_3COCHO and $CH_3CH_2CO_2H$. Identification of these compounds will decide whether the enol is (I) or (II), or both.

The type of tautomerism discussed above is known as known as the **keto-enol trial system**. (In this system a hydrogen atom migrates from atom 1 (*oxygen*) to atom 3 (*carbon*):

$$\overset{3}{\gt CH}-\overset{2}{C}=\overset{1}{O} \rightleftharpoons \overset{3}{\gt C}-\overset{2}{C}=\overset{1}{OH}$$

Enols resemble phenols in a number of ways; *e.g.*, both from soluble sodium salts; both give characteristic colorations with ferric chloride; and both couple with diazonium salts.

The keto-enol type of tautomerism is only one example of a triad system. The trial system is the most important class of tautomeric systems, and the following, which are exemplified in the text, are the commonest:

(*i*) *Three-carbon systems:*

$$CH-C=C \rightleftharpoons C=C-CH$$

(*ii*) *Nitroso-oximino system:*

$$CH-NO_2 \rightleftharpoons C=NOOH$$

(*iii*) *Nitroso-oximino system:*

$$CH-N=O \rightleftharpoons C=NOH$$

(*iv*) *Amidine system:*

$$NH-C=N \rightleftharpoons N=C-NH$$

(*v*) *Amido-imidol system:*

$$NH-C=O \rightleftharpoons N=C-OH$$

(*vi*) *Azo-hydozone system:*

$$N=N-CH \rightleftharpoons HN-N=C$$

(*vii*) *Diazo-amino (triazen) system:*

$$N=N-NH \rightleftharpoons NH-N=N$$

(*viii*) *Diazzo-nitrosamine system:*

$$Ar-N=NOH \rightleftharpoons Ar-NH-N=O$$

In addition to the 'open' systems of tautomerism, there is also, *e.g.*, *ring-chain tautomerism*.

Theories of Tautomerism

Ingold (1927) suggested the name *cationotropy* for all those cases of tautomerism which involve the separation of a *cation*; and the name *anionotropy* for those cases which involve the separation of an *anion*.

Lowry (1923) suggested the name *prototropy* for those cases in which a proton separates, and called such systems *prototropic* systems. Using Ingold's generalised classification of tuatomeric systems, it can be seen that prototropy is a special case of cationotropy. Braude and Jones (1944) have proposed the term *oxotropy* for anionotropic rearrangements involving only the migration of a *hydroxyl* group (see allylic rearrangement).

$$B + R_2CHCOR \longrightarrow BH^+ + R_2\bar{C}{-}\overset{\overset{\displaystyle O}{\|}}{C}R \longleftrightarrow R_2C{=}\overset{\overset{\displaystyle O^-}{|}}{C}R$$

(I)

Acid-catalysed enolisation involves the removal of a proton from the conjugate acid of the ketone, (II), and this process is dependent mainly on the hyperconjugation by the alkyl groups is the transition state for the formation of the carbon-carbon double bond.

$$H^+ + R_2\overset{\overset{\displaystyle H}{|}}{C}{-}\overset{\overset{\displaystyle O}{\|}}{C}{-}R \longrightarrow R_2\overset{\overset{\displaystyle H}{|}}{C}{-}\overset{\overset{\displaystyle OH}{|+}}{C}{-}R \longrightarrow H^+ + R_2C{=}\overset{\overset{\displaystyle OH}{|}}{C}{-}R$$

Evidence for these mechanisms is that alkyl groups depress base-catalysed reaction rates.

The mechanisms described above are step-wise mechanisms, and they appear to operate in certain cases. On the other hand, there is evidence that in other cases, acid-and base-catalysis of enolisation take place by a *concerted* mechanism, *i.e.*, the molecule undergoing change is attacked simultaneously at two places. Thus the enolisation of, *e.g.*, acetone, proceeds by, the simultaneous removal of a proton from a n α-carbon and the addition of a proton to the oxygen of the carbonyl group. This may be represented as follows (B is a base and HA is an acid):

$$\begin{array}{c}B\cdots\!\rightarrow H{-}CH_2\\ \quad\ \ C{=}O\leftarrow\cdots HA\\ \quad\ \ |\\ \quad\ \ CH_3\end{array} \longrightarrow \begin{array}{c}B\cdots H\cdots CH_2\\ \quad\ \ \vdots\\ \quad\ \ C{\cdots}O\cdots HA\\ \quad\ \ |\\ \quad\ \ CH_3\end{array} \longrightarrow \overset{+}{B}H + \begin{array}{c}CH_2\\ \|\\ C{-}OH\\ |\\ CH_3\end{array} + A^-$$

transition state

In this type of mechanism, the solvent water molecules are believed to be involved, one acting as a proton acceptor and other a proton donor. Furthermore, the above mechanism is termolecular, and Swain (1950) has presented evidence to show that both the acid- and base-catalysed enolisation of acetone are termoleular reactions. Emmons *et al.* (1956)

have obtained evidence supporting Swain's termolecular mechanism for ketone enolisation, but propose that the transition state for both acid- and base-catalysed enolisation may be represented by (III). Under these conditions in either acid-or base-catalysed reactions, the transition state will be very close to the enol and hence will be stabilised by hyperconjugation. In the acid-actalysed reaction, bond *b* is relatively tight and bond *a* is relatively loose. Hence in acid-catalysed enolisation, steric hindrance around bond *a* is not observed in most cases, and hyperconjugation is therefore a dominating factor in the enolisation of a ketone. In the base-catalysed reaction, bond *a* is relatively tight and bond *b* is relatively loose. Under these conditions a steric factor becomes important and is probably the reason why alkyl groups depress base-catalysed reaction rates.

The tremendous increase in speed by acid and base catalysts is well illustrated by the fact that the pure keto and enol forms of ethyl acetoacetate change very slowly into the equilibrium mixture (several weeks), whereas the latter is obtained rapidly (several minutes) by the addition of acid or base catalysts (lowry *et al.*, 1924).

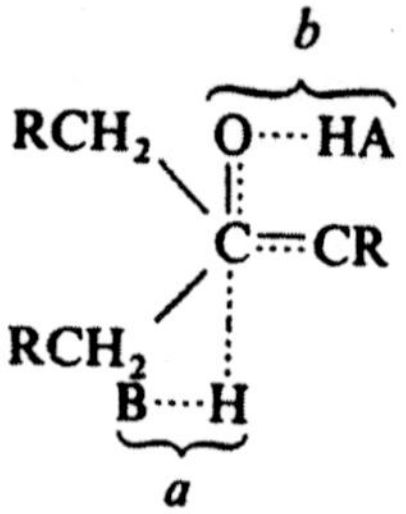

b
RCH_2 O····HA
C┈CR
RCH_2
B····H
a

(III)

Position of Equilibrium in Keto-enol Tautomerism

Since $\Delta G^{\ominus} = \Delta H^{\ominus} - T\Delta S^{\ominus} = -RT \ln K$

a knowledge of the values of the enthalpy and entropy changes would enable us to calculate $\Delta G^{\ominus}$, and from this, K, the equilibrium constant of the reaction:

$$\text{keto enol;} \rightleftharpoons K = [\text{enol}]/[\text{keto}]$$

The sum of the bond energies of the — CH_2CO— group is 1870 and that of the — CH═C(OH)— groups is 1820 KJ:

$$2C—H(828.4) + C — C(347.3) + C{=}O(694.5) = 1870 \text{ KJ}$$

$$C—H(414.2) + C{=} C(606.7) + C — O(334.7) + O — H(464.4) = 1820 \text{ KJ}$$

This means that the *enthalpy of formation* of the CH_2CO group is – 1870 KJ and that of the CH═C(OH) group is – 1820 KJ. Thus the keto from is more stable than the enol by 50 KJ mol^{-1}, and hence there must be some driving force to bring about enolisation.

If the equilibrium mixture contains 1 per cent enol, then.

$$\Delta G^{\ominus} = -5.7 \log 1/99 = 11.42 \text{ kJ}$$

If the equilibrium mixture contains 99 per cent enol, then

$$\Delta G^{\ominus} = -5.7 \log 99 = -11.42 \text{ kJ}$$

Thus a difference of 22.84 kJ mol^{-1} in ΔG will have large effects on the position of the equilibrium. Hence any attempt to calculate K from *estimated* values of ΔH and ΔS is almost certainly doomed to failure. However, a semi-quantitative approach is instructive.

Any factors hat affect $\Delta H^{\ominus}$ and $\Delta S^{\ominus}$ will affect the value of $\Delta G^{\ominus}$. Consider the following equilibria:

acetylacetone Me—C(=O)—CH_2—C(=O)—Me ⇌ MeC(—O—H···O=)CMe, with MeC=CH—CMe (enol, intramolecular H-bond)

ethyl acetoacetate Me—C(=O)—CH_2—C(=O)—OEt ⇌ MeC(—O—H···O=)COEt, with MeC=CH—COEt (enol, intramolecular H-bond)

ethyl malonate EtOC(=O)—CH_2—C(=O)—OEt ⇌ EtOC(—O—H···O=)COEt, with EtOC=CH—COEt (enol, intramolecular H-bond)

The enol form, in each case, is stabilised by intramolecular hydrogen bonding (say, 29.3 kJ mol^{-1}).

Thus, in the absence of any other factor affecting ΔH, we have:

$$\Delta H^{\ominus} = \Sigma H^{\ominus} \text{ (products)} - \Sigma H^{\ominus} \text{ (reactants)}$$

$$= -(1820 + 29.3) - (-1870) = +20.7 \text{ kJ}$$

$$\Delta G^{\ominus} = 20.7 - T\Delta S^{\ominus} = -RT \ln K$$

Now let us consider the resonance energies of crotonaldehyde (I), ethyl acetate (II) and ethyl

$$\underset{\text{(I) 10.04}}{\text{MeCH=CHCHO}} \qquad \underset{\text{(II) 75.31}}{\text{MeC(=O)—OEt}} \qquad \underset{\text{(III) 75.31}}{\text{CH}_2\text{=C(Me)—C(=O)—OEt}}$$

R.E. (kJ)

methacrylate (III). The important point here is that in (III) the ethylenic bond does not enter into resonance with the CO_2Et group. (III) contains crossed conjugation , and the contribution to resonance is only the path that leads to the greater R.E., which, in this case, is the carbethoxy group.

Inspection of the enol formulae shows that the ring contains a conjugated system:

$$\text{H—O—}\overset{|}{\text{C}}\text{=CH—}\overset{|}{\text{C}}\text{=O} \longleftrightarrow \text{H—}\overset{+}{\text{O}}\text{=}\overset{|}{\text{C}}\text{—CH=}\overset{|}{\text{C}}\text{—O}^-$$

The R.E. value of this is difficult to assess, but it can be expected to be considerably less than 73.31 kJ. Thus, in E.A.A., only the remaining group, HO—C=CH—, will make any contribution to the 'ring' resonance. It therefore follows that the enol form of E.A.A. is less stabilised by resonance than that of acetylacetone, and so will be present in smaller amount. It might also be noted here that if enolisation in E.A.A. occurred at the CO group of CO_2Et, then the R.E. of the system produced is far less. Thus there is no tendency to enolise in this direction. From this it can be seen that D.E.M. will have far less enol form than E.A.A. In the case of simple carbonyl compounds, *e.g.*, acetone, there is no intramolecular hydrogen bonding to stabilise the enol. Thus there is very little tendency for these carbonyl compounds to enolise.

Just as it is difficult to estimate $\Delta H^{\ominus}$, so it is more difficult to estimate $\Delta S^{\ominus}$. In the diketo form, the molecule is acyclic and consequently can assume various conformations. In the enol form, however, the presence of a double bond prevents change in conformation at this part of the molecule and, at the same time, conformational change is further restricted by hydrogen bonding, since this tends to maintaining structure. Thus both of these factors will tend to cause a *decrease* in entropy when the diketo compound enolises.

Some support for this argument is that Geo (1961) has found that cyclic monoketones contain more enol than the corresponding acyclic 2-one. In the cyclic ketones, change from keto to enol involves a relatively small change in strain in the ring due to the introduction of a double bond. For acyclic ketones, the introduction of the double bond has a much greater effect on the freedom to take up different conformations.

Another factor that determines the enol content is the steric factor. Let us consider the enol content of acetylacetone and α-methylacetylacetone; the values given are for the gas phase (Conant *et al.*, 1932). In the latter compound, thee is much greater steric repulsion due to the presence of the α-methyl group. Thus, the α-methyl enol form has greater internal strain than the enol of acetylacetone and so the former is less favourable energetically than the latter. Hence less enol will be present in the former equilibrium mixture than in the latter. A striking example for strain affecting the position of equilibrium is the case of triacylmethanes, (RCO)3CH. Nonhebel (1967) has shown, by means of i.r., u.v., and NMR studies, that as the size of R increases, the amount of enol content decreases, *e.g.*, $(Bu^tCO)_3CH$ is *completely* in the keto form.

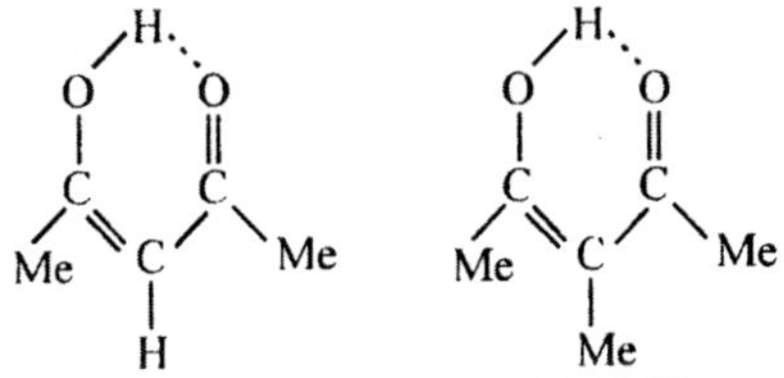

Enol of acetylacetone (Me, C–H)	Enol of α-methylacetylacetone (Me, C–Me)
91–93%	43.5–44.5%

Finally, let us consider the effect of the nature of the solvent on the enol content. If one form is more solvated than the other, then there may be considerable entropy effects connected with the nature of the solvent (*cf.* acids). Any solvent that can form hydrogen bonds with the carbonyl group of the keto form will cause a decrease in entropy. The enol form, however, since it forms an intramolecular bond, will be largely prevented from forming hydrogen bonds with the solvent, and consequently there will be a smaller decrease in entropy. Thus, the term ΔS becomes important in deciding the position of keto–enol equilibrium, *e.g.*, solvents such as water, methanol, acetic acid, etc., tend to reduce the enol content. On the other hand, in solvents such as hexane, benzene, etc., the enol content will be larger, *e.g.*, the enol content of acetylacetone in hexane is 92 per cent.

It has been shown that enol content of E.A.A. (pure liquid) decreases as the pressure increases (le Noble, 1960). From this it follows that the volume of the enol form is greater than that of the keto.

THE ACIDITY OF KETO COMPOUNDS

The proton transfer for these compounds may be written:

$$-CH_2-\overset{\overset{\displaystyle O}{\|}}{C}- + B \rightleftharpoons BH^+ + -\ddot{C}H-\overset{\overset{\displaystyle O}{\|}}{C}- \leftrightarrow -CH=\overset{\overset{\displaystyle O^-}{|}}{C}-$$

The carbanion produced is a resonance hybrid, and the transference of charge to the oxygen atom which has a higher electronegativity than the carbon atom will make this resonating structure a greater contributor to the resonance hybrid; the result is an ambident anion. The presence of the carbonyl group facilitates the release of a proton from the α-carbon atom, and the resonance hybrid nature of the conjugate base stabilises this with respect to its acid. Both factors are acting in the 'same' direction, and consequently such compounds will be acidic. However, the pK_a of the compound will depend on how strong these two driving forces are. As we have seen above, since enolisation involves a removal of a proton first, it follows that the acid strengths of these carbonyl compound, in a proton-accepting solvent, will run parallelled with the enol content of each compound, *e.g.*,

	E.A.A.	(MeCO)2CH2
pK_a	10·7	8·24

Infra-red Spectra of β-diketones

If β-diketones existed as such, one would expect to find absorption in the region of a single carbonyl group (1725–1700 cm^{-1}), but with greater intensity because two such groups are present. In practice, those β-diketones which exist partly (or wholly) in the enol from (as indicated by the coloration with ferric chloride) show the carbonyl absorption in the region 1640–1540 cm^{-1} (s); the enolic form of β-keto-esters absorbs in the region 1655-1635 cm^{-1} (s), α,β-Unsaturated ketones, *i.e.*, those containing the grouping C=C—C=O absorb in the region 1685–1665 cm^{-1} (s). Enols contain this grouping, but the shift that occurs in these compounds is attributed to hydrogen bonding as follows: —O···O=C— (see chelated structures of the enols described above). At the same time, the true alcoholic hydroxyl band near 3700 cm^{-1} is absent in enols, but there is a band near 2700 cm^{-1} (s) which is attributed to the *chelated* hydroxyl group. Thus, infra-red spectral studies of β-dicarbonyl compounds may be used to ascertain their structures.

NMR spectra studies have been very useful in the examination of keto -enol tautomeric mixtures, *e.g.*, Fig 2.12 is the NMR spectrum of ethyl acetoacetate (in CCl_4 at 60 MHz). and it can be seen that the CH_2 and =CH protons give separate signals. Furthermore, very careful integration will give the proton ratio of these two groups and so it is possible to estimate the amount of enol content.

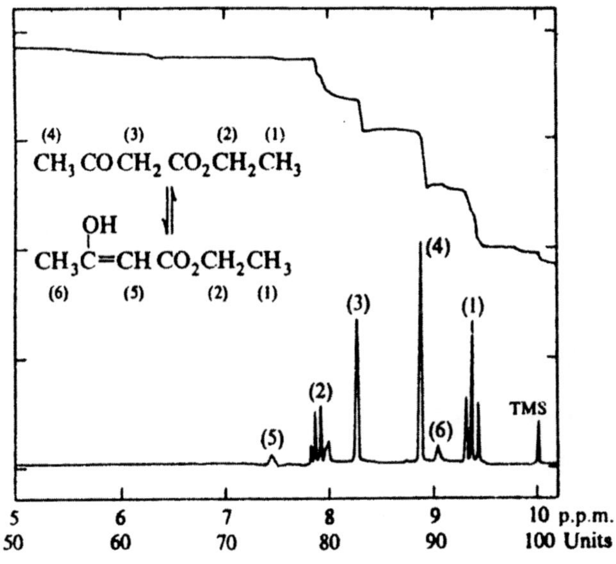

Fig. 2.12

The Preparation of Ethyl Acetoacetate

In the preparation of ethyl acetoacetate, condensation is effected between two molecules of the same ester. This is one example of the **Claisen condensation** (1887), in which a keto-ester is formed by the reaction between two molecules of an ester containing α-hydrogen atoms.

The Claisen condensation may also take place between an ester and a ketone to form a 1.3-diketone. The condensation is brought about by sodium ethoxide, sodamide, triphenylmethylsodium, etc.

Many mechanisms have been proposed for the formation of ethyl acetoacetate. The one most widely accepted is (*cf.* aldol condensation):

There is much evidence to support this mechanism, *e.g.*;

(*i*) Compounds containing an active methylene group undergo deuterium exchange with sodium ethoxide in the presence of EtOD (*inter alia*, Brown *et al.*, 1940); this can be explained by the *reversible* formation of a carbanion,

$$EtO^- + H{-}CH_2CO_2Et \rightleftharpoons EtOH + \bar{C}H_2CO_2Et$$

$$MeC(=O)(OEt) + \bar{C}H_2CO_2Et \rightleftharpoons MeC(O^-)(OEt){-}CH_2CO_2Et \rightleftharpoons$$

$$MeC(=O){-}CH(H)CO_2Et + EtO^- \rightleftharpoons EtOH + MeC(\bar{O}){=}CHCO_2Et \xrightarrow{MeCO_2H}$$

$$MeC(OH){=}CHCO_2Et \rightleftharpoons MeC(=O)CH_2CO_2Et \quad (28\text{–}29\%)$$

$$—CH_2CO— + EtO^- \rightleftharpoons EtOH + —\bar{C}HCO— \underset{}{\overset{EtOH}{\rightleftharpoons}} —CHDCO— + EtOH$$

(*ii*) Optically active esters of the type $R^1R^2CHCO_2Et$ are racemised by the ethoxide ion (Kenyon *et al.*, 1940); the intermediate carbanion would be expected to racemise.

The problem now is; Since the condensation is reversible, what factors favour shifting the overall equilibrium to the right? The accepted explanation is as follows. Because ethanol is a stronger acid than ethyl acetate, the equilibrium of the first stage lies on the left. However, E.A.A. is a stronger acid pK_a 10.68) than ethanol (pK_a 15.5). Thus, the former is present mainly as its anion, *i.e.*, conjugate base, in the last stage of the condensation, and so the equilibrium lies largely on the right. In this way, the E.A.A. anion is 'removed' from the reaction, and so the overall equilibrium is displaced to the right. This argument is supported by the fact that ethyl isobutyrate does *not* undergo the Claisen condensation. The β-keto ester that would be produced, $Me_2CHCOCMe_2CO_2Et$, does not contain an active α-hydrogen atom, and consequently the last reversible stage of the reaction given for ethyl acetate as substrate is now absent. Hence, for ethyl isobutyrate, the overall equilibrium lies almost completely on the left. If a very strong base such as the triphenylmethyl anion (as sodium salt), $Ph_3C^-Na^+$ (pK_a of Ph_3CH is about 33) is used, then the equilibrium of the first stage with ethyl isobutyrate lies on the right, and the Claisen condensation will proceed, since the position of equilibrium of the last stage will no longer matter.

Ethyl acetoacetate is also prepared industrially by polymerising ketone in acetone solution to diketone, which is then treated with ethanol. A possible mechanism is :

$$CH_2{=}C{=}O + CH_2{=}C{=}O \longrightarrow \underset{CH_2-C{=}O}{CH_2{=}C-O} \xrightarrow{EtOH} \underset{\underset{O^-}{|}}{\overset{CH_2{=}C-O}{CH_2-C-\overset{+}{O}(H)OEt}} \longrightarrow$$

$$\left[\overset{O^-\cdots H-\overset{+}{O}-Et}{CH_2{=}C-CH_2-C{=}O}\right] \longrightarrow \left[\overset{O-H\cdots O}{CH_2{=}C-CH_2-C-OEt}\right] \longrightarrow CH_3\overset{O}{\overset{\|}{C}}CH_2\overset{O}{\overset{\|}{C}}-OEt$$

In 'mixed ethyl acetoacetate condensations' between two different esters, a mixture of all four possible products is usually obtained; *e.g.*, with ethyl acetate and ethyl propionate:

$$CH_3CO\boxed{OC_2H_5+H}-CH_2CO_2C_2H_5 \xrightarrow{C_2H_5ONa} CH_3COCH_2CO_2C_2H_5+C_2H_5OH$$

$$CH_3CH_2CO\boxed{OC_2H_5+H}-\overset{CH_3}{\overset{|}{C}}H-CO_2C_2H_5 \xrightarrow{C_2H_5ONa} CH_3CH_2CO\overset{CH_3}{\overset{|}{C}}HCO_2C_2H_5+C_2H_5OH$$

$$CH_3CO\boxed{OC_2H_5+H}-\overset{CH_3}{\overset{|}{C}}HCO_2C_2H_5 \xrightarrow{C_2H_5ONa} CH_3CO\overset{CH_3}{\overset{|}{C}}HCO_2C_2H_5+C_2H_5OH$$

$$CH_3CH_2CO\boxed{OC_2H_5+H}-CH_2CO_2C_2H_5 \xrightarrow{C_2H_5ONa} CH_3CH_2COCH_2CO_2C_2H_5+C_2H_5OH$$

Esters with no α-hydrogen atoms will condense only 'one way' with compounds containing an active methylene group. Three important esters of this type are formic, benzoic and oxalic esters, and these give, respectively, the formyl, benzoyl and oxalyl derivatives Ethyl carbonate also does not contain any α-hydrogen atoms, and is very useful for introducing a carbethyoxyl group into compounds with an active methylene group.

β-Keto esters, other than E.A.A., may be prepared by the Claisen condensation between ketones and ethyl carbonate or ethyl oxalate (*cf.* decarbonylation of ethyl pyruvate):

$$RCOMe + EtO_2CCO_2Et \xrightarrow{EtONa} RCOCH_2COCO_2Et$$
$$\xrightarrow{heat} RCOCH_2CO_2Et + CO$$

PROPERTIES OF ETHYL ACETOACETATE

Ethyl acetoacetate is a colourless, pleasant-smelling liquid, p.b. 181°C (with slight decomposition), sparingly soluble in water, but miscible with most organic solvents. It is readily reduced by sodium amalgam, or catalytically to β-hydroxybutyric ester:

$$CH_3COCH_2COC_2H_5 \longrightarrow CH_3CH(OH)CH_2CO_2C_2H_5$$

This is also produced by using lithium aluminium hydride in pyridine. In the absence of pyridine, the product is butane-1,3-diol (Buchta *et al.*, 1951).

$$CH_3COCH_2COC_2H_5 \xrightarrow{LiAlH_4} CH_3CHOHCH_2CH_2OH\ (30\%)$$

Kent *et al.* (1963) have shown that sodium borohydride in boiling ethanol also produces the diol.

Ethyl acetoacetate is neutral to litmus, but is soluble in dilute sodium hydroxide solution: it is the enol form which dissolves to form the sodium salt. It does *not* give the haloform reaction. When ethyl acetoacetate is hydrolysed in the cold with dilute sodium hydroxide, the solution then acidified, extracted with ether and the ether removed in the cold under reduced pressure, free acetoacetic acid is obtained. Krueger (1952) has prepared acetoacetic acid as a crystalline solid, m.p. 36–37°C. As prepared in the usual way it is unstable, readily decomposing into acetone ad carbon dioxide:

$$CH_3COCH_2CO_2H \longrightarrow CH_3COCH_3 + CO_2$$

All β-ketonic acids readily decompose in this manner, but it is interesting to note that trifluoroacetoacetic acid, $CF_3COCH_2CO_2H$, is quite stable and can be distilled without much decomposition. Ethyl acetoacetate forms α-oximinoacetoacetate when treated with nitrous acid:

$$CH_3COCH_2CO_2C_2H_5 \xrightarrow{HNO_2} CH_3COC(=NOH)CO_2C_2H_5$$

On the other hand, monosubstituted derivatives of ethyl acetoacetate are split by nitrous acid; the nitroso-compound is formed first and then rearranges to the oximino form with fission of the molecule.

$$CH_3COCHRCO_2C_2H_5 \xrightarrow{HNO_2} CH_3CO\underset{\underset{NO}{|}}{C}RCO_2C_2H_5 \longrightarrow R\underset{\underset{NO}{||}}{C}CO_2C_2H_5$$

Ethyl acetoacetate forms a green copper compound which is soluble in organic solvents. This indicates that the copper compound is not an ionic but a chelate compound (the alternative is a π-delocalised structure). Bromination of E.A.A. is catalysed by the presence of cupric ions; this is due to the stabilisation of the enol form by chelate formation.

```
Me—C—O         O=C—OEt              Me—C⋯O     O⋯C—OEt
  //    \     ↙    \                   //    ⋰ ⋱     \\
 CH      Cu          CH       or      CH      Cu       CH
  \     ↗   \     //                   \\   ⋱ ⋰      //
EtO—C=O       O—C—Me               EtO—C⋯O     O⋯C—Me
```

E.A.A. reacts with ammonia (and primary amines) to form an enamine:

$$Me-\overset{O}{\overset{\|}{C}}-CH_2CO_2Et + NH_3 \longrightarrow Me-\underset{NH_2}{\underset{|}{\overset{OH}{\overset{|}{C}}}}-CH_2CO_2Et \xrightarrow{-H_2O} Me-\underset{NH_2}{\underset{|}{C}}=CHCO_2Et$$

Monoalkyl derivatives of E.A.A. also behave in a similar manner, but dialkyl derivatives form amides, $MeCOCR_2CONH_2$. In the latter, there is no α-hydrogen atom.

Ethyl acetoacetate reacts with hydoxylamine to form an isoxazolone.

$$CH_2COCH_2CO_2C_2H_5 + NH_2OH \longrightarrow CH_3\underset{NOH}{\underset{\|}{C}}CH_2CO_2C_2H_5 \longrightarrow \begin{array}{c} CH_3C—CH_2 \\ \| \quad\quad | \\ N \quad\; CO \\ \diagdown O \diagup \end{array} + C_2H_5OH$$

With phenylhydrazine, ethyl acetoacetate forms a pyrazolone.

Acetoacetic ester undergoes the Knoevengel reaction due to the presence of an 'active' methylene group, and also couples with diazonium salts. It reacts with Grignard reagents to form the hydrocarbon, which indicates that it reacts in the enol form, *e.g.*,

$$CH_3\overset{OH}{\overset{|}{C}}=CHCO_2C_2H_5 + CH_3MgI \longrightarrow CH_3\overset{OH}{\overset{|}{C}}=CHCO_2C_2H_5 + CH_4$$

$$\begin{array}{c} O \\ \| \\ C \\ \diagup \quad \diagdown \\ CH_2 \quad OEt \\ | \\ MeCO \end{array} + \underset{CO_2Et}{\underset{|}{\bar{C}HCOMe}} \longrightarrow \left[\begin{array}{c} \bar{O} \quad OEt \\ \diagdown | \diagup \\ C \\ \diagup \quad \diagdown \\ CH_2 \quad CHCOMe \\ | \qquad\quad | \\ MeCO \quad CO_2Et \end{array} \right]$$

$$\xrightarrow[\text{(enolisn)}]{-EtO^-} \left[\begin{array}{c} O \\ \| \\ C \\ \diagup \quad \diagdown \\ CH \qquad CHCOMe \\ \| \qquad\quad | \quad O \\ MeC \qquad\quad C \diagup \\ \diagdown O—H \quad \diagdown OEt \end{array} \right] \xrightarrow{-EtOH} \text{(3-acetyl-6-methyl-pyran-2,4-dione; Me, O, COMe, =O)} \rightleftharpoons \text{(4-hydroxy-3-acetyl-6-methyl-2-pyrone; OH, COMe, Me, O, =O)} \; (53\%)$$

The *O*-methyl derivative is also produced by the action of diazomethane on ethyl acetoacetate. Thus the latter compound reacts in the enol form.

When acetoacetic ester is heated under reflux with a trace of sodium hydrogen carbonate, *dehydroacetic acid*, m.p. 109°C, is obtained.

Dehydroacetic acid may also be prepared by the polymerisation of diketen:

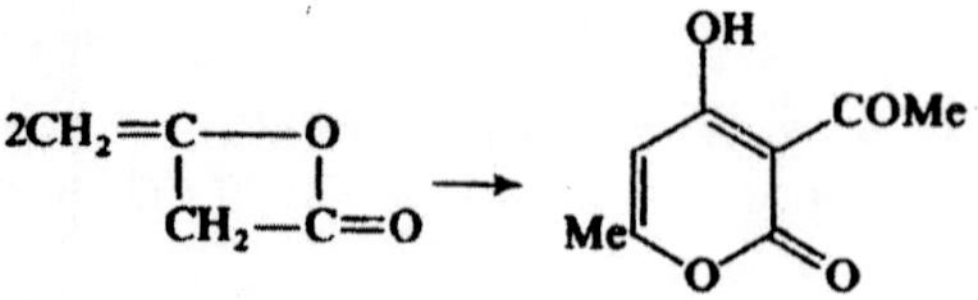

$$2CH_2{=}C{-}O,\ CH_2{-}C{=}O \longrightarrow \text{(pyrone ring with OH, COMe, Me, =O)}$$

When E.A.A. is allowed to stand in concentrated sulphuric acid (5–6 days), *isodehydroacetic* acid, m.p. 155°C, and its ester are formed:

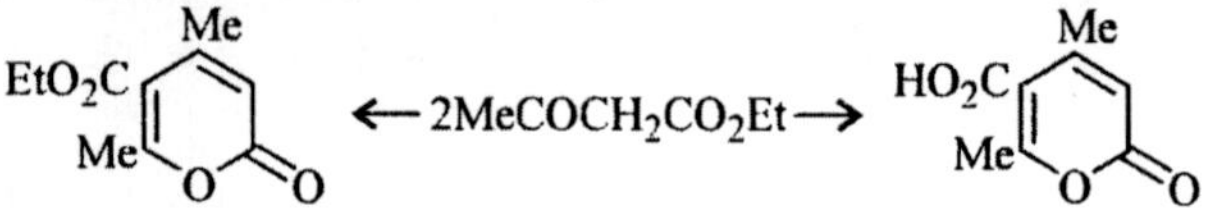

$$\text{(EtO}_2\text{C, Me, Me pyrone)} \longleftarrow 2MeCOCH_2CO_2Et \longrightarrow \text{(HO}_2\text{C, Me, Me pyrone)}$$

(27–36%) (22–27%)

Both acids are α-pyrone deivatives.

The Use of Acetoacetic Ester in the Synthesis of Ketones and Monocarboxylic Acids

The synthetic use of acetoacetic ester depends on two chemical properties:

1. (*a*) When treated with sodium ethoxide, acetoacetic ester forms sodioacetoacetic ester:

$$CH_3C(OH){=}CHCO_2C_2H_5 + C_2H_5ONa \longrightarrow$$

$$\left[CH_3C(O^-){=}CHCO_2C_2H_5\right]Na + C_2H_5OH\ (ex.)$$

(*b*) So acetoacetic ester readily reacts with primary and secondary alkyl halides (vinyl and aryl halides do *not* react) to produce alkyl derivatives of acetoacetic ester in which the alkyl group is attached to carbon. This fact has given rise to considerable speculation regarding the mechanism of the alkylation process. The problem is still not settled, but a highly favoured theory is that the negative ionis a resonance hybrid.

$$\underset{\text{(I)}}{CH_3\overset{\overset{O^-}{|}}{C}=CHCO_2C_2H_5} \longleftrightarrow \underset{\text{(II)}}{CH_3\overset{\overset{O^-}{\|}}{C}-\overset{..}{C}HCO_2C_2H_5}$$

i.e., this ion is ambident. Kornblum *et al.* (1955) have pointed out that the tendency for alkylation at the more electronegative atom of an ambident anionusually increases with the S_N1 character of the reaction. As far as the alkylation of E.A.A. is concerned, the evidence is in favour of an S_N2 mechanism and consequently attack occurs at the carbon atom.

$$\begin{array}{c} CH_3-C=O \\ | \\ H\overset{..}{C}:^- \; R-X \\ | \\ CO_2C_2H_5 \end{array} \longrightarrow \begin{array}{c} CH_3-C=O \\ | \\ HC-R + X^- \\ | \\ CO_2C_2H_5 \end{array}$$

For the sake of simplicity the alkylation of sodioacetoacetic ester will, in future, be represented as:

$$[CH_3COCHCO_2C2H_5]^-Na^+ + RX \longrightarrow CH_3COCHRCO_2C_2H_5 + NaX (v.g.)$$

After one alkyl group has been introduced, the whole process may then be repeated to give the dialkyl derivative of acetoacetic ester:

$$CH_3COCHR^1CO_2C_2H5 \xrightarrow{C_2H_5ONa} [CH_3COCR^1CO_2C_2H_5]^-Na^+$$

$$\xrightarrow{R^2X} CH_3COCR^1R_2CO_2C_2H_5 \; (g.\text{–}v.g.)$$

Formerly, it was not considered possible to prepare *disubsituted* derivatives of acetoacetic ester in *one* step. Sandberg (1957), however, has now carried out this one-step reaction in the preparation of ethyl β-acetotricarballylate from acetoacetic ester (0.5 mole), ethyl bromoacetate (1.2 mole) and sodium hydride (1.2 mole) in benzene solution.

$$CH_3COCH_2CO_2C_2H_5 + 2BrCH_2CO_2C_2H_5 \xrightarrow{2NaH}$$

$$CH_3COC\begin{cases} CH_2CO_2C_2H_5 \\ CO_2C_2H_5 \\ CH_2CO_2C_2H_5 \end{cases} \quad (77\%)$$

2. Acetoacetic ester and its alkyl derivatives can undergo two types of hydrolysis with potassium hydroxide:

(*a*) **Ketonic hydrolysis.** Ketonic hydrolysis, so called because a ketone is the chief product, is carried out by boiling with *dilute* aqueous or ethanolic potassium hydroxide solution, and is usually followed by acidification (dilute sulphuric acid) and warming:

$$CH_3COCHRCO_2C_2H_5 \xrightarrow[\text{(ii) } H_2SO_4]{\text{(i) KOH}} [CH_3COCHRCO_2H] \xrightarrow{-CO_2} CH_3COCH_2R \ (g.)$$

The ketone obtained is acetone or its derivatives, and the latter *alway contain the goup* CH_3CO—.

Dehn and Jackson (1933) found that 85 per cent phosphoric acid was a very good catalyst for the ketonic hydrolysis of acetoacetic ester and its alkyl derivatives, the yield of ketone reaching 95 per cent.

The mechanism of the decarboxylation of E.A.A. has been the subject of much discussion. Evidence obtained by several workers has led to the proposal that decarboxylation occurs through a cyclic transition state, and that the product is originally produced in the enol form.

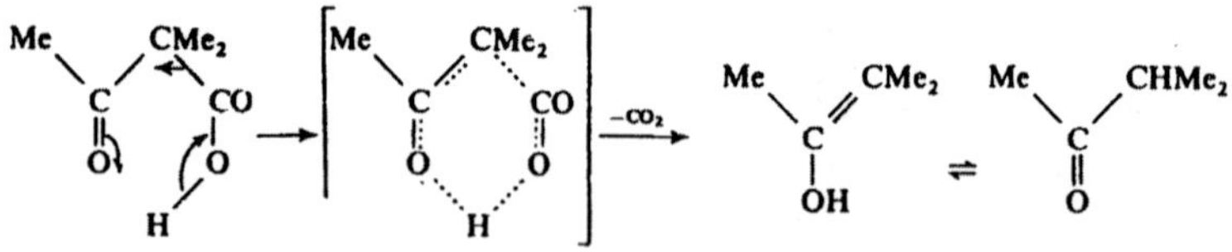

Me, CMe₂, C, CO, O, O, H → [Me, CMe₂, C, CO, O, O, H] $\xrightarrow{-CO_2}$ Me, CMe₂, C, OH ⇌ Me, CHMe₂, C, O

(*b*) **Acid hydrolysis.** Acid hydrolysis, so called because an acid is the chief product, is carried out by boiling with *concentrated* ethanolic potassium hydroxide solution. *e.g.*,

$$CH_3COCH_2CO_2C_2H_5{+}2KOH \longrightarrow CH_3CO_2K{+}CH_3CO_2K{+}C_2H_5OH \ (f.g.)$$

$$CH_3COCH_2CO_2C_2H_5{+}2KOH \longrightarrow CH_3CO_2K{+}R_2CHCO_2K{+}C_2H_5OH \ (f.g.)$$

The acid obtained is acetic acid or its derivatives as the potassium salt. Form this the free acid is readily obtained by treatment with inorganic acid.

The mechansim of this cleavage is possibly:

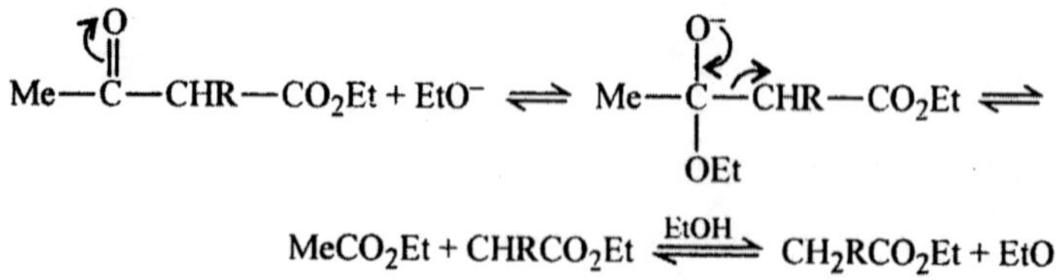

$$Me{-}\overset{O}{\overset{\|}{C}}{-}CHR{-}CO_2Et + EtO^- \rightleftharpoons Me{-}\overset{O^-}{\overset{|}{\underset{OEt}{\underset{|}{C}}}}{-}CHR{-}CO_2Et \rightleftharpoons$$

$$MeCO_2Et + CHRCO_2Et \overset{EtOH}{\rightleftharpoons} CH_2RCO_2Et + EtO$$

This is the reverse of the Claisen condensation. However, the esters formed are hydrolysed by the potassium hydroxide to their salts, and since do not undergo the condensation, the final equilibrium is driven completely to the right.

EXAMPLES OF THE SYNTHESIS OF KETONES

The formula of the ketone is written down, and *provided it contains the group* CH_3CO—, the ketone can be synthesised via acetoacetic ester as follows. The acetone nucleus is picked out, and the alkyl groups attached to it are then introduced into acetoacetic ester one at a time; this is followed by ketonic hydrolysis. It is usually better to introduce the larger group before the smaller (steric effect):

(*i*) $\boxed{CH_3COCH_2}CH_3$

$$CH_3COCH_2CO_2C_2H_5 \xrightarrow{C_2H_5ONa} [CH_3COCHCO_2C_2H_5]^-Na^+$$

$$\xrightarrow{CH_3I} CH_3COCH(CH_3)CO_2C_2H_5 \xrightarrow[\text{(ii) } H^+]{\text{(i) dil. KOH}} CH_3COCH_2CH_3$$

(*ii*) 3-*Methylentan-2-one.* $\boxed{CH_3COCH(CH_3)}CH_2CH_3$

$$CH_3COCH_2CO_2C_2H_5 \xrightarrow{C_2H_5ONa} [CH_3COCHCO_2C_2H_5]^-Na^+$$

$$\xrightarrow{C_2H_5I} CH_3COCH(C_2H_5)CO_2C_2H_5 \xrightarrow{C_2H_5ONa}$$

$$[CH_3COCH(C_2H_5)CO_2C_2H_5]^- Na^+ \xrightarrow{CH_3I} CH_3COC(CH_3)(C_2H_5)CO_2C_2H_5 \xrightarrow[\text{(ii) } H^+]{\text{(i) dil. KOH}} CH_3COCH(CH_3)CH_2CH_3$$

Examples of the Synthesis of Monocarboxylic Acids

The approach is similar to that for ketones except that the acetic acid nucleus is picked out, and the acetoacetic ester derivative is subjected to 'acid' hydrolysis. The acetoacetic ester acid synthesis is usually confined to the preparation of straight-chain acids or brached-chain acids where the branching occurs on the α-carbon atom:

(*i*) n-*Butyric acid.* $CH_3CH_2\boxed{CH_2CO_2H}$

$$CH_3COCH_2CO_2C_2H_5 \xrightarrow{C_2H_5ONa} [CH_3COCHCO_2C_2H_5]^-Na^+$$

$$\xrightarrow{C_2H_5I} CH_3COCH(C_2H_5)CO_2C_2H_5$$

$$\xrightarrow[\text{(ii) } H^+]{\text{(i) c. KOH}} CH_3CH_2CH_2CO_2H$$

(*ii*) a-*Methyl-n-valeric acid.* $CH_3CH_2CH_2\boxed{CH(CH_3)CO_2H}$

$$CH_3COCH_2CO_2C_2H_5 \xrightarrow{C_2H_5ONa} [CH_3COCHCO_2C_2H_5]^-Na^+$$

$$\xrightarrow{C_3H_7Br} CH_3COCH(C_3H_7)CO_2C_2H_5 \xrightarrow{C_2H_5ONa}$$

$$[CH_3COC(C_3H_7)CO_2C_2H_5]^- Na^+ \xrightarrow{CH_3I} CH_3COC(CH_3)(C_3H_7)CO_2C_2H_5 \xrightarrow[\text{(ii) } H^+]{\text{(i) c. KOH}} CH_3CH_2CH_2CH(CH_3)CO_2H$$

It has been found that 'ketonic' hydrolysis and 'acid' hydrolysis of acetoacetic ester *always take place simultaneously*, but one or the other can be made to predominate by adjusting the concentration of the potassium hydroxide. Thus, in the preparation of acids, there will always be some ketone formed as a by-product; and vice versa. For this reason it is better to use ethyl malonate to synthesise acids since the yields are greater. However, Ritter *et al.* (1962) have improved the preparation of acids via E.A.A. by heating the mono-alkyl derivatives with catalytic amounts of alkoxide in excess of absolute ethanol:

$$MeCOCHRCO_2Et + EtOH \xrightarrow{OEt^-} MeCO_2Et + RCH_2CO2Et (70\text{-}90\%)$$

In addition to the acetoacetic ester ketone synthesis, ketones may be prepared:

(*i*) by the alkylation of simpler ketones (sodamide followed by RI). The difficulty with this method is that over-alkylation usually occurs. This synthesis has been modified to prepare t-carboxylic acids from acetophenone.

$$CH_3COC_6H_5 \xrightarrow[\text{(ii) } R^1I]{\text{(i) } NaNH_2} R^1CH_2COC_6H_5 \xrightarrow[\text{(ii) } R^2I]{\text{(i) } NaNH_2} (R^1)(R^2)CHCOC_6H_5 \xrightarrow[\text{(ii) } R^3I]{\text{(i) } NaNH_2}$$

$$(R^1)(R^2)(R^3)CCOC_6H_5 \xrightarrow{NaNH_2} (R^1)(R^2)(R^3)CCONH_2 \xrightarrow{HNO_2} (R^1)(R^2)(R^3)CCO_2H$$

(*ii*) Stork *et al.* (1963) have introduced a new method of alkylating aldehydes and ketones, and so this affords a means of preparing more complicated aldehydes and ketones from simpler ones. An aliphatic primary amine is condensed with an aldehyde or ketone containing at least one active a-hydrogen atom, and the imine produced is treated with ethylmagnesium bromide in boiling tetrahydrofuran. The magneisum salt so produced readily reacts with alkyl halides, *e.g.*, a-benzyliso-butyraldehyde:

$$Me_2CHCHO \xrightarrow{Me_3CNH_2} Me_2CHCH{=}NCMe_3 \xrightarrow{3EtMgBr} Me_2C{=}CHN\begin{matrix}\diagup MgBr \\ \diagdown CMe_3\end{matrix}$$

$$\xrightarrow{PhCH_2Cl} \underset{\displaystyle CH_2Ph}{Me_2\overset{}{C}}CH{=}NCMe_3 \xrightarrow{HCl} PhCH_2CMe_2CHO \quad (80\%)$$

(*iii*) β-Keto-acids, $R^1COCHR^2CO_2H$, can be prepared by α-carboxylation of ketones, $R^1COCH_2R^2$, with magensium methyl carbonate. The intermediate magnesium salt can be alkylated *in situ* and then decarboxylated to give ketones, $R^1COCHR^2R^2$ (Stiles *et al.*, 1959).

Sodioacetoacetic ester reacts with many other halogen compounds besides alkyl halides, and so may be used to synthesise a variety of compounds.

(*i*) 1,3–*Diketones*. In the synthesis of 1,3–diketones the halogen compound used is an acid chloride. Since acid chlorides react with ethanol, the reaction cannot be carried out in this solvent in the usual way. The reaction, however, may conveniently be carried out by treating acetoacetic ester in benzene solution with magnesium and the acid chloride, *e.g.*, pentane-2,4-dione may be obtained by the ketonic hydrolysis of the intermediate product ethyl diacetylacetate:

$$CH_3COCH_2CO_2C_2H_5 + CH_3COCl \xrightarrow{Mg} (CH_3CO_2)CHCO_2C_2H_5$$

$$\xrightarrow[\text{(ii) } H^+]{\text{(i) dil. KOH}} CH_3COCH_2CH_3$$

If sodioacetoacetic ester or acetoacetic ester itself is treated with acetyl chloride in pyridine as solvent, the *O-acetyl* derivative of acetoacetic ester, acetoxycrotonic ester, is obtained, and not the carbon-linked compound (as above):

$$[CH_3COCHCO2C_2H_5]^- Na^+ + CH_3COCl \xrightarrow{\text{pyridine}} \underset{\displaystyle CH_3C{=}CHCO_2C_2H_5}{\overset{\displaystyle OCOCH_3}{|}} + NaCl$$

If we use Kornblum's generalisation with ambident ions, then the first reaction presumably occurs by the S_N2 mechanism and the latter by the S_N1.

(*ii*) *Dicarboxylic acids*. Dicarboxylic acids may be prepared by interaction of sodioacetoacetic ester and a halogen derivative of an ester, *e.g.*, succinic acid from ethyl chloroacetate:

$$[CH_3COCHCO_2C_2H_5]^-Na^+ + ClCH_2CO_2C_2H_5 \longrightarrow$$

$$\begin{array}{c} CH_2CO_2C_2H_5 \\ | \\ CH_3COCHCO_2C_2H_5 \end{array} \xrightarrow[\text{(ii) } H^+]{\text{(i) c.KOH}} \begin{array}{c} CH_2CO_2H \\ | \\ CH_2CO_2H \end{array}$$

Ketonic hydrolysis of this acetoacetic ester derivative gives the γ-keto-acid ester (ethyl ester of laevulic acid):

$$\begin{array}{c} CH_2CO_2C_2H_5 \\ | \\ CH_3COCHCO_2C_2H_5 \end{array} \xrightarrow[\text{(ii) } H^+]{\text{(i) dil.KOH}} CH_3COCH_2CH_2CO_2C_2H_5$$

(*iii*) *Long-chain monocarboxlic acids*. This method involves a combination of methods (*i*) and (*ii*) described above (Mrs. Robinson, 1930):

$$[CH_3COCHCO_2C_2H_5]^- Na^+ \xrightarrow{Br(CH_2)_xCO_2C_2H_5} \begin{array}{c} (CH_2)_xCO_2C_2H_5 \\ | \\ CH_3COCHCO_2C_2H_5 \end{array}$$

$$\xrightarrow[\text{(ii) } CH_3(CH_2)_yCOCl]{\text{(i) } C_2H_5ONa} \begin{array}{c} (CH_2)_xCO_2C_2H_5 \\ | \\ CH_3COCHCO_2C_2H_5 \\ | \\ CO(CH_2)_yCH_3 \end{array}$$

$$\xrightarrow[\text{hydrolysis}]{\text{graded}} CH_3(CH_2)_yCOCH_2(CH_2)_xCO_2H$$

These keto-acids are readily reduced by means of the Clemmensen reduction. Improved modifications of this method have now been developed.

(*iv*) The acetyl group in acetoacetic ester may be replaced by another acyl group under suitable conditions, *e.g.*, ethyl benzoylacetate:

$$MeCOCH_2CO_2Et + PhCOCl \xrightarrow{NaOH} PhCOCH(COMe)CO_2Et$$

$$\xrightarrow{aq.\ NH_4Cl} PhCOCH_2CO_2Et + MeCO_2NH_4$$

(v) Alicyclic compounds and heterocyclic compounds may be prepared by the use of E.A.A.

Malonic Ester Syntheses

Malonic ester, $CH_2(CO_2C_2H_2)_2$, which is the diethyl ester of malonic acid, $CH_2(CO_2H)_2$, is prepared by dissolving potassium cyanoacetate in ethanol, adding concentrated hydrochloric acid, and warming the mixture on the water-bath:

$$NCCH_2{-}CO_2K + 2C_2H_5OH + 2HCl \longrightarrow CH_2(CO_2C_2H_5)_2 + KCl{+}NH_4Cl\ (v.g.)$$

It is a pleasant-smelling liquid, b.p. 199°C. Its use as a synthetic reagent depends on two chemical properties.

(1) With sodium ethoxide it forms a sodium derivative, sodiomalonic ester, which reacts with compounds containing a reactive halogen atom, *e.g.*, alkyl halides, acid chlorides, halogensubstituted esters, etc. (*cf.* acetoacetic ester). In all cases the yields are *g.–v.g.*

$$C_2H_5O{-}\overset{\overset{O}{\|}}{C}{-}CH_2{-}\overset{\overset{O}{\|}}{C}{-}OC_2H_5 \xrightarrow{C_2H_5ONa}$$

$$\left[C_2H_5O{-}\overset{\overset{\bar{O}}{|}}{C}{=}CH{-}\overset{\overset{O}{\|}}{C}{-}OC_2H_5\right]Na^+ \longleftrightarrow$$

$$\left[C_2H_5O{-}\overset{\overset{O}{\|}}{C}{-}\bar{C}H{-}\overset{\overset{O}{\|}}{C}{-}OC_2H_5\right]Na^+ \xrightarrow{RX}$$

$$C_2H_5O{-}\overset{\overset{O}{\|}}{C}{-}\bar{C}HR{-}\overset{\overset{O}{\|}}{C}{-}OC_2H_5 + NaX$$

The process may then be repeated to produce the disubstituted derivative of malonic ester:

$$R^1CH(CO_2C_2H_5)_2 \xrightarrow{C_2H_5ONa} [R^1C(CO_2C_2H_5)_2]^-Na^+$$
$$\xrightarrow{R^2X} R^1R^2C(CO_2C_2H_5)_2$$

These disubstituted derivatives of malonic ester can readily be prepared in *one* step by treating the ester with *two* equivalents of sodium ethoxide and then with *two* equivalents of alkyl halide. This procedure is only used if it is required to introduce two identical alkyl groups.

$$CH_2(CO_2C_2H_5)_2 \xrightarrow[\text{(ii) 2RX}]{\text{(i) 2EtONa}} R_2C(CO_2C_2H_5)_2 + NaX$$

(2) Malonic acid and its derivatives eliminate a molecule of carbon dioxide when heated just above the melting point of the acid (usually between 150° and 200°C) to form acetic acid or its derivatives (the yields are *v.g.-ex.*):

$$HO_2CCH_2CO_2H \longrightarrow CH_3CO_2H + CO_2$$

$$HO_2CCHRCO_2H \longrightarrow RCH_3CO_2H + CO_2$$

Decarboxylation may also be effected by refluxing malonic acid or its derivatives in sulphuric acid solution.

The mechanism of decarboxylation is piobably similar to that of açetoacetic acid:

$$R_2C(COOH)(CO_2H) \xrightarrow{-CO_2} \left[R_2C{=}C\begin{matrix}OH\\OH\end{matrix}\right] \longrightarrow R_2CHCO_2H$$

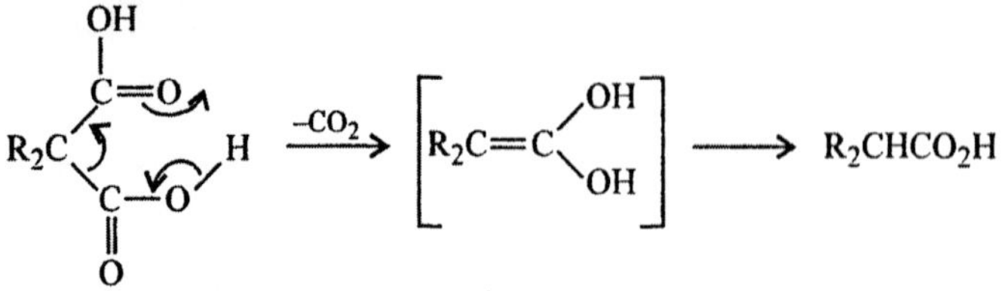

In a number of cases ethyl cyanoacetate (b.p. 207°C) may be used instead of malonic ester in many syntheses. The cyano-group is readily converted into the carboxyl group on hydrolysis.

SYNTHESIS OF MONOCARBOXYLIC ACIDS

Malonic ester is preferable to acetaocetic ester in synthesising acids, and should be used wherever possible. The structural formula of the acid required is written down, the acetic acid nucleus picked out, and the required alkyl groups introduced into sodiomalonic ester. The substituted ester is then refluxed with potassium hydroxide solution, acidified with hydrochloric acid, and the precipitated salt may be refluxed with sulphuric acid:

(*i*) n-*Valeric acid.* $\boxed{CH_3CH_2CH_2}CH_2CO_2H$

$$CH_2(CO_2C_2H_5)_2 \xrightarrow{C_2H_5ONa} [CH(CO_2C_2H_5)_2]^-Na^+ \xrightarrow{C_3H_7Br} C_3H_7CH(CO_2C_2H_5)_2 \xrightarrow{KOH}$$

$$C_3H_7CH(CO_2K_5)_2 \xrightarrow{HCl} C_3H_7CH(CO_2H)_2 \xrightarrow{150-200°C} CH_3CH_2CH_2CH_2CO_2H$$

(*ii*) *Dinethylacetic acid.* $CH_3\boxed{\overset{CH_3}{\overset{|}{C}}HCO_2H}$

$$CH_2(CO_2C_2H_5)_2 \xrightarrow{C_2H_5ONa} [CH(CO_2C_2H_5)_2]^-Na^+ \xrightarrow{CH_3I}$$

$$CH_3CH(CO_2C_2H_5)_2 \xrightarrow{C_2H_5ONa} [CH_3C(CO_2C_2H_5)_2]^-Na^+ \xrightarrow{CH_3I}$$

$$(CH_3)_2C(CO_2C_2H_5)_2 \xrightarrow{KOH} (CH_3)_2C(CO_2K)_2 \xrightarrow{HCl}$$

$$(CH_3)_2C(CO_2H)_2 \xrightarrow{150-200°C} (CH_3)_2CHCO_2H$$

Since two methyl groups are required for this synthesis, their introduction an be carried out in one step (see above):

$$CH_2(CO_2C_2H_5)_2 \xrightarrow[2CH_3I]{2C_2H_5ONa} (CH_3)_2C(CO_2C_2H_5)_2 \xrightarrow{KOH, etc.} (CH_3)_2CHCO_2H$$

Since aryl groups cannot be introduced directly into malonic ester, aryl-substitued derivatives must be prepared indirectly. A very good method uses the Claisen condensation followed by decarboxylation *e.g.*, ethyl phenylmalonate:

$$PhCH_2CO_2Et + (CO_2Et)_2 \xrightarrow{EtONa} EtOH +$$

$$PhCH(CO_2Et)COCO_2Et \xrightarrow{heat} PhCH(CO_2Et)_2 + CO \ (80\text{-}85\%)$$

This method may also be used of the preparation of long-chain alkylmalonic esters; the ester of a longchain monocarboxylic acid with two α-hydrogen atoms is used instead of the ester of phenylacetic acid given above.

Synthesis of Dicarboxylic Acids

Dicaboxylic acids of the type $R^1R^2C(CO_2H)_2$ are readily prepared from malonic ester as shown above.

Dicarboxylic acids of the type $HO_2C(CH_2)N_nCO_2H$ are very important, and the malonic ester synthesis is particularly useful for their preparation. The actual procedure depends mainly on the value of *n*, and the following examples illustrate this point:

(*i*) *Adipic acid.* $\boxed{HO_2CCH_2}CH_2CH_2\,CH_2CO_2H$

The acetic acid is blocked off at *one* end, and the remaining fragment is considered from the point of view of accessibility. In this example the fragment required is g-bromobutyric ester. This is not readily accessible, and so the procedure now is to block off the acetic acid nucleus at the *other* end of the adipic acid molecule, and then to use *two* molecules of malonic ester and the a, w-*dihalide* of the polymethylene fragment that joins together the two acetic acid nuclei. In this case it is ethylene dibromide, and since this is readily accessible, the synthesis of adipic acid may be carried out as follows:

$$[(C_2H_5O_2C)_2CH]^-Na^+ + BrCH_2CH_2Br + Na^+[CH(CO_2C_2H_5)_2]^- \longrightarrow (C_2H_5O_2C)_2CHCH_2CH_2CH(CO_2C_2H_5)_2 \xrightarrow[\text{(ii) HCl}]{\text{(i) KOH}}$$

$$(HO_2C)_2CHCH_2CH_2CH(CO_2H)_2 \xrightarrow{150\text{–}200°C} HO_2CCH_2CH_2CH_2CH_2CO_2H$$

Should the fragments required by either route be inaccessible, then a method not involving the use of malonic ester may be more satisfactory.

(*ii*) *Succinic acid.* $\boxed{HO_2CCH_2}CH_2CO_2H$

If *one* acetic acid nucleus is blocked off, the fragment required is ethyl chloroacetate. This is readily accessible and hence succinic acid may be synthesised from *one* molecule of malonic ester as follows:

$$[CH(CO_2C_2H_5)_2]^-Na^+ + ClCH_2CO_2C_2H_5 \longrightarrow$$

$$(C_2H_5O_2C)_2CHCH_2CO_2C_2H_5 \xrightarrow[\text{(ii) HCl}]{\text{(i) KOH}} (HO_2C)_2CHCH_2CO_2H$$

$$\xrightarrow{150\text{–}200°C} HO_2CCH_2CH_2CO_2H$$

If the *other* acetic acid nucleus in succinic acid is blocked off, there is no intervening fragment left. However, the union of two malonic ester molecules may be effected by means of iodine.

$$[(C_2H_5O_2C)_2CH]^-Na^+ + I_2 + Na^+[CH(CO_2C_2H_5)_2]^- \longrightarrow$$

$$(C_2H_5O_2C)_2CHCH(CO_2C_2H_5)_2 \xrightarrow[\text{(ii) HCl}]{\text{(i) KOH}} (HO_2C)_2CHCH(CO_2H)_2$$

$$\xrightarrow{150-200°C} HO_2CCH_2CH_2CO_2H$$

Synthesis of Ketones

Bowman *et al.* (1952) have introduced the following general synthesis of ketones and b-keto-esters (R^2= tetrahydropyran-2-ol):

$$R^1CH(CO_2R^2)_2 \xrightarrow[\text{(ii) } R^3COCl]{\text{(i) NaH}} R^3COCR^1(CO_2R^2)_2 \xrightarrow[\text{reflux}]{CH_3CO_2H}$$

$$R^3COCH_2R^1 \text{ (90-92\%)}$$

Johnson *et al.* (1952) have used t-butyl esters of malonic acid (R^2 = t-butyl):

$$R^1CH(CO_2R^2)_2 \xrightarrow[\text{(ii) } R^3COCl]{\text{(i) NaH}} R^3COCR^1\ (CO_2R^2)_2 \xrightarrow[CH_3CO_2H]{CH_3-C_6H_4-SO_3H}$$

$$R^3COCH_2R^1 \text{ (56–85\%)}$$

Good yields of ketone are obtained by the above method only when anhydrous conditions are used throughout the preparation. Consequetly, insead of the usual method of hydrolysis (aqueous alkali), acidolysis is used.

Acetic acid is satisifactory for this purpose. especially in the presence of a small amount of sulphuric acid or *p*-toluenesulphonic acid. In this way, the ester group is transferred from the malonic ester to the acetic acid.

However, for the transfer to be successful, the ester group must be very sensitive to acidolysis; the above two esters are particularly sensitive in this way.

Synthesis of higher ketonic acids. Sodiomalonic ester is treated with the acid chloride-ester derivative of a dicarboxylic acid, *e.g.*, e-ketoheptanoic acid:

$$[(C_2H_5O_2C)_2CH]^-Na^+ + ClCO(CH_2)_4CO_2C_2H_5 \longrightarrow$$

$$(C_2H_5O_2C)_2CHCH(CH_2)_4CO_2C_2H_5 \xrightarrow[\text{(ii) HCl}]{\text{(i) KOH}}$$

$$(HO_2C)_2CHCO(CH_2)_4CO_2H \xrightarrow{150-200°C} [HO_2CCH_2CO-$$

$$(CH_2)_4CO_2H] \xrightarrow{-CO_2} CH_3CO(CH_2)_4CO_2H$$

The intermediate b-keto-dicarboxylic acid is unstable and is readily decarboxylated This might have been anticipated, since it may be regarded as a derivative of acetoacetic acid, which is readily decarboxylated on warming.

Synthesis of Polycarboxylic Acids

It has been pointed out previously that the monoalkyl derivatives of malonic acid are readily brominated, and that these bromo-derivative may be used to synthesise polycarboxylic acid, *e.g.*,

$$CH_2(CO_2C_2H_5)_2 \xrightarrow{Br_2} CHBr(CO_2C_2H_5)_2$$

$$CHBr(CO_2C_2H_2)_2 + [CH(CO_2C_2H_5)_2]^-Na^+$$

$$\longrightarrow NaBr + (C_2H_5O_2C)_2CHCH(CO_2C_2H_5)_2 \xrightarrow{Br_2}$$

$$\begin{array}{l} (C_2H_5O_2C)_2C-Br \\ \quad\quad\quad\quad\;\; | \\ (C_2H_5O_2C)_2C-Br \end{array} \xrightarrow{2[CH(CO_2C_2H_5)_2]^-Na^+} \begin{array}{l} (C_2H_5O_2C)_2C-CH(CO_2C_2H_5)_2 \\ \quad\quad\quad\quad\;\; | \\ (C_2H_5O_2C)_2C-CH(CO_2C_2H_5)_2 \end{array}$$

Malonic ester may also be used for the preparation of unsaturated acids (Knoevenagel reaction), alicyclic compounds and heterocyclic compounds.

HYDROXYALDEHYDES AND HYDROXYKETONES

When naming a compound which contains more than one functional group, it is necessary to choose one as the principal function. The compound is then named by using the suffix of the principal function and the prefixes of the other functions.

The carboxylic and the sulphonic acid group are *always* chosen as principal functions, and the usual order for choosing the principal function is:

Carboxylic, sulphonic, acid halide, amide, imdie, aldehyde, cyanide, isocyanide, ketone, alcohol, phenol, thioalcohol, amine, imine.

Examples. Acetoacetic acid contains a ketonic group and a carboxyl group, and since the latter is the principal function, acetoacetic acid is named as a ketone acid, *viz.*, β-ketobutyric acid, 3-ketobutanoic acid, or 3-oxobutanoic acid.

Acetoacetic acid may also be named propan-2-one-1-carboxylic acid—each functional group is indicated by its appropriate suffix.

Table 2.2

Function	*Prefix*	*Suffix*
Acid	carboxy	carboxylic or -oic
Acid amide	carbamoyl	amide or carbonamide
Acid chloride	chloroformyl	oyl or carbonyl
Alcohol	hydroxy	ol
Aldehyde	oxo, aldo (for aldehydic O) or formyl (for CHO)	al
Amine	amino	amine
Azo-derivative	azo	—
Azoxy-derivative	azoxy	—
Carbonitrile (nitrile)	cynao	carbonitrile
Double bond	—	ene
Ester	alk(yl)oxycarbonyl	carboxylate
Ether	alkoxy	—
Ethylene oxide, etc.	epoxy	—
Halogenide (halide)	halogeno (halo)	—
Hydrazine	hydrazino	hydrazine
Ketone	oxo or keto	one
Nitro-derivative	nitro	—
Nitroso-derivative	nitroso	—
'Quinquevalent' nitrogen	—	onium, inium
Sulphide	sulphonyl	—
Sulphonic derivative	sulpho	sulphonic
Sulphoxide	sulphinyl	—
Sulphonic derivative	sulpho	sulphonic
Sulphoxide	sulphinyl	—
Thioalcohol (Mercaptan)	mercapto	thiol
Triple bond	—	yne
Urea	ureido	urea

CH_3OHCH_2CHO. This is both an alcohol and an aldehyde, so that, bearing in mind the order of preference, the name of the compound will be β-hydroxypropionaldehyde or 3-hydroxypropanl.

Table 2.2 indicates the prefixes and suffixes used for designating the functions (in alphabetical order).

The simplest hydroxyladehyde is **glycoladehyde** (*hydroxyethanal*). It may be prepared by the oxidation of glycol with *Fenton's reagent* (hydrogen peroxide and ferrous sulphate):

$$HOCH_2CH_2OH + H_2O_2 \xrightarrow{FeSO_4} HOCH_2CHO + 2H_2O$$

It may also be prepared by ozonolysis of allyl alcohol:

$$CH_2{=}CHCH_2OH \xrightarrow{O_3} \begin{array}{c} CH_2{-}O{-}CHCH_2OH \\ | \qquad\qquad | \\ O{-}{-}{-}{-}{-}O \end{array} \xrightarrow[Zn]{H_2O} OCHCH_2OH$$

The most convenient means of preparing glycoladehyde, however, is by heating dihydroxymaleic acid, which is obtained from the oxidation of tartaric acid (*q.v.*):

$$\begin{array}{c} HO_2C{-}C{-}OH \\ \| \\ HO_2C{-}C{-}OH \end{array} \longrightarrow 2CO_2 + \left[\begin{array}{c} H{-}C{-}OH \\ \| \\ H{-}C{-}OH \end{array}\right] \longrightarrow \begin{array}{c} CHO \\ | \\ CH_2OH \end{array}$$

Glycolaldehyde exists in the solid form as the dimer, m.p. 96°C; but in aqueous solution it exists as the monomer which forms the stable hydrate, $CH(OH)_2CH_2OH$ (*cf.* chloral). Careful oxidation of glycolaldehyde with bromine water produces glycollic acid:

$$HOCH_2CHO + [O] \longrightarrow HOCH_2CO_2H$$

Glycolaldehyde is a powerful reducing agent, reducing ammoniacal silver nitrate and Fehling's solution at room temperature. With phenhylhydrazine it forms the *osazone*. This osazone is identical with that formed from glyoxal.

Glycolaldehyde undergoes the aldol condensation in the presence of alkali; with sodium hydroxide solution a tetrose sugar is formed, and with sodium carbonate solution, a hexose sugar:

$$2HOCH_2CHO \xrightarrow{NaOH} HOCH_2CHOHCHOHCHO$$

$$3HOCH_2CHO \xrightarrow{Na_2OH_3}$$
$$HOCH_2CHOHCHOHCHOHCHOHCHO$$

Glycolaldehyde is formed in small amounts with formaldehyde is allowed to stand in the presence of calcium carbonate:

$$2HCHO \xrightarrow{CaCO_3} HOCH_2CHO$$

Aldol (*acetaldol, β-hydroxybutyraldehyde*, 3-*hydroxybutanal*), p.b. 83°C/20 mm., may be prepared by the aldol condensation of acetaldehyde:

$$2CH_3CHO \xrightarrow{NaOH} CH_3CHOHCH_2CHO$$

When heated, aldol is dehydrated to crotonaldehyde:

$$CH_3CHOHCH_2CHO \longrightarrow CH_3CH = CHCHO + H_2O$$

There is some doubt about the structure of aldol. Recent work suggests it is an equilibrium mixture of β-hydroxybutyraldehyde and the cyclic hemiacetal:

$$CH_3CHOHCH_2CHO \rightleftharpoons CH_3\overset{\ulcorner O \urcorner}{CHCH_2CH}OH$$

This is an example of ring-chain tautomerism.

The simplest hydroxyketone is **hydroxyacetone** (*acetol, pyruvic alcohol*), b.p. 145°C. The best method of preparation is by heating bromoacetone with potassium hydroxide in methanolic solution, and adding ethyl formate:

$$HCO_2C_2H_5 + KOH \longrightarrow HCO_2K + C_2H_5OH$$

$$CH_3COCH_2Br + HCO_2K \longrightarrow CH_3COCH_2OOCH + KBr$$

$$CH_3COCH_2OOCH + CH_3COCH_2OH \longrightarrow CH_3COCH2OH + HCO_2CH_3 \ (54\text{-}88\%)$$

This is an example of alcoholysis, hydroxyacetone being replaced by methanol.

Hydroxyacetone reduces ammoniacal silver nitrate, thereby being oxidised to DL-lactic acid, and reduces Fehling's solution, thereby being oxidised to a mixture of formic and acetic acids. Although ketones do not normally reduce ammoniacal silver nitrate and Fehling's solution, α-hydroxyketones are exceptions. Similarly, ketones form phenylhydrazones with phenylhydrazine, but α-hydroxyketones form osazones; thus, hydroxyacetone gives the same osazone as methylglyoxal (*cf.* glycolaldehyde):

$$\begin{matrix} CH_3 \\ | \\ CO \\ | \\ CH_2OH \end{matrix} \xrightarrow{C_6H_5NHNH_2} \begin{matrix} CH_3 \\ | \\ C{=}NNHC_6H_5 \\ | \\ CH{=}NNHC_6H_5 \end{matrix}$$

Hydroxyacetone appears to exist as an equilibrium mixture of hydroxyketone and cyclic hemiacetal forms (ring-chain tautomerism).

$$CH_3COCH_2OH \rightleftharpoons CH_3\underset{\underset{OH}{|}}{C}\!\!-\!\!\!\underset{\diagdown O \diagup}{}\!\!\!-\!\!CH_2$$

Acyloins are α-hydroxyketones, and may be readily prepared by reduction of esters. Acetoin, p.b. 148°C, the simplest member, may be prepared from ethyl acetate.

Diacetone alcohol (4-*hydroxy*-4-*methylpentan*-2-*one*), p.b. 164°C, may be prepared by the condensation of acetone in the presence of barium hydroxide (p. 218):

$$2CH_3COCH_3 \xrightarrow{Ba(OH)_2} (CH_3)_2C(OH)CH_2COCH_3$$

When heated with a trace of acid or iodine, diacetone alcohol eliminates a molecule of water to form mesityl oxide:

$$(CH_3)_2C(OH)CH_2COCH_3 \xrightarrow{I_2} (CH_3)_2C = CHCOCH_3 + H_2O$$

Diacetone alcohol is oxidised by sodium hypobromite to β-hydroxyisovaleric acid (*cf.* haloform reaction):

$$(CH_3)_2C(OH)CH_2COCH_3 \xrightarrow{[O]} (CH_3)_2C(OH)CH_2CO_2H$$

DIALDEHYDES, KETOALDEHYDES AND DIKETONES

Dialdehydes

The simplest dialdehyde is **glyoxal** (*oxaldehyde, ethanedial*). It may be obtained by the oxidation of ethanol, acetaldehyde or glycol with nitric acid, but the yields are poor. It is most convenietly prepared, as the bisulphite compound, by refluxing a mixture of paraldehyde, 50 per cent aqueous acetic acid, dioxan and selenous acid (yield 72-74 per cent on the selenous acid):

$$(CH_3CHO)_3 + 3H_2SEeO_3 \longrightarrow 3OCHCHO + 3Se + 6H_2O$$

$$\xrightarrow{NaHSO_3} OCHCHO{\cdot}2NaHSO_3{\cdot}H_2O$$

Glyoxal is manufactured by the vapour-phase oxidation of glycol with air 250-300°C in the presence of copper as actalyst.

Glyoxal exists as a colourless polymer giving, on distillation, the monomer. This is a green vapour, which condenses to yellow crystals (m.p. 15°C) which polymerise on standing to a colourless solid of unknown molecular weight. Glyoxal exists in aqueous solution as the dihydrate, $CH(OH)_2CH(OH)_2$, which has been isolated (*cf.* chloral hydrate).

Glyoxal undergoes many of the reactions of a dialdehyde. *e.g.*, it reduces ammoniacal silver nitrate, and forms addition compounds with two molecules of hydrogen cyanide and sodium hydrogen sulphite. It does *not* reduce Fehling's solution. Since it has no α-hydrogen atom, glyoxal undergoes an intramolecuelar Cannizzaro reaction in the presence of alkali to form glycollic acid.

$$H{-}C(=O){-}C(=O)H + OH^- \rightleftharpoons H{-}C(=O){-}C(OH)(H){-}O^- \xrightarrow[\text{Shift}]{H^-} H{-}C(O^-)(H){-}C(OH){=}O \longrightarrow H{-}C(OH)(H){-}CO_2^-$$

With phenylhydrazine, glyoxal forms the osazone (this is identical with that from glycoladehyde):

$$\begin{matrix}CHO\\|\\CHO\end{matrix} + 2C_6H_5NHNH_2 \longrightarrow \begin{matrix}CH{=}NNHC_6H_5\\|\\CH{=}NNHC_6H_5\end{matrix} + 2H_2O$$

It also combines with *o*-phenylenediamines of form *quinoxaline* (heterocyclic compounds); *e.g.*, with *o*-phenylenediamine it forms quinoxaline itself:

$$C_6H_4(NH_2)_2 + \begin{matrix}O{=}CH\\|\\O{=}CH\end{matrix} \longrightarrow \text{quinoxaline} + 2H_2O$$

Methylglyoxal (pyruvaldehyde) may be prepared by the oxidiation of hydroxyacetone with hydrogen peroxide, or of acetone with selenium dixoide, or by the hydrolysis of oximinoacetone:

$$\left.\begin{matrix}CH_3COCH_2OH \xrightarrow{H_2O_2}\\ CH_3COCH_3 \xrightarrow{SeO_2}\\ CH_3COCH{=}NOH \xrightarrow{H_3O^+}\end{matrix}\right] \longrightarrow CH_3COCHO$$

It is manufactured by the vapour-phase oxidation of propene glycol with air at 250-300°C in the presence of copper as catalyst:

$$CH_3CHOHCH_2OH \xrightarrow[Cu]{O_2} CH_3COCHO + 2H_2O$$

It is a yellow oil with a pungent odour, and beings to boil at 72°C to give a light green vapour.

The liquid form of methylglyoxal is the dimer, and dimer, and this slowly polymerises at room temperature to form a glassy mass of unknown molecular weight (*cf.* glyoxal).

Malondialdehyde is prepared commercially as its tetra-acetal, 1,1,3,3-tetaethoxypropane, $(EtO)_2CHCH_2CH(OEt)_2$, p.b. 100°C, is formed when ethyl formate undergoes the Claisen condensation with acetone.

$$CH_3COCH_3 + HCO_2Et \xrightarrow{EtONa} CH_3COCH_2CH$$

Fomylacetone is strongly acidic, *e.g.*, it dissolves in sodium hydroxide solution to form the sodium salt, and gives a violet colour with ferric chloride. Thus it is enolised. Moreover, infra-red spectra studies by Boileau (1954) have shown that it is the aldehyde group that has enolised almost 100 per cent *i.e.*, formylacetone is hydroxymethyleneacetone, MeCOCH = CHOH.

It appears that all β-ketoaldehydes behave in this way. The interesting point about their direction of enolisation is that they were the first exceptions to the *Erlenmeyer rule* (1880, 1892) that a terminal = CHOH group (in acyclic compounds) is so unstable that it always rearranges to the aldehyde (*cf.* 'vinyl alochol').

More recently, Garbisch, Jr. (1963) has shown, from his NMR spectra studies of cyclic β-keto-aldehydes, that the direction of enolisation depends on their structure.

Succinaldehyde (succindialdehyde, butanedial) may be prepared by the ozonolysis of hexa-1,5-diene:

$$CH_2=CHCH_2CH_2CH=CH=CH_2 \xrightarrow{\text{ozonolysis}} OCHCH_2CH_2CHO$$

Another method of preparation is to allow pyrrole and hydroxylamine to interact, and to treat the product, succinaldoxime, with aqueous nitrous acid:

$$\begin{array}{l} CH=CH\diagdown \\ | \qquad\qquad NH \\ CH=CH\diagup \end{array} \xrightarrow{NH_2OH} \begin{array}{l} CH_2CH=NOH \\ | \\ CH_2CH=NOH \end{array} \xrightarrow{HNO_2} \begin{array}{l} CH_2CHO \\ | \\ CH_2CHO \end{array}$$

Barton *et al.* (1964) have obtained succinaldehye (yield: 80 per cent) from butane-1,4-diol via the alkyl chloroformate.

Succinaldehyde is a colourless oil, p.b. 170°C, which readily polymerises. It undergoes the usual reactions of a (di)aldehyde; *e.g.*, it reduces ammoniacal silver nitrate and Fehling's solution, forms addition compounds with two equivalents of hydrogen cyanide ad sodium hydrogen sulphite; etc. When treated with phosphorus pentoxide, ammonia and phosphorus pentasulphide, it gives respectively furan, pyrrole and thiophen:

$$\begin{array}{l} CH_2CHO \\ | \\ CH_2CHO \end{array} \begin{cases} \xrightarrow{P_2O_5} \text{furan (ring with O)} \\ \xrightarrow{NH_3} \text{pyrrole (ring with NH)} \\ \xrightarrow{P_2S_5} \text{thiophene (ring with S)} \end{cases}$$

Diketones

These are classified as α, β, γ . . . diketones according as the two carbonyl groups are in the 1,2, 1,3, 1,4, . . . positions, respectively.

The infra-red absorption region of α-diketones is 1730-1710cm^{-1} (s), and that of γ-diketones is 1725-1705 cm^{-1}(s). Both of these are in the absorption region of acyclic monoketones (1725-1700 cm^{-1}) but of greater intensity, and since they are completely different from the region of enolisable β-diketones, it therefore means that α-and γ-diketones are like acetone, *i.e.*, are in the keto form.

Butane-2,3-dione, dimethylglyoxal (*diacetyl*), is the simplest α-diketone. It may be prepared: (*i*) By the hydrolysis of oximinobutan-2-one formed by the action of nitrous acid on the ketone:

$$CH_3COCH_2CH_3 \xrightarrow{HNO_3} CH_3COC(=NOH)CH_3$$
$$\xrightarrow{HCl} CH_3COCOCH_3$$

(ii) By the oxidation of butan-2-one with selenium dioxide:

$$CH_3COCH_2CH_3 + SeO_2 \longrightarrow CH_3COCOCH_3 + Se + H_2O$$

(iii) Acetoin may be readily oxidised by bismuth trioxide in acetic acid; the yield of dimethylglyoxal is almost quantitative (Rigby, 1951). This reagent is specific for this reaction, *i.e.*, for the oxidation of acyloins to α-diketones.

Butane-2,3-dione is a yellow oil, b.p. 88°C. It gives the usual reactions of a diketone; *e.g.*, it forms the monoxime and di-oxime with hydroxylamine, the bisulphite compound with sodium hydrogen sulphite, etc. It is oxidised by hydrogen peroxide to acetic acid:

$$CH_3COCOCH_3 \xrightarrow{H_2O_2} 2CH_3CO_2H$$

Reduction with lithium aluminium hydride produces the corresponding diol.

The dioxime of butane-2,3-dione, *i.e.*, *dimethylglyoxime*, $CH_3C(=NOH)C(=NOH)CH_3$, forms chelate compounds with many metals. and is used to estimate nickel, whose salts produce a red precipitate when treated with dimethylglyoxime. According to Brady *et al.* (1930), the structure of bisdimethylglyoxime nickel is (I). The planer configuration round the nickel ion has been demonstrated (Sugden *et al.*, 1935), and confirmed by X-ray analysis (Rundle *et al.*, 1953).

```
            O···H—O
          ↗        \
CH3C=N              N=CCH3
        \          ↙
           Ni
        ↗          \
CH3C=N              N=CCH3
          ↘        /
            O···H—O
```

(I)

Pentane-2,4-dione, acetylacetone, is the simplest β-diketone. It may be prepared:

(*i*) By means of the Claisen condensation between ethyl acetate and acetone (yield 38-45 per cent on the acetone):

$$CH_3CO_2C_2H_5 + CH_3COCH_3 \xrightarrow{C_2H_5ONa} CH_3COCH_2COCH_3$$

Bloomfield (1962) has shown that β-diketones may be obtained in very good yields by this method by using dimethyl sulphoxide as solvent and sodium hydride as the base.

$$CH_3COCH_2CO_2C_2H_5 \xrightarrow[Mg]{CH_3COCl} CH_3CO\underset{}{\overset{COCH_3}{\overset{|}{C}}}H_2C_2H_5 \xrightarrow[(ii)\ H^+]{(i)\ dil.KOH} CH_3COCH_2COCH_3$$

(*iii*) By the condensation of acetic anhydride with acetone in the presence of boron trifluoride as catalyst (yield 80-85 per cent on the acetone):

$$CH_3COCH_3 + (CH_3CO)_2O \xrightarrow{BF_3} CH_3COCH_2COCH_3 + CH_3CO_2H$$

Pentane-2,4-dione is a colourless liquid, p.b. 139°C/746 mm. It exhibits tautomerism, *e.g.*, it gives a red colour with ferric chloride.

$$CH_3COCH_2COCH_3 \rightleftharpoons CH_3COCH{=}C(OH)CH_3$$

It is readily oxidised to acetic acid, and is converted into a mixture of acetone and acetic acid when heated with potassium hydroxide solution (*cf.* 'acid hydrolysis' of acetoacetic ester). It forms pyrazoles when treated with hydrazine or its derivatives; *e.g.*, with phenylhydrazine it forms 3,5-dimethyl-1-penylpyrazole:

$$CH_3COCH_2COCH_3 + PhNHNH_2 \longrightarrow \text{3,5-dimethyl-1-phenylpyrazole} + 2H_2O$$

Pentane-2,4-dione forms chelated compounds with various metals, *e.g.*, iron, aluminium, copper, etc. (*cf.* E.A.A.):

$$(CH_3COCHCOCH_3)_2Cu \quad or \quad \text{(delocalised chelate structure)}$$

All β-diketones are split by aqueous alkali, and when they are unsymmetrical, fission can occur in two different ways:

$$R^1COCH_2-C(O^-)(OH)R^2 \xleftarrow{OH^-} R^1COCH_2COR^2 \xrightarrow{OH^-} R^1C(O^-)(OH)-CH_2COR^2$$

$$R^1COCH_3 + R^2CO_2^- \qquad\qquad R^1CO_2^- + CH_3COR^2$$

In general, fission occurs in the direction that produces the stronger acid in greater yield.

Pentane-2,4-done is oxidised by selenium dioxide to pentane-2,3,4-trione:

$$MeCOCH_2COMe \xrightarrow{SeO_2} MeCOCOCOMe\ (29\%)$$

Hexane-2,5-dione, acetonylacetone, p.b. 188°C, is the simplest γ-diketone. It may be prepared by menas of the acetoacetic ester synthesis:

$$[CH_3COCHCO_2C_2H_5]^-Na^+ + BrCH_2COCH_3 \longrightarrow$$

$$CH_3COCH(CH_2COCH_3)CO_2C_2H_5 \xrightarrow[\text{(ii) } H^+]{\text{(i) dil.KOH}} CH_3COCH_2CH_2COCH_3$$

An alternative and more convenient acetoacetic ester synthesis is as follows:

$$2[CH_3COCHCO_2C_2H_5]^-Na^+ \xrightarrow{I_2} \begin{matrix} CH_3COCHCO_2C_2H_5 \\ | \\ CH_3COCHCO_2C_2H_5 \end{matrix} \xrightarrow[\text{(ii) HCl}]{\text{(i) KOH}}$$

$$\begin{array}{l} CH_3COCHCO_2H \\ \quad | \\ CH_3COCHCO_2H \end{array} \xrightarrow{heat} \begin{array}{l} CH_3COCH_2 \\ \quad | \\ CH_3COCH_2 \end{array} + 2CO_2$$

It readily forms five-membered rings: with phosphorus pentoxide, 2,5-dimethylfuran; with ammonia, 2,5-dimethylpyrrole; and with phosphorus pentasulphide, 2,5-dimethylfuran; with ammonia, 2,5-dimethylpyrrole; and with phosphorus pentasulphide, 2,5-dimethylthophen (*cf.* succinaldehyde).

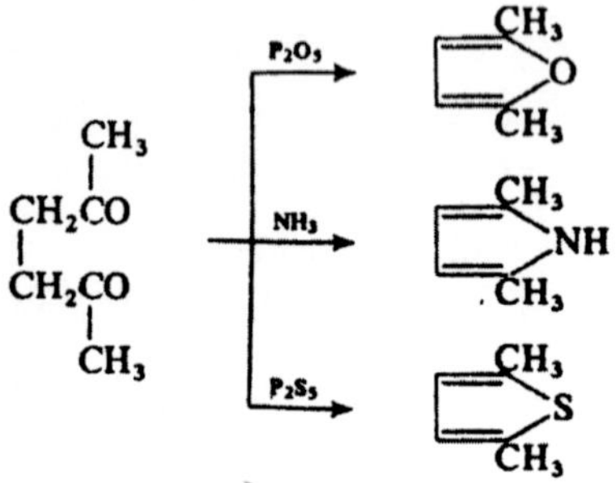

$CH_3COCH_2CH_2COCH_3 \xrightarrow{P_2O_5}$ 2,5-dimethylfuran; $\xrightarrow{NH_3}$ 2,5-dimethylpyrrole; $\xrightarrow{P_2S_5}$ 2,5-dimethylthiophen

Fig. 2.13

Hexane-2,5-dione is oxidised by selenium dioxide to hex-3-ene-2,5-dione (*cf.* pentane-2,4-dione):

$$MeCOCH_2CH_2COMe \xrightarrow{SeO_2} MeCOCH{=}CHCOMe$$

Aldehydic and Ketonic Acids

Glyoxylic acid (*glyoxalic acid, oxoethanoic acid*) is the simplest aldehydic acid. It occurs in unripe fruits, *e.g.*, gooseberries, and disappears during ripening; it also occurs in animal tissues and fluids. It may be prepared by the oxidation of ethanol, glycol or glycollic acid with nitric aid (yields are poor). It may also be prepared by the reduction of oxalic acid either electrolytically or by means of magnesium and sulphuric acid, but it is most conveniently prepared by the hydrolysis of dichloroacetic acid with water:

$$CHCl_2CO_2H + H_2O \longrightarrow OCHCO_2H + 2HCl$$

Glyoxylic acid crystallises from water with one molecule of water which is combined as water of constitution, $CH(OH)_2CO_2H$, *i.e.*, dihydroxyacetic acid (*cf.* chloral hydrate). The anhydrous acid may be obtained as a thick syrup by evaporating the aqueous solution over phosphorus pentoxide *in vacuo*.

Glyoxylic acid gives all the reactions of aldehyde and an acid; *e.g.*, it reduces maniacal silver nitrate, forms a bisulphite compound, etc. Since it has no α-hydrogen atoms, glyoxylic acid undergoes the Cannizzaro reaction to form glycollic and oxalic acids:

$$2OCHCO_2H \xrightarrow{NaOH} HOCH_2CO_2Na + (CO_2Na)_2$$

When glyoxylic acid is reduced by means of metal and acid, tartaric acid is obtained as well as glycollic acid:

$$OCHCO_2H \xrightarrow{e;\,H^+} HOCH_2CO_2H + \begin{array}{l} CH(OH)CO_2H \\ | \\ CH(OH)CO_2H \end{array}$$

Pyruvic acid (*acetylformic acid, pyroracemic acid, α-ketopropionic acid, 2-oxopropanoic acid*), p.b. 165°C, is the simplest keto-acid. It may be prepared:

(*i*) By heating tartaric acid alone, or better with potassium hydrogen sulphate at 210-220°C.

The reaction is believed to take place via the formation of hydroxyamleic acid (I), which rearranges to oxalacetic acid, (II):

$$\begin{array}{l} CH(OH)CO_2H \\ | \\ CH(OH)CO_2H \end{array} \xrightarrow{-H_2O} \begin{array}{l} CHCO_2H \\ \| \\ C(OH)CO_2H \end{array} \longrightarrow \begin{array}{l} CH_2CO_2H \\ | \\ COCO_2H \end{array} \longrightarrow$$

$$CH_3COCO_2H + CO_2 \quad (50 - 55\%)$$

This is the best method for preparing pyruvic acid, and it was this method which gave rise to the name pyroracemic acid:

(*ii*) Other methods of preparation are the oxidation of lactic (α-hydroxypropionic) acid, the hydrolysis of α, α-dibromopropionic acid or acetyl cyanide, *e.g.*,

$$CH_3COCl \xrightarrow{KCN} CH_3COCN \xrightarrow{H^+} CH_3COCO_2H$$

A general method of preparing α-keto-acids uses the Claisen condensation between ethyl oxalate and monocarboxylic esters:

$$EtO_2CCO_2Et + RCH_2CO_2Et \xrightarrow{EtONa} EtO_2CCOCHRCO_2Et$$

$$\xrightarrow{10\%\ H_2SO_4} [HO_2CCOCHRCO_2H] \xrightarrow{-CO_2} RCH_2COCO_2H(94\%)$$

Pyruvic acid behaves as a ketone and as an acid: it forms an oxime, hydrazone, etc. It is reduced by sodium amalgam to alctic acid and dimethyltartaric acid (*cf.* glyoxylic acid, above):

$$3CH_3COCO_2H \xrightarrow{e;\ H^+} CH_3CH(OH)CO_2H + \begin{array}{c} CH_3C(OH)CO_2H \\ | \\ CH_3C(OH)CO_2H \end{array}$$

Pyruvic acid reduces ammoniacal silver nitrate, itself being oxidised to acetic acid:

$$CH_3COCO_2H + [O] \longrightarrow CH_3CO_2H + CO_2$$

It is oxidised by warm nitric acid to oxalic acid:

$$CH_3COCO_2H + [O] \longrightarrow (CO_2H)_2 + CO_2 + H_2O$$

Pyruvic acid is easily decarboxylated with warm dilute sulphuric acid to give acetaldehyde:

$$CH_3COCO_2H \longrightarrow CH_3CHO + CO_2$$

The mechanism of this reaction is uncertain, but since it occurs only with α-keto-acids, it suggests that the -I effect of the α-acrbonyl group plays an important part (*cf.* trichloroacetic acid). On this basis, the following mechanism could be postulated, involving a carbene intermediate.

$$CH_3-\overset{O}{\overset{\|}{C}}-C\begin{smallmatrix} \nearrow O \\ \searrow OH \end{smallmatrix} \xrightleftharpoons{H^+} CH_3-\overset{\overset{+}{O}H}{\overset{\|}{C}}-C\begin{smallmatrix} \nearrow O \\ \searrow OH \end{smallmatrix} \longleftrightarrow$$

$$CH_3-\overset{OH}{\overset{|}{C^+}}-C\begin{smallmatrix} \nearrow O \\ \searrow O-H \end{smallmatrix} \longrightarrow CO_2 + H^+ + CH_3-\ddot{C}-OH$$

$$\xrightarrow{H^+} CH_3-\overset{+}{C}H-O-H \xrightarrow{-H^+} CH_3CHO$$

On the other hand, when warmed with *concentrated* sulphuric acid, pyruvic acid undergoes *decarboylation, i.e.*, eliminates a molecule of carbon monoxide, to form acetic acid:

$$CH_3\overset{O}{\overset{\|}{C}}-C\begin{smallmatrix} \nearrow O \\ \searrow OH \end{smallmatrix} \xrightarrow{+2H^+} CH_3\overset{\overset{+}{O}H}{\overset{\|}{C}}-\overset{+}{C}\begin{smallmatrix} \nearrow OH \\ \searrow OH \end{smallmatrix} \xrightarrow[-H^+]{-H_2O}$$

$$CH_3\overset{O}{\overset{\|}{C}}-\overset{+}{C}=O \longrightarrow CO + CH_3\overset{+}{C}=O \xrightarrow[(ii)\ -H^-]{(i)\ H_2O} CH_3CO_2H$$

In this case, the possible mechanism proceeds though an intermediate acylium ion (*cf.* the A_{AC} mechanism).

Both of these reactions with sulphuric acid are characteristic of α-keto-acids. α-Keto-esters are also decarbonylated when heated, and Calvin *et al.* (1940), using ethyl pyruvate labelled with ^{14}C, showed that the carbon monoxide came from the carbethoxyl group. In this case, the results may be explained by he formation of an epoxide intermediate:

$$CH_3-\overset{O}{\overset{\|}{C}}-\underset{:OEt}{\overset{O}{\overset{\|}{C}}} \xrightarrow{110^\circ C} CH_3-\overset{O}{\overset{\|}{C}}-\overset{O}{\overset{\|}{C}}\ (\text{epoxide}, \overset{+}{O}Et) \longrightarrow CH_3-\underset{OEt}{\overset{O}{\overset{\|}{C}}} + CO$$

The methyl group is made reactive the adjacent carbonyl group, and so pyruvic acid undergoes many condensation reactions characteristic of a compound containing an active methylene group; *e.g.*, in the presence of dry hydrogen chloride, pyruvic acid forms α-keto-γ-valerolactone-γ-carboxylic acid:

$$CH_3\overset{CO_2H}{\overset{|}{C}}=O + CH_3COCO_2H \xrightarrow{HCl} \left[CH_3\underset{OH}{\overset{CO_2H}{\overset{|}{\underset{|}{C}}}}-CH_2-COCO_2H\right]$$

$$\xrightarrow{-H_2O} CH_3\overset{CO_2H}{\overset{|}{C}}-CH_2-COCO \quad (\text{C and CO joined through } O)$$

Pyruvic acid is a very important substance biologically, since it is an intermediate product in the metabolism of carbohydrates and proteins.

Laevulic acid (β-*acetylpropionic acid, γ-keovaleric acid, butan*-3-*one-1-carboxylic acid*), m.p. 34°C, is the simplest γ-keto-acid. It may be prepared via acetoacetic ester as follows:

$$[CH_3COCHCO_2C_2H_5]^- Na^+ + ClCH_2CO_2C_2H_5 \longrightarrow$$

$$NaCl + CH_3CO\overset{CH_2CO_2C_2H_5}{\overset{|}{C}}HCO_2C_2H_5 \xrightarrow[\text{(ii) } H^+]{\text{(ii) dil. KOH}} CH_3COCH_2CH_2CO_2H$$

It may also be prepared by heating a hexose sugar, particularly laevulose, with concentrated hydrochloric acid:

$$C_6H_{12}O_6 \longrightarrow CH_3COCH_2CH_2CO_2H + HCO_2H + H_2O$$

In practice it is customary to use cane-sugar as the starting material:

$$C_{12}H_{22}O_{11} \longrightarrow 2CH_3COCH_2CH_2CO_2H + HCO_2H + H_2O \ (21\text{-}22\%)$$

By heating dilute solutions of sucrose sunder pressure with hydrochloric acid, the yield is increased to 50 per cent.

Laevulic acid behaves as a ketone (forms oxime, etc.) and as an acid (forms esters, etc.). On the other hand, many of the reactions of laevulic acid indicate that it exists in the *lactol* form, *i.e.*, hydroxylactone (this is an example of ring-chain tautomerism), *e.g.*,

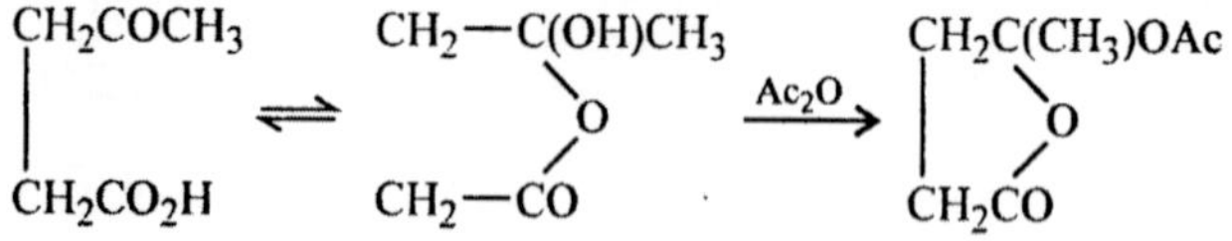

$$\begin{matrix} CH_2COCH_3 \\ | \\ CH_2CO_2H \end{matrix} \rightleftharpoons \begin{matrix} CH_2-C(OH)CH_3 \\ \quad\quad \diagdown O \\ \quad\quad \diagup \\ CH_2-CO \end{matrix} \xrightarrow{Ac_2O} \begin{matrix} CH_2C(CH_3)OAc \\ | \quad \diagdown O \\ | \quad \diagup \\ CH_2CO \end{matrix}$$

When heated for some time, laevulic acid is converted into α-and β-*angelica* lactones (derived from the lactol form):

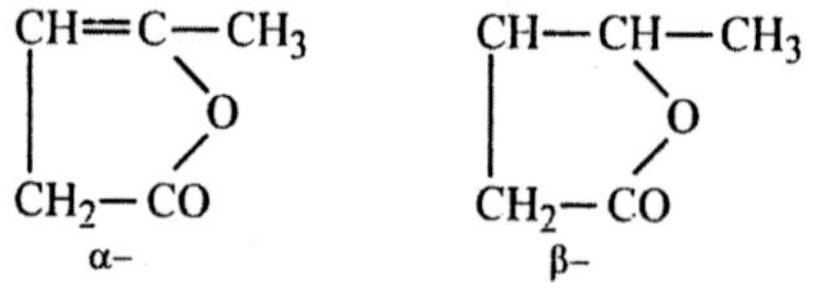

$$\underset{\alpha-}{\begin{matrix} CH{=}C-CH_3 \\ | \quad \diagdown O \\ | \quad \diagup \\ CH_2-CO \end{matrix}} \qquad \underset{\beta-}{\begin{matrix} CH-CH-CH_3 \\ | \quad \diagdown O \\ | \quad \diagup \\ CH_2-CO \end{matrix}}$$

A general method for preparing γ-keto-acids is to treat a ketone with lithium amide and lithium bromoacetate; the ketone undergoes α-carboxymehylation (Puterbaugh *et al.*, 1959):

$$R^1COCH_2R^2 \xrightarrow[CH_2BrCO_2Li]{LiNH_2} R^1COCHR^2CH_2CO_2H$$

γ-Keto-acids may also be synthesised by other methods.

Keto-acids with the oxo group in the δ- or a higher position are very conveniently prepared by the action of a dialkylcadmium on the half-ester acid chloride. On the other hand, δ-keto-acids may be prepared as follows from cyclohexane-1,3-dione (Stetter *et al.*, 1952); *e.g.*, 5-oxoheptanoic acid:

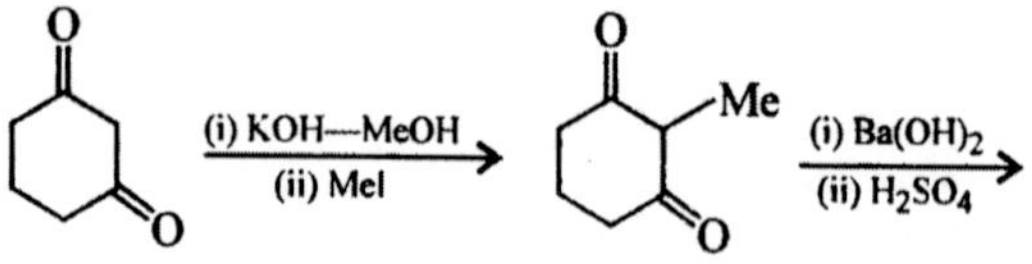

$$\text{cyclohexane-1,3-dione} \xrightarrow[\text{(ii) MeI}]{\text{(i) KOH—MeOH}} \text{2-methylcyclohexane-1,3-dione (Me)} \xrightarrow[\text{(ii) } H_2SO_4]{\text{(i) } Ba(OH)_2} MeCH_2CO(CH_2)_3CO_2H$$

The cyclic dione, since it contains an active methylene group, forms the potassium salt which can be alkylated, and the product readily undergoes fission, thus behaving like the acyclic β-diketones. The keto-acids may be reduced by the Wolff-Kishner method to the corresponding saturated acid; in the above example, to heptanoic acid. Alternatively, the alkylated ring may be opened and the carbonyl group reduced in one step step by refluxing with hydrazine in alkaline glycol solution.

EXERCISES

1. Write out the structures and I.U.P.A.C. names of isomeric secondary heptanols containing one methyl side-chain.
2. The dehydration of butan-2-ol with acid gives a mixture of 1- and 2-butenes. Write out the equations, state which predominates, and give your reasons. Draw the energy profiles of the reactions.
3. Convert ethanol into (i) n-propanol; (ii) n-butanol.
4. Arrange FtOH, CF_3CH_2OH, and $CC_{13}CH_2OH$ in order of increasing strength as acids and give your reasons.
5. The dehydration of n-BuOH with acid gives 1- and 2-butene. Explain this, state which product predominates, and why.
6. Convert propene into (i) n-PrOH; (ii) isoPrOH.
7. How could you show the existence of the following equilibrium?

$$ROH + H^+ \underset{}{\overset{H_2SO_4—H_2O}{\rightleftharpoons}} R^+ + H_2O$$

8. What are the disadvantages of preparing alkenes by:

   (*i*) RX + base; ROH + acid?

9. 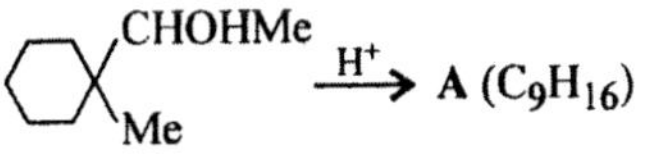 (cyclohexane bearing CHOHMe and Me on the same carbon) $\xrightarrow{H^+}$ **A** (C_9H_{16})

A, on ozonolysis, gives nonane-2,8-dione. What is **A** and how is it formed?

10. Complete the following equation and comment:

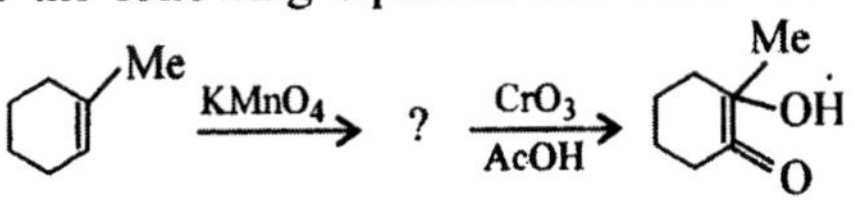

(1-methylcyclohexene) $\xrightarrow{KMnO_4}$? $\xrightarrow[AcOH]{CrO_3}$ (2-hydroxy-2-methylcyclohexanone: ring bearing Me, OH and =O)

11. Convert (i) n-C4H9OH into MeC≡CMe; (iii) n-PrOH into *cis*-hex-2-ene.
12. A concentrated aqueous solution of HBr reacts with EtOH to give EtBr, but a concentrated aqueous solution of NaBr does not. Explain.

13. What are A to E in the following reactions?

$$(\mathbf{A})\ C_4H_{10}O \xrightarrow{SOCl_2} C_4H_9Cl\ (\mathbf{B})$$

$$C_4H_{10}O \xrightarrow{H_2SO_4} C_4H_8(\mathbf{C})$$

$$(\mathbf{D})\ C_3H_6O_2 \xleftarrow[\text{(ii) } H_2O_2]{\text{(i) } O_3} C_4H_8(\mathbf{C}) \xrightarrow{HCl} C_4H_9Cl(\mathbf{E})$$

14. Complete the following equations:

(i) $n\text{-}C_3H_7CO_2H \xrightarrow{?} n\text{-}C_4H_9OH$

(ii) $Me_2CO + EtMgI \longrightarrow ? \xrightarrow{H^+} ?$

(iii) $EtCO_2Et + 2MeMgI \longrightarrow ? \xrightarrow{H_2O} ?$

(iv) (cyclohexylidene)=$CHCO_2Et \xrightarrow{?}$ (cyclohexylidene)=$CHCH_2OH \xrightarrow{MnO_2} ?$

(v) $CH_3CH_2CH_2OH \xleftarrow{?} BrCH_2CH_2COCl \xrightarrow{?} BrCH_2CH_2OH$

15. How many NMR signals would be given by the following alcohols in the presence of a trace of acid? Indicate all signals other than singlets by (m).

(i) $CH_3CH_2CH_2OH$;

(ii) Me_2CHCH_2OH;

(iii) Me_3COH;

(iv) Me_3CCH_2OH;

(v) $MeCH_2CMe_2OH$.

16. The table lists some peaks in the mass spectra of the four isomeric butyl alcohols, A, B, C, D. Assign each alcohol to a letter, and give your reasons.

m/e	18	29	31	41	43	45	56	59	74
A(%)	—	34	100	63	60	7	86	—	1
B(%)	4	24	67	55	100	52	3	5	9
C(%)	—	15	22	11	10	100	1	18	0.2
D(%)	5	13	34	20	15	—	1.4	100	0.1

17. Complete the following equations and comment where necessary:

(i) $Me(C \equiv C)_3COCH = CH_2 \xrightarrow{NaBH_4} ? \xrightarrow{H^+} ?$

(ii) $\overset{\diagup O \diagdown}{CH_2—CH_2}$?

(iii) $CH_2 = CHCl + Cl_2 \longrightarrow ? \xrightarrow[300°C]{Al_2O_3} ?$

(iv) $CH_2 = CHCN + CtOH \xrightarrow{H_2SO_4 - H_2O} ?$

(v) $CH_2 = CHCOCH_3 \xrightarrow{NaOCl} ?$

(vi) $MeCH = CHCOEt \xrightarrow[H_2(1\ atm)]{Pd-c} ?$

(vii) $MeCH = CHCOEt \xrightarrow[press.]{Ni;H_2} ?$

(viii) $MeCH = CHCO_2Et + HCN \longrightarrow ?$

(ix) $CH_2 = CHCO_2Et + EtOH \xrightarrow{EtONa} ?$

(x) $CH_2 = CHCOMe \xrightarrow{H_2O_2;OH^-} ?$

(xi) $CH \equiv CNa + Cl(CH_2)_3I \xrightarrow[NH_3]{liq.} ? \xrightarrow[(ii)\ H^+]{(i)\ NaCN} ?$

(xii) $MeCH = CHCOMe + NH_3 \longrightarrow ?$

(xiii) $MeCH = CHCH_2Cl + ? \xrightarrow{?} MeCH = CHCH_2I$?

(xiv) $MeCH = CHCO_2Et + CH_3NO_2 \xrightarrow{EtONa} ?$

(xv) $MeCH = CHCHO + HCN \xrightarrow{OH^-} ?$

(xvi) $MeCH = CHCHO \xrightarrow{LAH} ?$

(xvii) $PhCHO + ? \xrightarrow{pyridine} PhCH = CHCO_2H$?

(xvii) $EtCHO + 2CH_2(CO_2)Et_2 \xrightarrow{piperidine} ?$

(xix) $EtCHI + CH_3(CO_2H)_2 \xrightarrow{pyridine} ?$

18. A compound has structure $MeCH = CHCH_2CHO$ (A) or $MeCH_2CH = CHCHO$ (B). Distinhuish betweebn the two by (i) a chemical method, (ii) i.r., (iii) NMR

19. Explain the following.

    (i) μ of acraldehyde > μ of propionaldehyde, (ii) μ of ethyl bromide > μ of vinyl bromide.

20. Complete the following equation for.

    (i) $B_{AC}2$, (ii) B_{AL} 1, (iii) A_{AC} 1 mechanism.

$$CH_2 = CHCHMeOOCCH_3 \xrightarrow{OH^-\ or\ H^+}$$

21. How could you demonstrate the S_N1' mechanism in the reaction

$$CH_2 = CHCH_2OH + HCl \longrightarrow CH_2 = CHCH_2Cl\ ?$$

22. Calculate $\Delta H^{\oplus}$ for the 1,2 and 1,4-additions of HZ to C = C – C = O. Which addition is favoured ?

23. Draw the M.O for the group C = C – C = O, and suggest a reason for the 1,4-addition properties.

24. Convert MeCH = $CHCO_2H$ into $MeCH(CH_2C_2Et)_2$.

25. Synthesise $EtO_2CCH = CHCH_2COCO_2Et$ starting from compounds not containing ore than *four* carbon atoms.

26. Aldol may undergo dehydration in two possible ways.

$$CH_2 = CHCH_2CHO \xleftarrow{-H_2O} MeCHOCCH_2CHO$$

$$\xrightarrow{-H_2O} MeCH = CHCHO$$

On of these products is obtained almost exclusively. Which one and why ?

27. 2,2-Dimethylbut-3-enoic acid, on warming, gives trimethylethylene. Suggest a mechanism.

28. Suggest a synthesis for.

(i) $\begin{array}{c} Me_2CCH_2CO_2H \\ | \\ Me_2CCH_2CO_2H \end{array}$, (i) $Me_2C(CH_2CO_2H)_2$

(iii) $CH_3CH = CHCH_2CH_2OH$.

29. Draw the energy profiles for tautomerism between A and B and for resonance between A and B.

30. Explain the following equation:

$$MeCOCH_2CO_2Et + MeCOCHR^1CO_2Et \xrightarrow[(ii)\,1\,R^2X]{(i)\,1\,EtONa}$$

$$MeCOCHR^2CO_2Et + MeCOCHR^1CO_2Et$$

31. How could you differentiate, by i.r. and NMR spectroscopy, between

(A) $MeCOCH_2CH_2COMe$ and

(B) $MeCOCH_2COCH_2CH_3$?

32. Starting with E.A.A. or D.E.M., give one synthesis (using equations) for:

    (i) 4-methylpentan-2-one;

    (ii) 2,3-dimethylbutanoic acid;

    (iii) methylpentanedioic acid;

    (iv) cyclopentate-1,2-dicarboxylic acid;

    (v) $PhCOCH_2CO_2Et$;

    (vi) 3-carboxy-2-methylpentanedioic acid.

33. The b-keto-acid shown in decarboxylated with difficulty. Suggest an explanation.

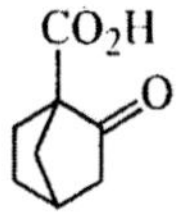

    CO_2H

    O

34. What two compounds will condense to give each product shown (after acidification)?

    (i) $A + B \xrightarrow{EtONa} MeCOCH_2COCO_2Et$

    (ii) $C + D \xrightarrow{EtONa} PhCH(CN)COCO_2Et$

    (iii) $E + F \xrightarrow{EtONa} PhCH(CN)COMe$

35. Starting with E.A.A., synthesise 3-methylpentan-2-ol.

36. Complete the equations:

    (i) $CH_2(CO_2Me)_2 + ? \xrightarrow[(ii)\ AcOH]{(i)\ Na} CH(CO_2Me)_3$

    (ii) $MeCH(CO_2Et)COCO_2Et \xrightarrow{heat} ?$

    (iii) $MeCH_2CO_2Et + ? \xrightarrow[(ii)\ AcOH]{(i)\ EtONa} MeCH(CHO)CO_2Et$

    (iv) $MeCH_2CO_2Et + (CO_2Et)_2 \xrightarrow[(ii)\ AcOH]{(i)\ EtONa} ? \xrightarrow{heat} ?$

37. Suggest a synthesis for ethyl 4-keto-3-mehylpentanoate.

38. A compound $C_4H_6O_2$, shows a very strong i.r. band at 1721 cm^{-1} and its N.M.R. spectrum shows only one singlet signal. What is the compound?

3

HYDROXYACIDS, STEREOCHEMISTRY, UNSATURATED DICARBOXYLIC ACID

MONOBASIC HYDROXYACIDS

Monobasic hydroxyacids are monocarboxylic acids which contain one or more hydroxyl groups in the carbon chain.

Nomenclature

The usual method is to name the hydroxyacid as a derivative of the parent acid (named according to the trivial system), the position of the hydroxyl group being indicated by a Greek letter, *e.g.*,

$CH_3CHOHCH_2CO_2H$ β-hydroxybutyric acid

According to the I.U.P.A.C. system of nomenclature, the position of the hydroxyl group is indicated by a number, *e.g.*,

$CH_3CH_2CHOHCO_2H$ 2-hydroxybutanoic acid or

1-hydroxypropane-1 -carboxylic acid

Many hydroxyacids which occur in nature are given special names indicating the source, *e.g.*, $CH_3CHOHCO_2H$, lactic acid, occurs in sour milk (Latin: *lac*, milk).

General Methods of Preparation

The methods used depend on the position of the hydroxyl group in the chain.

α-Hydroxyacids may be prepared by the hydrolysis of aldehyde or ketone cyanohydrins or the hydrolysis or α-bromo-acids.

Less useful methods are the controlled oxidation of 1,2-glycols with dilute nitric acid or the deamination of α-amino-acids.

β-Hydroxyacids are usually best prepared by means of the Reformatsky reaction or by the oxidation of aldose with Tollens' reagent. The hydrolysis of ethylene cyanohydrin is useful for the preparation of β-hydroxypropionic acid.

α-, β-, γ-, or δ-Hydroxy acids may be prepared by the catalytic reduction of the corresponding keto-esters, *e.g.*,

$$CH_3COCH_2COEt \xrightarrow[\text{Press:100 °C}]{H_2-Ni} CH_3CHOCH_2CO_2Et$$

$$\xrightarrow[(ii)\,H_3O^+]{(i)\,KOH} CH_3CHOHCH_2CO_2H$$

Alternatively, hydrolysis of the corresponding halogen-acid gives the hydroxyacid or lactone (see below). The reduction of half-esters of α, ω-dicarboxylic acids is particularly useful for preparing ω-hydroxyacids.

General Properties and Reactions

Glycollic acid, the first member of the series, is a solid; the higher members are liquids. All are soluble in water, generally more so than are either the corresponding acid or alcohol. This is to be expected, since hydroxyacids have *two* functional groups which can form hydrogen bonds with water.

1. Hydroxyacids behave both as acids and alcohols; in many reactions the hydroxyl and carboxyl groups do not interfere with each other, particularly when they are far apart. Further-more, by esterifying the carboxyl group, the ester then behaves predominantly as a hydroxy- compound, *i.e.*, esterification masks, to a large extent, the presence of the carboxyl group. The carboxyl group may be converted into the ester, amide, nitrile, acyl chloride, etc. The hydroxyl group (when the carboxyl group has been esterified) may be converted into the ester, ether, etc.; *e.g.*, glycollic acid reacts with acetyl chloride to form acetyiglycollic acid (behaving as a hydroxy-compound):

$$HOCH_2CO_2H \longrightarrow CH_3COOCH_2CO_2H + HCl$$

Glycollic acid reacts with phosphorus pentachloride or thionyl chloride to form chloroacetyl chloride (behaving both as an alcohol and an acid); the chloroacetyl chloride is readily hydrolysed by water to chloroacetic acid:

$$HOCH_2CO_2H \xrightarrow{PCl_5} ClCH_2COCl \xrightarrow{H_2O} ClCH_2CO_2H$$

2. When heated with dilute sulphuric acid or acid permanganate, a-hydroxyacids are converted into aldehydes or ketones:

$$RCHOHCO_2H \xrightarrow{H_2SO_4} RCHIO + HCO_2H \quad (v.g.)$$

$$R_2C(OH)CO_2H + [O] \xrightarrow{KMnO_4} R_2CO + CO_2 + H_2O \ (v.g.)$$

These reactions offer a very good means of stepping down the monocarboxylic acid series (via the, H.V.Z. reaction), one carbon atom at a time.

β-Hydroxyacids, on oxidation with alkaline permanganate, give ketones:

$$RCHOHCH_2CO_2H \longrightarrow [RCOCH_2CO_2H] \xrightarrow{-CO_2} RCOCH_3$$

3. When heated with concentrated hydriodic acid, hydroxyacids are reduced to the corresponding carboxylic acid:

$$CH_3CHOHCO_2H + {}_2HI \longrightarrow CH_3CH_2CO_2H + H_2O + I_2$$

4. When hydroxyacids are heated, the product formed depends on the relative positions of the hydroxyl and carboxyl groups.

(i) α-Hydroxyacids form **lactides**; these are six-membered ring compounds formed by reaction between two molecules of the hydroxyacid, and are named systematically as 3,6-dialkyl-1, 4-dioxan-2, 5-dione:

```
RCHO[H  HO]OC                 RCH—O—CO
|           |        ——→      |      |     + 2H2O
CO[OH    H]OCHR               CO—O—CHR
```

The tendency to form lactides is very pronounced, in many cases the lactide being formed by allowing the α-hydroxyacid to stand in adesiccator over concentrated sulphuric acid. Lactides are readily hydrolysed to the acid by alkali.

The distillation of α-hydroxyacids produces aldehydes via the lactide:

```
2RCHOHCO2H ——→ RCHOCO ——→ 2RCHO + 2CO
                |   |
               COOCHR
```

When heated with a trace of zinc chloride, lactides are converted into linear polyesters HO—(—CHRCOO—)$_n$—H, which regenerate the lactide on distillation under reduced pressure.

(ii) When β-hydroxyacids are heated, they eliminate a molecule of water to form mainly the α, β-unsaturated acid and a very small amount of the β, γ-unsaturated acid. The reaction is best carried out by refluxing with dilute sulphuric acid or dilute alkali.

$$RCHOHCH_2CO_2H \longrightarrow RCH = CHCO_2H + H_2O$$

(iii) On heating, γ- and δ-hydroxyacids readily form *internal* esters known as **lactones:**

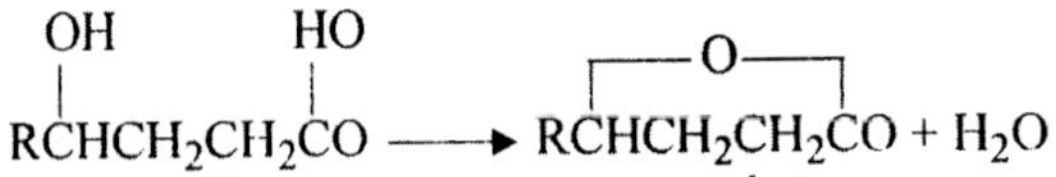

$$\underset{\gamma\text{-acid}}{R\overset{OH}{\overset{|}{C}}HCH_2CH_2\overset{HO}{\overset{|}{C}}O} \longrightarrow \underset{\gamma\text{-lactone}}{RCHCH_2CH_2CO\ (\text{ring via }O)} + H_2O$$

$$\underset{\delta\text{-acid}}{R\overset{OH}{\overset{|}{C}}HCH_2CH_2\overset{HO}{\overset{|}{C}}O} \longrightarrow \underset{\delta\text{-lactone}}{RCHCH_2CH_2CO\ (\text{ring via }O)} + H_2O$$

Lactone formation often occurs spontaneously when the sodium salts of γ- or δ-hydroxyacids are acidified, particularly with the former.

γ- and δ-Lactones are also formed when γ, δ- or δ, ε-unsaturated acids are treated with concentrated sulphuric acid:

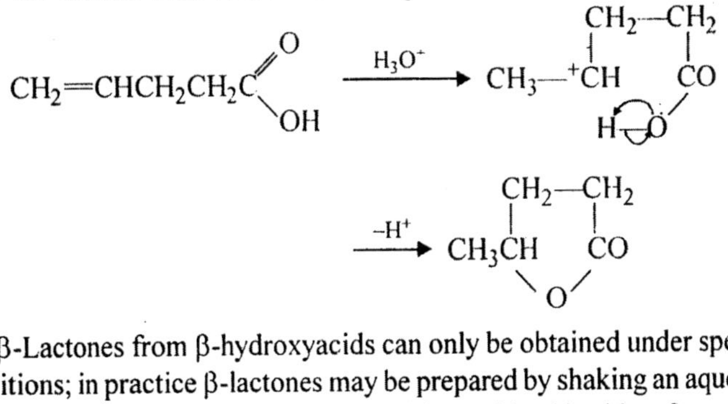

$$CH_2{=}CHCH_2CH_2C(=O)OH \xrightarrow{H_3O^+} CH_3{-}{}^{+}CH(CH_2{-}CH_2{-}CO{-}\ddot{O}{-}H)$$

$$\xrightarrow{-H^+} CH_3CH(CH_2{-}CH_2{-}CO{-}O) \ (\text{ring})$$

β-Lactones from β-hydroxyacids can only be obtained under special conditions; in practice β-lactones may be prepared by shaking an aqueous solution of the solidum salt of the β-chloroacid with chloroform:

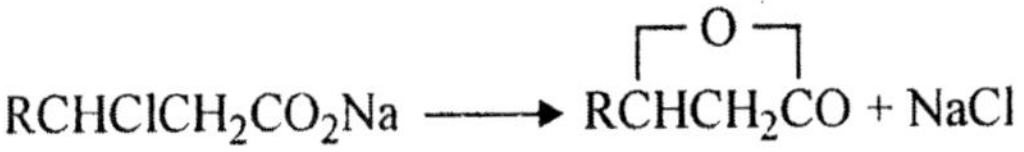

$$RCHClCH_2CO_2Na \longrightarrow RCHCH_2CO\ (\text{ring via }O) + NaCl$$

They may, however, be prepared more readily by reactions between keten and a carbonyl compound; *e.g.*, β-Droniolactone from keten and formaldehyde. β-Lactones are more reactive than other types of lactones; *e.g.*, β-propiolactone readily reacts with nucleophilic reagents to form β-substituted propionic acid derivatives.

$$CH_2{=}C{=}O + HCHO \longrightarrow \overbrace{CH_2CH_2CO}^{O}$$

According to the I.U.P.A.C. system of nomenclature, lactones are known as -olides, *e.g.*, $\overbrace{CH_2CH_2CH_2CH_2CO}^{O}$ δ-valerolactone or 1,5-pentanolide. The systematic name of this lactone is 4-*hydroxybutane-1 -carboxylic acid lactone.*

γ-Butyrolactone is manufactured by the oxidation of tetramethylene glycol over a copper catalyst:

$$HOCH_2CH_2CH_2OH \xrightarrow[Cu]{O_2} \overbrace{CH_2CH_2CH_2CO}^{O}$$

It is an important intermediate in the preparation of polyamides.

Lactones are converted into alkali salts when refluxed with excess of alkali:

$$\overbrace{RCHCH_2CH_2CO}^{O} + NaOH \longrightarrow RCHOHCH_2CH_2CO_2Na$$

They are reduced by sodium amalgam in acid solution to the corresponding acid:

$$\overbrace{RCHCH_2CH_2CH_2CO}^{O} \xrightarrow{e;\ H^+} RCH_2CH_2CH_2CH_2CO_2H$$

On the other hand, lactones are reduced to diols by LAH or by catalytic hydrogenation.

When treated with concentrated halogen acid, lactones form the corresponding halogen-acid:

$$\overbrace{RCHCH_2CH_2CO}^{O} + HX \rightleftharpoons RCHXCH_2CH_2CO_2H$$

With concentrated ammonium hydroxide solution, the hydroxyamide is formed; in some cases this product may be' the y-lactone:

$$\overbrace{RCH(CH_2)_2CO}^{O} \xrightarrow{NH_3} RCHOH(CH_2)_2CONH_2 \text{ or } \overbrace{RCH(CH_2)_2CO}^{NH}$$

Lactones readily react with potassium cyanide to form cyano-acids, *e.g.*, γ-butyrolacton forms glutaric acid as follows:

$$\underset{\text{O—CO—CH}_2}{\overset{\text{CH}_2\text{—CH}_2}{|}} \xrightarrow{\text{KCN}} \text{KO}_2\text{C(CH}_2)_3\text{CN} \xrightarrow{\text{conc. HCl}} \text{HO}_2\text{C(CH}_2)_3\text{CO}_2\text{H} \quad (79\text{–}83\%)$$

Lactones also react with Grignard reagents to form diols, *i.e.*, they behave as esters; *e.g.*,

$$\underset{\text{O—CO—CH}_2}{\overset{\text{CH}_2\text{—CH}_2}{|}} \xrightarrow{\text{2MeMgI}} \underset{\text{OMgI}}{\text{CH}_2}(\text{CH}_2)_2\underset{\text{OMgI}}{\text{CMe}_2}$$

$$\xrightarrow{\text{H}_3\text{O}^+} \text{HOCH}_2(\text{CH}_2)_2\text{CMe}_2\text{OH}$$

The infra-red absorption region of the carbonyl group (C=O stretch) in γ-lactones is 1780-1760 cm^{-1}(s), whereas that for δ-lactones is 1750-1735 cm^{-1}(s). Thus the two may be distinguished, but the latter absorb in the same region as esters. The molecular formula should differentiate between saturated esters and lactones.

(iv) ε-Hydroxyacids, in certain cases, may form the lactone on heating. Usually they either eliminate a molecule of water to form two unsaturated acids, δ, ε- and ε, ξ-, or form linear polyesters.

(v) Hydroxyacids with the hydroxyl group further removed than the ε-position, on heating, either eliminate a molecule of water to form unsaturated acids (of two types; cf. ε-hydroxy-acids), or form linear polyesters.

Large ring lactones (fourteen- to eighteen-membered rings) have been prepared by the oxidation of cyclic ketones with Caro's acid (Ruzicka and Stoll, 1928). This is an example of the Baeyer-Villiger oxidation:

$$\text{CH}_2\begin{matrix}\diagup(\text{CH}_2)_n\diagdown\\ \diagdown(\text{CH}_2)_n\diagup\end{matrix}\text{CO} + [\text{O}] \xrightarrow{\text{H}_2\text{SO}_5} \text{CH}_2\begin{matrix}\diagup(\text{CH}_2)_n\text{—CO}\\ \diagdown(\text{CH}_2)_n\text{—O}\end{matrix}\Big|$$

By using the **high dilution principle** of Ruggli (1912), large ring lactones have also been prepared from hydroxyacids in which the hydroxyl group is far removed from the carboxyl group. According to this principle, by using sufficiently dilute solutions of a hydroxyacid, the distance between *different* molecules can be made greater than the distance between the hydroxyl and carboxyl groups of the same molecule.

Thus, the cyclic compound (lactone) is formed instead of lineal condensation taking place; *e.g.*, Stoll and his co-workers (1934) found that (ω-hydroxypenta-decanoic acid gave a high yield of lactone in very dilute solution.

Hunsdiecker and Eribach (1947) have also prepared large ring lactones by the dilution principle. These workers cyclised (u-bromo-aliphatic acids by boiling dilute solutions in butanone with excess of potassium carbonate; they obtained .lactones in yields varying from 56.3 to 96.8 per cent, the yield increasing with the size of the ring.

Glycollic acid (*hydroxyacetic acid, hydroxyethanoic acid*), m.p. 80°C, is the simplest hydroxy-acid, and occurs in the juice of beet and sugar-cane, and in unripe grapes. It may be prepared by refluxing an aqueous solution of potassium chloroacetate with sodium carbonate;

$$ClCH_2CO_2K + H_2O \xrightarrow[\text{(ii) HCl}]{\text{(i) } Na_2CO_3} HOCH_2CO_2H + KCl$$

(80%)

It may also be prepared by warming a solution of formalin with potassium cyanide:

$$HCHO + KCN + 2H_2O \longrightarrow HOCH_2CO_2K + NH_3$$

$$\xrightarrow{HCl} HOCH_2CO_2H \quad (70\%)$$

Glycollic acid is prepared industrially by the electrolytic reduction of oxalic acid and by heating at 160-170°C and under pressure, a mixture of formaldehyde, carbon .monoxide and water in acetic acid with sulphuric acid as catalyst:

$$HCHO + CO + H_2O \longrightarrow HOCH_2CO_2H$$

If methanol is used instead of water, methyl glycollate is obtained.

Glycollic acid is oxidised to oxalic acid by nitric acid. Its lactide is known as *glycollide*.

Lactic acid (α-*hydroxypropionic acid*, 2-*hydroxypropanoic acid*) may be prepared by any of the general methods used for α-hydroxyacids, *e.g.*, the hydrolysis of acetaldehyde cyanohydrin, or β-heating α-bromopropionic acid with aqueous sodium hydroxide.

One industrial preparation of lactic acid is the fermentation of lactose (the sugar in milk) by *Bacillus acidic lactili*:

$$C_{12}H_{22}O_{11} + H_2O \longrightarrow 4CH_3CHOHCO_2H$$

Another method is the fermentation of sucrose by *Rhizopus oryza*.

Lactic acid undergoes all the general reactions of α-hydroxyacids; oxidation with Fenton's reagent converts it into pyruvic acid:

$$CH_3CHOHCO_2H \xrightarrow{H_2O_2/Fe^{2+}} CH_3COCO_2H$$

It is oxidised to acetic acid by permanganate. Lactic acid is used in the tanning and dyeing industries, and ethyl lactate is used as a solvent for cellulose nitrate.

Since lactic acid contains one asymmetric carbon atom, it can (theoretically) exhibit optical activity . Three forms are known: L(+)-lactic acid, m.p. 26°C; D(—)-lactic acid, m.p. 26°C; and DL or (±)-lactic acid, m.p. 18°C. The lactic acid prepared by the above methods is (–)-(+)-Lactic acid may be obtained from meat extract; this acid is also known as *sarcolactic acid* (Greek: *sarkos*, flesh). (–)-Lactic acid may be obtained by the fermentation of sucrose by *Bacillus acid lavolactiti*.

STEREOCHEMISTRY

A description of conformational analysis and brief accounts of optical and geometrical isomerism have already been given, and these should be revised before reading the following discussion.

Isomerism consists of three types:

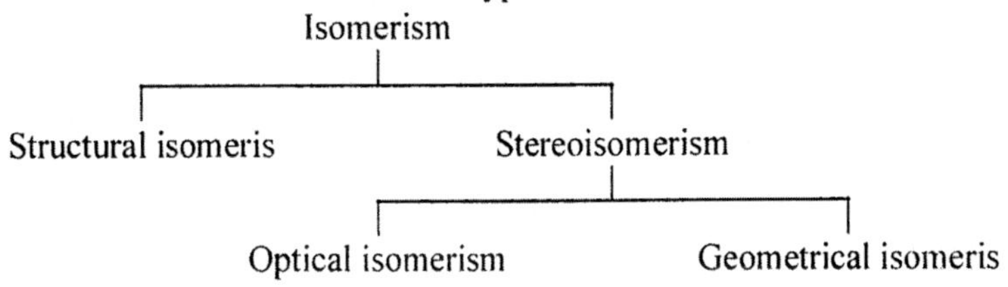

Structural isomerism is due to the difference in structure, and is exhibited in three different ways.

(i) *Chain or nuclear isomerism* is exhibited by compounds which differ in the arrangement of the carbon atoms, *e.g.*, n- and isobutane.

(ii) *Position isomerism* is exhibited by compounds having the *same* carbon skeleton but differing in the position occupied by a substituent group, *e.g.*, n- and isopropyl alcohols; α-, β- and γ-hydroxybutyric acids: *ortho-*, *meta-* and *para*-nitrophenols.

(iii) *Functional group isomerism* is exhibited by compounds having different functional groups, *i.e.*, compounds with the same molecular formula but belonging to different homologous series, *e.g.*, ethanol and dimethyl ether; acetone and propionaldehyde.

Tautomerism may be regarded as a special case of functional group isomerism.

Stereoisomerism is exhibited by isomers having the same structure but different *configurations*. Different configurations are possible because carbon forms mainly covalent bonds and these have direction in space.

Optical isomerism is characterised by compounds having the same structure but different configurations, and because of their *molecular asymmetry* these compounds are optically active.

Geometrical isomerism or *cis-trans isomerism* is characterised by compounds having the same *structure but different configurations*, and because of their *molecular symmetry* these compounds are *not* optically active. They can also exhibit optical isomerism if the structure of the molecule, apart from giving rise to geometrical isomerism, also satisfies the requirements for optical isomerism.

The instrument used for measuring the rotatory power of a substance is the *polarimeter* Essentially it consists of two Nicol prisms, one the polariser (P) and the other, the analyser (A), and between them a tube (T) which contains the substance (a liquid or a solution) to be examined (Fig. 3.1). S is a source of monochromatic light.

Fig. 3.1

If the substance rotates the plane of polarisation to the right, *i.e.*, if the analyser has to be turned to the right (clockwise) to restore the original field, the substance is said to be dextrorotatory, if to the left (anticlockwise), *lavorotatory*.

It has been found that the amount of the rotation depends, for a given substance, on a number of factors, *e.g.*, (i) the thickness of the layer traversed, (ii) the nature of the solvent (if in solution), (iii) the temperature and (iv) the wavelength of the light used. If [α] represents the *specific rotation*, *l* the thickness of the layer in decimetres, c the number of grams of substance (per 100 ml of solution), and the determination is carried out at temperature l°C using sodium light (the D line), then if a is the *observed* rotation (+ or –),

$$[\alpha]t_D = \frac{100\alpha}{lc}$$

Since the value of the rotation depends on the solvent, this should also be stated.

Designation of optical isomers. Originally, the dextrorotatory (+) isomer was prefixed by and the lavorotatory (—) isomer by *l*. It soon became apparent that slereochemical relationships were far more important than the actual direction of rotation, and so the symbols D and L are now used to indicate relationships between configurations. If such relationships are unknown, the symbols (+) and (–) are used to indicate the *sign* of rotation. The prefixes *dextro* and *laevo* (with hyphens) are also used.

In order to develop a system of stereochemical relationships it is necessary to choose a standard to which all compounds may be referred. Rosanoff (1906) chose the (+)- and (–)-forms of glyceraldehyde as the arbitrary standard and proposed that the formula of (+)-glyceraldehyde should be written as shown in Fig. 3.2(a).

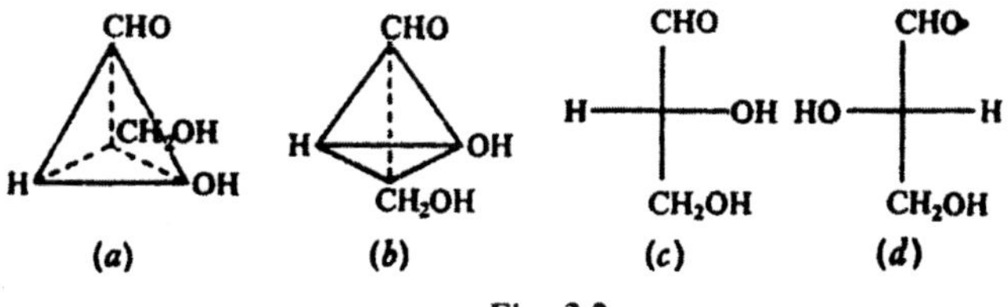

Fig. 3.2

The tetrahedron is drawn so that three corners are imagined to be *above* the plane of the paper, and the fourth *below* this plane. Furthermore, the spatial arrangement of the four groups joined to the central carbon atom *must be placed as shown in* (*a*), *i.e.*, the accepted convention for drawing D (+)-glyceraldehyde places the *hydrogen atom at the left and the hydroxyl group at the right, with the aldehyde group at the top corner*. Now imagine the tetrahedron to rotate about the horizontal line joining H and OH until it takes up position (*b*).

This is the conventional position for a tetrahedron, groups joined to *full horizontal* lines being *above* the plane of the paper, and those joined to *broken vertical* lines being *below* the plane of the paper. The *conventional plane-diagram* is obtained by drawing the full horizontal and broken vertical lines of (b) as full lines, placing the groups as they appear in (b), and taking the asymmetric carbon atom to be at the point where the lines cross. Although (c) is a plane-diagram, it is most important to remember that horizontal lines represent groups above the plane, and vertical lines groups below the plane of the paper. Figure (d) represents the plane-diagram formula of L(—)-glyceraldehyde; here *the hydrogen atom is to the right and the hydroxyl group to the left*. Another way of drawing (c) and (d) is to use broken vertical lines (instead of the full lines shown). Thus, any compound that can be prepared from, or converted into, D(+)-glyceraldehyde will belong to the D-series. Similarly, any compound that can be prepared from, or converted into, L(—)-glyceraldehyde will belong to the L-series. When representing relative configurational relationships of molecules containing more than one asymmetric carbon atom, *the asymmetric carbon atom of glyceraldehyde is always drawn at the bottom*, the rest of the molecule being built up from this unit.

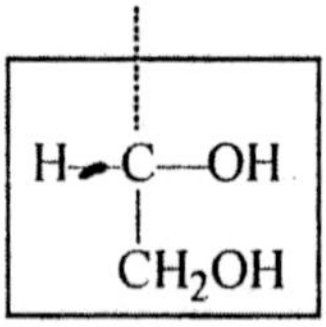

H—C—OH CH_2OH	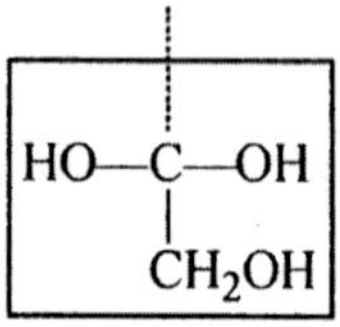 HO—C—OH CH_2OH	CO_2H H — — OH HO — — H CO_2H
D-series	L-series	(+)-tartaric acid

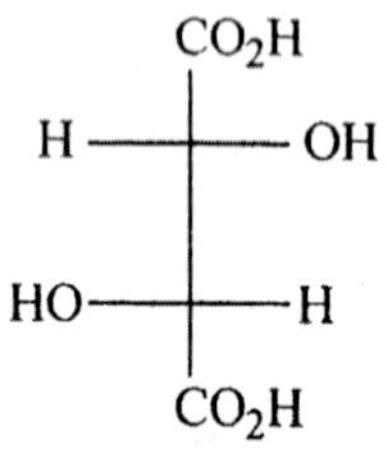

Thus we have a scheme of classification of *relative* configurations based on D(+)-glyceral-dehyde as *arbitrary* standard. Until recently there was no way of determining, with certainty, the absolute configuration of molecules.

Arbitrary choice makes the configuration of D(+)- glyceraldehyde have the hydrogen to the left and the hydroxyl to the right. Bijvoet *et al.* (1951), however, have shown by X-ray analysis that dextrorotatory tartaric acid has the configuration assigned to it by E. Fischer. Hence tartaric acid can be used as an absolute standard.

In 1848. Pasteur separated sodium ammonium racemate into two kinds of crystals by hand, and found that the specific rotation of each kind of crystal was the same, but one was dextrorotatory and the other, laevorotatory. Pasteur was able to separate the crystals by hand because he observed that they had hernihedral facets, one set of crystals being the mirror images of the other set. Such sets of crystals are said to be *enantiomorphous*.

It has been found that only those structures, crystalline or molecular, which are not super- impossible on their mirror images, are optically active. Such structures may be *asymmetric*, or *dissymmetric*. Asymmetric means completely devoid of the elements of symmetry. Dissymmetric means not completely devoid of symmetry, but possessing so few elements of symmetry as still to be capable of existing in two form's (one the mirror image of the other) which are not super- imposable. To avoid unnecessary complications, we shall use the term asymmetric to cover both cases (of asymmetry and dissymmetry). If a compound is asymmetric, then it is to be expected that the original molecule and its mirror image might differ in some properties although their structures-are identical. Experience shows that the most marked difference is their action on polarised light.

Optical activity may be due entirely to the *crystal structure being asymmetric*, *e.g.*, quartz. In such cases the substance is optically active only so long as it remains solid, the optical activity being lost when the solid is fused or dissolved in a solvent. Quartz crystals exist in hernihedral forms. Hernihedral faces are those faces not symmetrically placed with respect to other faces; they occur only in *half* the positions where they might be expected to occur, and thus give the crystal an asymmetric structure (actually dissymmetric). The (+)- and (–)-forms of quartz are mirror images; but it should be noticed that many optically active crystals do not possess hemihedral faces.

On the other hand, optical activity may be due entirely to molecular structure. In this case the *molecular structure* is *asymmetric*, *i.e.*, the compounds have molecules in which the *atoms* are arranged spatially so that the original molecule is not superimposable on its mirror image. Such compounds are optically active in the solid, fused, dissolved or gaseous state, *e.g.*, sucrose, lactic acid, limonene, etc.

A molecule and its mirror image, when they are not superimposable, are known as *enantiomorphs* (this name taken from crystallography), *enantiomers* or *optical antipodes*. It appears that enantiomers are identical physically except (i) in their manner of rotating polarised light; the rotations are equal but opposite; (ii) they absorb dextro- and laevo-circularly polarised light unequally (circular dichroism). The crystal forms of enantiomers may be mirror images of each other, *i.e.*, the crystals themselves may be enantiomorphous, but this is unusual. Enantiomers are similar chemically, but their rates of reaction with other optically active substances are usually different. They may also be different physiologically, *e.g.*, (+)-histidine is sweet, (–)- tasteless; (–)-nicotine is more poisonous than (+)-; L(+)-ascorbic acid (vitamin C) is more efficient than D(–)-.

In 1874, van't Hoff and Le Bel, independently, gave the solution to the problem of optical isomerism. Van't Hoff proposed the theory that if the four valances of the carbon atom are

arranged tetrahedrally with the carbon atom at the centre, then all the cases of isomerism known are accounted for. Le Bel's theory is substantially the same as van't Hoff's, but differs in that whereas van't Hoff believed that the valency distribution was definitely tetrahedral and fixed as such, Le Bel believed that the valency directions were not rigidly fixed, and did not specify the tetrahedral arrangement, but thought that *whatever* the spatial arrangement, the molecule *Cabde* would be *asymmetric*. Van't Hoff's theory is more in keeping with later work, *e.g.*, in recent years physico-chemical evidence—X-ray and dipole moment studies—has shown that saturated carbon compounds exhibit a tetrahedral structure and that the carbon atom is situated inside the tetrahedron at the centre. Before the tetrahedral theory was suggested, it was believed that the four valencies of carbon were planar.

The simplest type of asymmetric structure is that which contains one carbon atom joined to four different atoms or groups, *i.e.*, a molecule of the type *Cabde*, in which the groups, a, b, d ore may or may not contain carbon. The carbon atom in *Cabde* is said to be asymmetric (actually, of course, it is the group which is asymmetric; a carbon atom cannot be asymmetric). Up to the present time, compounds of the type Ca_4 (*e.g.*, CCl_4), Ca_3b (*e.g.*, $CHCl_3$), Ca_2b_2 (*e.g.*, CH_2Cl_2), Ca_2bd (*e.g.*, $HOCH_2CO_2H$) have never been observed to exist as optical isomers. Only one form of each is known. This agrees with the tetrahedral configuration, *e.g.*, Ca_2bd (Fig. 3.3) (II), the mirror image of (1), is superimposable on (1); however the four groups are arranged in the tetrahedron, (II) is always superimposable on (1) (see footnote). Thus there is only one form of Ca_2bd. Similarly, there is only one form of Ca_4, Ca_3b or Ca_2b_2. On the other hand, the tetrahedral structure of *Cabde* gives two forms (no more), one related to the other as object and mirror image, which are *not* superimposable (Fig. 3.3). Thus a molecule of the type *Cabde* should exist in two forms; and these should be detectable if the difference between them (physical or chemical) is sufficiently great. This would account for the structural identity and optical activity of molecules of the type *Cabde*, *e.g.*, the lactic acids, $CH_3CHOHCO_2H$ (Fig. 3.3). (III) and (IV) are mirror images and cannot be superimposed. Further evidence that optical activity is due to this arrangement is shown by the fact that if lactic acid is reduced to propionic acid, optical activity disappears; propionic acid, $CH_3CH_2CO_2H$, is a molecule of the type Ca_2bd, which is superimposable on its mirror image.

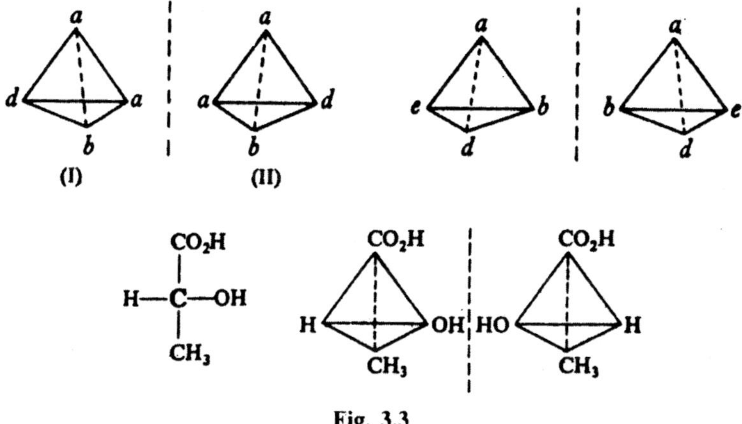

Fig. 3.3

Groups a, b, d, e are all different, but two or more may be *structural* isomers, *e.g.*, isopropylpropyl-methanol is optically active. The substitution of hydrogen by deuterium has also been investigated in recent years to ascertain whether these two atoms are sufficiently different to give rise to optical iso-merism. The earlier work gave conflicting results, but later work is conclusive in favour of optical activity, *e.g.*, Eliel (1949) prepared optically active methylphenyideuteromethane, $CH_3CHDC6H_5$, by reducing optically active methylphenylmethyl chloride with lithium aluminium deuteride.

If we examine an equimolecular mixture of a pair of enantiomers, we shall find that the mixture is optically inactive. This is to be expected, since enantiomers have equal but opposite rotatory power. Such a mixture (of equimolecular amounts) is said to be *optically inactive external compensation,* and is known as a *racemic modification.* A racemic modification may be a purely mechanical mixture, a compound, or a solid solution, and is designated by the prefixes (–)-or DL-, *e.g.* (–)- or DL-lactic acid. Now let us examine the stereochemistry of a molecule containing two asymmetric carbon atoms. First let us consider the case of a compound containing two structurally dissimilar carbon atoms, *i.e.*, compounds of the type *CabdCabe*, *e.g.*, $CH_3CHBrCHBrCO_2H$. There are four possible spatial arrangements for this type of structure (Fig. 3.4). (I) and (II) are enantiomers, and an equimolecular mixture of them forms a racemic modification; similarly for (III) and (IV). Thus there are four optically active forms. In general, a compound containing n different asymmetric carbon atoms exists in 2^n optically active.

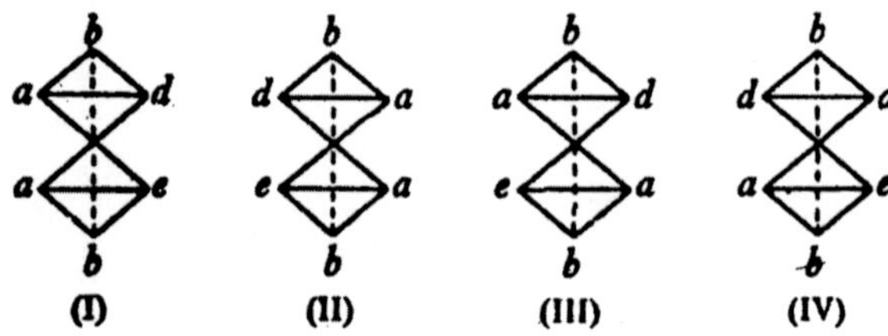

Fig. 3.4

(I) and (III) are not identical in configuration and are not mirror images; they are known as **diastereoisomers**, *i.e.*, they are optical isomers but not mirror images (not enantiomers). Thus a compound of the type *CabdCabe* exists in six forms: two pairs of enantiomers, and two racemic modifications. Diastereoisomers differ in physical properties, such as melting point, solubility, specific rotation, etc. Chemically they are similar, but their rates of reaction with other optically active compounds are different (geometrical isomerism).

The plane-diagrams of the molecules (I-IV) in Fig. 17.6 willbe (V-VIII), respectively (see also below).

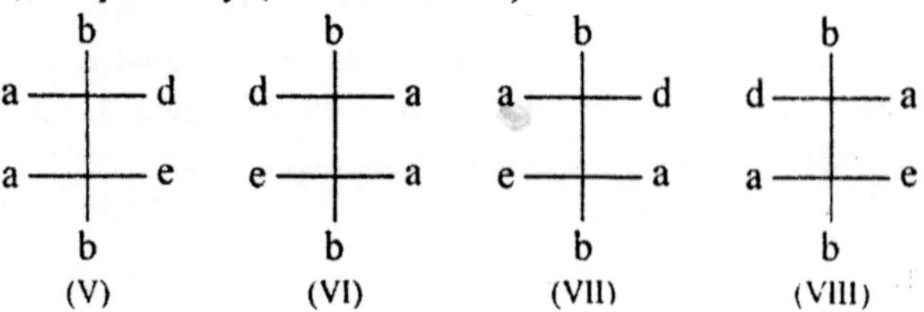

Fig. 3.5

Pairs of enantiomers of the type *CabdCabe* are distinguished by the prefixes *erythro* and *three*. In the former, the two identical groups a and the two identical groups b can eclipse each other (a, a and b,b) in one conformation, whereas in the latter this cannot be done. Thus (I) and (II) are the *erythro*-forms, and (III) and (IV) are the *threo*-forms. These names are derived from erythrose and threose.

Instead of writing down all the possible configurations, the number of optical isomers for a compound of the type *CabdCabe* may be obtained by indicating the configuration of each asymmetric carbon atom by the symbol + or –, or by D or L; thus:

Cabd	+ –	+ –	D_1 L_2	D_1 L_1
|				
Cabe	+ –	+ –	D_2 L_2	L_2 D_2
	(±)	(±)	DL	DL

Now let us consider the case of a compound containing two asymmetric carbon atoms which are structurally the same *i.e.*, compounds of the type *CabdCabd*, *e.g.*, tartaric acid. $HO_2CCHOHCHOHCO_2H$. In compounds of this type, it is obvious that D = D_2. If we ignore the problem of the different conformations possible for this molecule (but see below). We can proceed as follows.

Cabd	D	L	D	L
|				
Cabe	L	L	L	D
	(IX)	(X)	(XI)	(XII)

In molecules (IX) and (X), the upper and lower halves reinforce each other; hence (IX), as a whole, has the dextro-, and (X), the laevo-configuration. *i.e.*, (IX) and (X) are optically active, and enantiomeric. On the other hand, in (XI) the two halves are in opposition, and hence the molecule, *as a whole* will not show optical activity. It is also obvious that (XI) and (XII) are identical, *i.e.*, there is only we optically inactive form of *CabdCabd*. Molecule (XI) is said to be *optically inactive by internal compensation*; it is known as the *meso*-form, and is a diastereoisomer of (IX) and (X). Thus there are four possible forms: D-, L-, DL- and *meso*-.

Molecule (XI) is an example of a *compound having two asymmetric carbon atoms, but is optically inactive* (by internal compensation). It is therefore obvious that inspection of the usual structural formula, which contains two (or more) asymmetric carbon atoms, is not sufficient to decide whether the molecule is optically active or not. *The molecule as a whole must be asymmetric.* The test of superimposing the original formula (tetrahedral) on its mirror image definitely indicates whether the molecule is symmetrical or not. The most satisfactory way in which superimposability may be ascertained is to buildup models of the molecule and its mirror image. Usually, this is not convenient. A simple device to decide whether a molecule is symmetrical or not is to ascertain whether it contains a *plane of symmetry*, a *centre of symmetry*, or an *alternating axis of symmetry*. If any one of these is present the molecule is symmetrical, *i.e.*, superimposable on its mirror image.

A *plane of symmetry* divides a molecule in such a way that points (atoms or groups of atoms) on the one side of the plane, form mirror images of those on the other side. This test may be applied to both solid and plane-formulae, *e.g.*, the plane-formula of the meso form of *CabdCabd* possesses a plane of symmetry; the other two, D- and L-, do not:

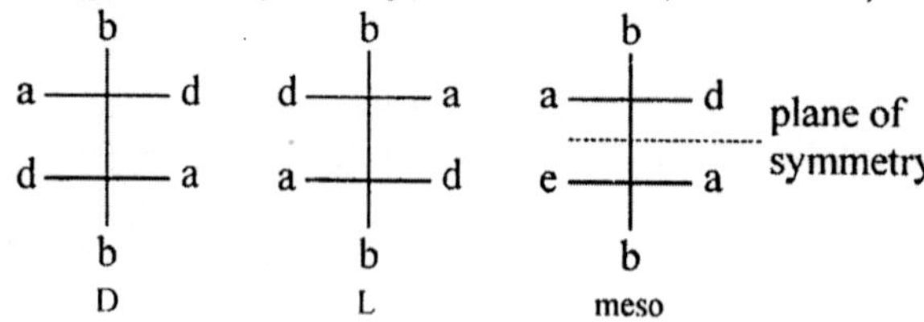

Fig. 3.6

A centre of symmetry is a point from which lines, when drawn on one side and produced an equal distance on the other side, will meet identical points in the molecule. This test can be satisfactorily applied only to three-dimensional formulae, particularly those of ring systems, *e.g.*,

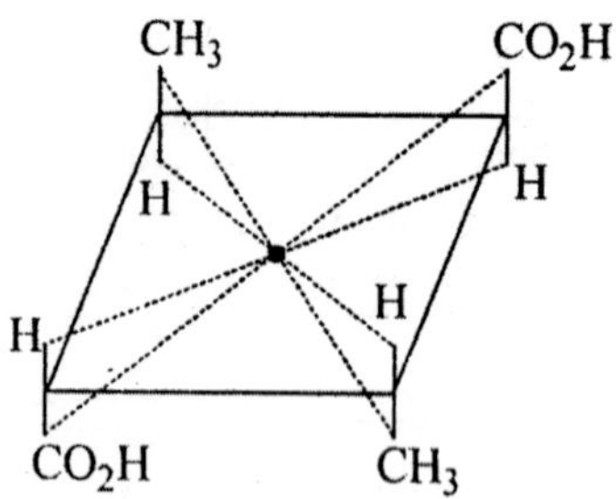

Fig. 3.7

2,4-dimethylcyclobutane-1,3-dicarboxylic acid. The form shown possesses a centre of symmetry which is the centre of the ring. This form is therefore optically inactive.

(ii) Dimethyidiketopiperazine exists in two geometrical isomeric forms, cix and trans:

```
CH3              CH3              CH3              H
 |                |                |               |
 | ,CO—NH,        |                | ,CO—NH,       |
 |/       \       |                |/       \      |
 C         \      C                C         \     C
 |\NH—CO/         |                |\NH—CO/        |
 |                |                |               |
 H                H                H              CH3
          cis                              trans
```

The *cis* isomer has neither a plane nor a centre of symmetry. It can therefore exist in two enantiomeric forms; both are known. The *trans* isomer has a centre of symmetry and is therefore optically inactive.

It is important to note that only *even-numbered* rings can possibly possess a centre of symmetry.

Up to the present, all optically active *natural* compounds have been found to consist of molecules which owe their asymmetry to the absence of both a plane and centre of symmetry. It is possible, however, for these elements of symmetry to be *absent* and the molecule to be superimposable on its mirror image and hence not be optically active. Such a molecule will possess an alternating axis of symmetry. This may be defined as follows: a molecule possesses an n-fold **alternating** axis of **symmetry** if, when rotated through an angle of 360° about this axis and then followed by reflection in a plane perpendicular to the axis, the molecule is the same as it was in the starting position.

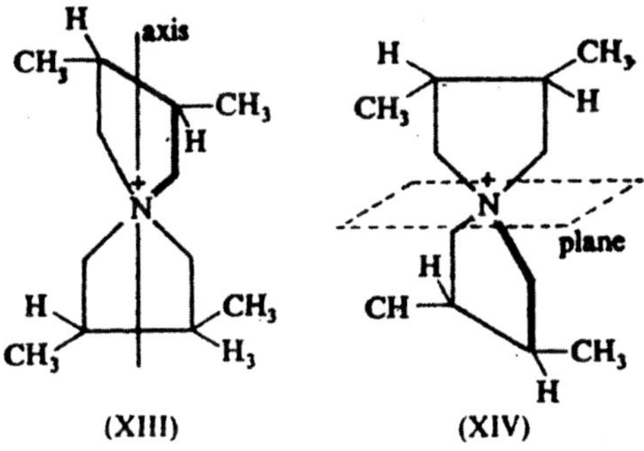

Fig. 3.8

Mc Casland and Proskow (1956) synthesised a compound which owes its symmetry to the presence of an alternating axis of symmetry only (this compound possesses neither a plane nor a centre of symmetry). It is the N-spiro-compound (XIII) (as the -toluenesulphonate). If (XIII) is rotated through 90° about the co-axis of the two rings, (XIV) is obtained. Reflection of (XIV) through a central plane perpendicular to this axis gives a molecule identical and coincident with (XIII).

The application of the elements of symmetry is always to the Fischer plane projection formulae, *which are always those of the eclipsed conformations*. These molecules, however, may have other conformations which are devoid of any element of symmetry. Nevertheless, such compounds are not optically active. Although such conformations are *individually* optical; active, statistically it can be expected that for every optically active conformation present, so will also be present its mirror image conformation, *i.e.*, each conformation will be present as pair of enantiomers. Thus, the whole collection of molecules will be optically inactive by 'external compensation'. It therefore follows that when a molecule can exist in a number of conformations then provided that at least one conformation (whether preferred or not) is superimposable on it: mirror image, the compound will be optically inactive by 'external compensation'. Such 'racemic modifications' cannot be resolved because of the low energy of interconversion of the enantiomers.

Chirality. As we have seen, the structures of enantiomers differ only in 'handedness', one being left-handed and the other being right-handed. Any molecule which is nor superimposable on it mirror image is said to possess **chirality** (from the Greek *kheir*, hand). Thus, the term chirality means 'having handedness'. This term was first introduced by Kelvin (1884), and has been used by Cahn et. al. in their system for the specification of absolute configuration (see below) Chirality expresses the necessary and sufficient condition for the existence of enantiomers (*i.e.*, chirality is now equivalent to asymmetry or dissymmetry). The adjective chiral equivalent to being left- or right-handed, and so a **chiral centre** is one which can be left- fight-handed. On the other hand. when a molecule is superimposable on its mirror image, that molecule does not possess 'handedness' (it is not asymmetric), and is said to be achiral.

The commonest cause of optical activity is the presence of one or more *achiral* centres, which in organic chemistry are usually asymmetric carbon atoms.

Specification of absolute configurations. Since the configuration of (+)-tartaric acid has been related to that of (+)-glyceraldehyde, and since the *absolute* configuration of (+)- tartaric acid has been determined, it is now possible to assign absolute configurations to many compounds whose relative configurations to (+)-glyceraldehyde are known. This raised the problem of using one system of specifying *absolute configurations*. Cahn el al. (1956. 1964) have proposed such a system and this is now widely used. Let us first consider the procedure for a molecule containing one asymmetric carbon atpm, *i.e.*, containing one chiral centre.

(1) The four groups are first ordered according to the *sequence rule*. According to this rule, the groups are arranged in *decreasing atomic number* of the atoms by which they are bound to the asymmetric carbon atom. If two or more of these atoms have the same atomic number, then the relative priority of the groups is determined by a similar comparison of the atomic numbers of the *next* atoms in the groups. If this fails, then the next atoms in the (groups are considered. Thus one works *outwards* from the asymmetric carbon atom until a selection can be made for the sequence of the groups.

(2) It is next determined whether the sequence describes a right- or left-handed pattern on the molecular model as viewed according to the *conversion rule*. When the four groups in the molecule *Cabcd* have been ordered in the priority a, b, c, d, the conversion rule states that their spatial pattern shall be described as right- or left-handed according as the sequence a → b → c is clockwise or anticlockwise when viewed from an external point on the side *remote* from d (the atom with the lowest priority), *e.g.*, (1) shows a right-handed (*i.e.*, clockwise) arrangement.

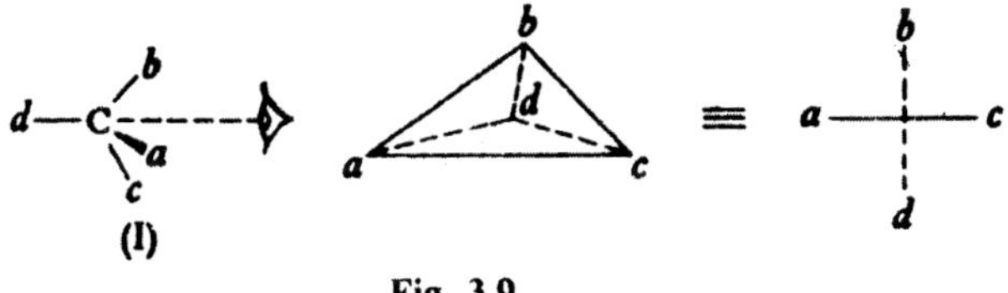

Fig. 3.9

(3) Absolute-configuration labels are then assigned. The asymmetry leading under the sequence and conversion rules to a right- and left-handed pattern, is indicated by R and S respectively (R, *rectus*, right; S, *sinister*, left).

```
        b                      Cl
        ⋮                      ⋮
  a ────┼──── c      Br ───────┼──── CO₂H
        ⋮                      ⋮
        d                      H
```

Consider bromochloroacetic acid (II). Priority of the groups according to the sequence rule is Br (a), Cl (b), C of CO_2H (c), and H (d). Hence by the conversion rule, (II) is the *R*-form.

When multiple bonds or rings are present, the procedure for determining priority is as follows. Both atoms attached to the multiple bond are considered to be duplicated (for a double bond) or triplicated (for triple bond), *e.g.*,

```
                                            C        C
                                            |        |
 —C=CH—  ≡  —CH—CH—                       —C=O  ≡  C—O
              |   |                                  |  |
              C   C                                  O  C

     H                                               H  C
     |                                               |  |
   C=O   ≡  —C—O              —C≡N   ≡             —C—N
              |  |                                   |  |
              O  C                                   N  C

                  CH₂—
                 /
 (cyclohexane) ≡ —CH                —(benzene)
                 \
                  CH₂—

                                                   C   C
                                                   |   |
                                                   C—C—
                                                  /
      ↔  —(benzene)            ≡              —C—C
                                                  \
                                                   C—
```

The priority sequence is then determined by the consideration of the duplicated or triplicated 'structure' in which there are *phantom atoms*, *e.g.*, —CHO is —CH(O)—O(C) and $—CH(OH)_2$ is —CHOH—OH (the phantom atoms in -the former are those in parentheses). Both groups contain a carbon atom joined to two oxygen atoms, but since C precedes H, —CHO precedes $—CH(OH)_2$ in priority.

Ring systems arc treated as branched chains, and if unsaturated, then duplication is used for a double bond or triplication for a triple bond. By using these rules, it can be shown that the order of priority sequence (for some of the common substituents) is: I, Br, Cl, SO, H, SH, F, OCOR, OR, OH, NO_2, NR_2, NHR, NH_2, CO_2R, CO_2H, COR, CHO, CH_2OH, CN, Ph, CR_3, CHR_2, CH_2R, CH_3, D, H.

CHO
H — — OH
CH_2OH
D(+)
(III)

CHO
CH_2OH — — OH
H
L(–)
(IV)

CHO
H — — CH_2OH
HO
D(+)
(V)

Now let us consider D(+)-glyceraldehyde. By *convention* it is drawn as (III); (this is also the *absolute* configuration). Reference to the sequence list gives the priority sequence: OH (a), CHO (b), CH,OH (c), and H(d). Since the interchanging of two groups inverts the configuration, the sequence (III) → (IV) → (V) gives the *original* configuration, and because (V) corresponds to a → b → c in a clockwise fashion, D(+)-glyceraldehyde is therefore (R)-glyceraldehyde.

When a molecule contains two or more asymmetric carbon atoms, each asymmetric carbon atom (chiral centre) is assigned a configuration according to the sequence and conversion rules and is then specified with R or S, *e.g.*, (+)-tartaric acid.

CO_2H
H — 2 — OH
1 $CHOHCO_2H$

CO_2H
H — 2 — OH
HO — 1 — H
CO_2H

≡

2 $CHOHCO_2H$
HO — 1 — H
CO_2H

≡ (two interchanges)

CO_2H
HO — 2 — $CHOHCO_2H$
1 H

b
a — — c
d

≡ (two interchanges)

CO_2H
HO — 1 — $CHOHCO_2H$
H

Thus, (+)-tartaric acid is (RR)-tartaric acid.

The sequence rule was designed to relate the symbols D and L with the symbols R and S. How ever, D and L are obtained by means of chemical transformations, whereas R and 5 are derived from geometrical models and are independent of correlations. Because of this, R and S must be applied only to compounds whose absolute stereochemistry has been determined; they do not necessarily correlate chemical families, *e.g.*, (+)-tartaric acid, whether it be D or L (according to the method of correlation,), has the absolute configuration specified by (RR).

Resolution of racemic modifications

When optically active compounds are prepared by synthetic methods, the usual result is a racemic modification; *e.g.*, bromination of propionic acid results in the formation of DL-α- bromopropionic acid:

$$CH_3CH_2CO_2H \xrightarrow{Br_2/P} CH_3CHBrCO_2H$$

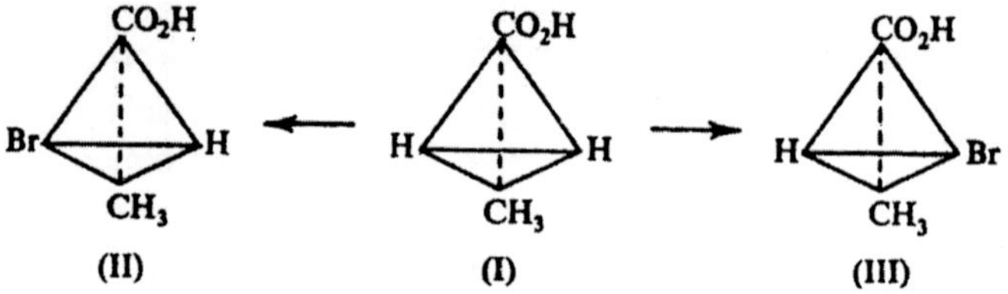

CO_2H / Br, H / CH_3	←	CO_2H / H, H / CH_3	→	CO_2H / H, Br / CH_3
(II)		(I)		(III)

Fig. 3.10

(II) and (III) (Fig. 3.10) are enantiomers, and since molecule (1) is symmetrical about its vertical axis, it can be anticipated from the theory of probability that either hydrogen atom should be replaced equally well to give DL-α-bromopropionic acid. This actually does occur in practice.

From what has been said above, it would appear that the two hydrogen atoms in (1) are alike. This is true for their behaviour with bromine, but the point to note is that replacement of one or other hydrogen does not produce the *same* molecule; a pair of enantiomers is produced.

These two hydrogens are therefore said to be *enantiotopic*. This term may be defined as follows: Two atoms or groups in a molecule are enantiotopic if replacement of each in turn by some other group leads to a pair of enantiomers. It should be noted that corresponding groups in a pair of *enantiomers* are also said to be enantiotopic, *e.g.*, the methyl groups or bromine atoms in (II) and (III) are enantiotopic.

If separate replacements produce the same molecule, the groups are said to be *homotopic*.

Enantiotopic groups react with non-asymmetric (achiral) reagents at the same rate, but react with asymmetric (chiral) reagents at different rates (cf. bromination, above).

In addition to referring to groups as being enantiotopic, *faces of double bonds* are also said to be enantiotopic if stereoisomers are produced by addition reactions, *e.g.*, the addition of hydrogen cyanide to the carbonyl double bond (attack at the front face and at the back face).

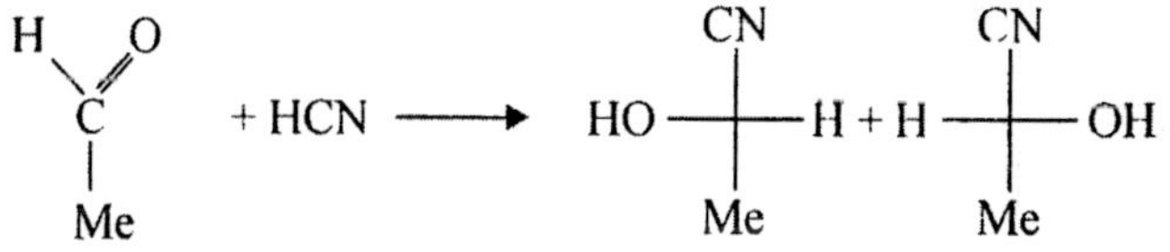

$$\text{MeCHO} + \text{HCN} \longrightarrow \text{HO–C(CN)(Me)–H} \;(\textit{front}) + \text{H–C(CN)(Me)–OH} \;(\textit{back})$$

If we now consider the bromination of (IV), it can be seen that the two α-hydrogen atoms are *not* symmetrically placed with respect to the lower half (which is asymmetric). Hence it might be expected that the two hydrogen atoms would be replaced by bromine (an achiral reagent) at *different rates*. This is the case in practice; the product is a pair of diastereoisomers in *unequal* amounts. These two hydrogens are therefore said to be **diastereotopic**. This term may be defined as follows: Two atoms or groups in a molecule are diastereotopic if replacement of each in turn by some other group leads to a pair of diastereoisomers. Also corresponding groups in a *pair of diastereoisomers* are said to be diastereotopic (*cf.* enantiotopic groups above).

The term *diastereotopic faces* may be used with respect to the *faces of a double bond* when one of the groups attached to the unsaturated carbon atom contains a chiral centre.

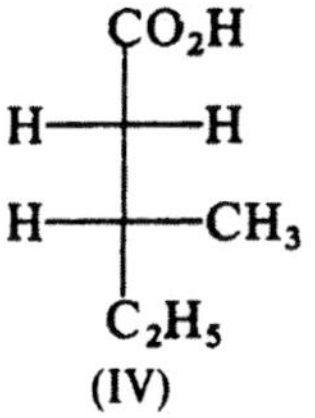

CO_2H / H–C–H / H–C–CH_3 / C_2H_5

(IV)

Fig. 3.11

The process of separating a racemic modification into its enantiomers is known as *resolution*. Various methods have been introduced.

1. **Mechanical separation** (Pasteur, 1848). In this method the crystals are, if sufficiently well defined, actually separated by hand; the crystals must be enantiomorphous. This method is applicable only to racemic mixtures, and is mainly of historical interest.

2. **Biochemical separation** (Pasteur, 1858). Certain bacteria and moulds, when they grow in a dilute solution of a racemic modification, destroy one enantiomer more rapidly than the

other; *e.g.*, *Penicillium glaucum* (a mould), when grown in a solution of ammonium racemate, attacks the D-form leaving the L-.

3. **By means of salt-formation** (Pasteur, 1858). This method, which is the best of all the methods of resolution, consists in converting the enantiomers in a racemic modification into *diastereo- isomers*, *e.g.*, if an optically active base is combined with a racemic acid, two diastereoisomers are obtained:

$$(D_{base} + L_{acid}) + 2D_{base} \longrightarrow (D_{acid}D_{base}) + (L_{acid}D_{base})$$

Because of their different solubilities, these diastereoisomers may be separated by fractional crystallisation. After separation, the acids may be regenerated by hydrolysis with inorganic acids or with alkalis.

Bases which are used for the resolution of racemic acids are mainly alkaloids, *e.g.*, quinine, strychnine, brucine, cinchonine, morphine, etc., and acids which are used for racemic bases are, *e.g.*, tartaric acid, camphorsulphonic acid, bromocamphorsulphonic acid.

The method of salt-formation has been extended to compounds other than acids and bases, *e.g.*,

(i) Racemic alcohols are converted into the acid ester derivative with phthalic anhydride:

$$C_6H_4(CO)_2O + ROH \longrightarrow C_6H_4(CO_2R)(CO_2H)$$

Fig. 3.12

The acid ester, consisting of equimolecular amounts of thE D- and L-forms, may now resolved by the method used for acids.

(ii) Racemic aldehydes and racemic ketones may be resolved by means of optically active derivatives othydrazine, *e.g.*, (–)-menthylhydrazine.

Resolution of racemic modifications by means of salt formation may be complicated by I phenomenon of *assymmetric transformation.* This is exhibited by compounds that are optically unseat *i.e.*, the enantiomers are readily interconvertible:

$$(+)\text{-C} \rightleftharpoons (-)\text{-C}$$

It is possible to get *complete* conversion of C into the form (as salt) that crystallises from solution. The form may be (+) or (–), depending on the nature of the base (used for resolving acids) and the solvent.

4. **Selective adsorption.** Optically active substances may be selectively adsorbed by some optically active adsorbent, *e.g.*, Henderson and Rule (1939) resolved a racemic modification of a camphor derivative on D-lactose as adsorbent; Prolog and Wieland (1944) resolved Troger's base on D-lactose.

5. **Channel complex formation**. This method is possible because the complexes of each enantioroer have different solubilities; *e.g.*, Schlenk (1952) has resolved (±)-2-chloro-octane by means of channel complex formation with urea.

Racemisation

By using suitable conditions, it is possible to cause most optically active compounds to lose their optical activity without changing their structure. This means that the (+)- and (–)-forms of most optically active compounds are convertible one into the other, the final result being a racemic modification. Such a transformation is known as racemisation.

The method of effecting racemisation depends on the nature of the compound in question; generally, heat, light or chemical reagents may be used. Thus, if the starting material is the (+)-form, then after treatment half will be converted into the (–)-form; similarly, when starting with the (–)-form, half will be converted into the (+)-form.

In some cases racemisation occurs spontaneously at room temperature; it is then known as *autoracemisation*, *e.g.*, dimethyl bromosuccinate autoracemises.

Many different types of compounds can racemise, and a number of mechanisms have been proposed, each mechanism applying to a particular type of compound.

One important type of compound that readily racemises is that in which the asymmetric carbon atom is joined to a hydrogen atom and can undergo tautomeric change; the mechanism proposed for this racemisation is one via enolisation, *e.g.*, the acid-catalysed racemisation of 2-butyl phenyl ketone:

$$\underset{(+)}{\mathrm{Ph}\overset{\mathrm{O}}{\overset{\|}{\mathrm{C}}}-\overset{\mathrm{H}}{\overset{|}{\mathrm{C}}}\mathrm{MeEt}} \underset{-\mathrm{H}^+}{\overset{+\mathrm{H}^+}{\rightleftharpoons}} \mathrm{Ph}\overset{\mathrm{OH}}{\overset{\|}{\mathrm{C}^+}}-\overset{\mathrm{H}}{\overset{|}{\mathrm{C}}}\mathrm{MeEt}$$

$$-\mathrm{H}^+ \Downarrow\Uparrow +\mathrm{H}^+$$

$$\underset{(+)}{\mathrm{Ph}\overset{\mathrm{O}}{\overset{\|}{\mathrm{C}}}-\underset{\mathrm{H}}{\underset{|}{\mathrm{C}}}\mathrm{MeEt}} \underset{\text{steps}}{\overset{\text{two}}{\rightleftharpoons}} \underset{\text{(loss of asymmetry)}}{\mathrm{Ph}\overset{\mathrm{OH}}{\overset{|}{\mathrm{C}}}=\mathrm{CMeEt}}$$

Hence, whether the starting material is the (+)- or (–)-enantiomer, the final product is (±)-2-butyl phenyl ketone, since both enantiomers have equal energy content and so both an formed in equal amounts.

Another example is the base catalysed racemisation of (–)-lactic acid. This may be formulated via a carbanion intermediate as follows:

$$\mathrm{Me}\overset{\mathrm{OH}}{\overset{|}{\underset{\mathrm{H}}{\underset{|}{\mathrm{C}}}}}-\mathrm{C}\begin{matrix}\nearrow\!\!\!\!/ \mathrm{O}\\ \searrow \mathrm{O}^-\end{matrix} + \mathrm{OH}^- \rightleftharpoons + \mathrm{H_2O} + \left[\mathrm{Me}\overset{\mathrm{OH}}{\overset{|}{\overset{-}{\mathrm{C}}}}-\mathrm{C}\begin{matrix}\nearrow\!\!\!\!/ \mathrm{O}\\ \searrow \mathrm{O}^-\end{matrix} \longleftrightarrow \;\rangle\mathrm{C}=\mathrm{C}\langle\;\right]$$

(loss of asymmetry)

$$\rightleftharpoons \mathrm{Me}\overset{\mathrm{OH}}{\overset{|}{\underset{\mathrm{H}}{\underset{|}{\mathrm{C}}}}}-\mathrm{C}\begin{matrix}\nearrow\!\!\!\!/ \mathrm{O}\\ \searrow \mathrm{O}^-\end{matrix} + \mathrm{OH}^-$$

Here again, whether we start with the (–)-or (+)-enantiomer, the final product is (±)-lactic acid

Some compounds which cannot undergo tautomeric change can, nevertheless, be racemised *e.g.*, (–)-limonene, some biphenyl compounds. The mechanism of some of these racemisation is uncertain.

The Walden Inversion (Optical Inversion)

By a series of reactions, Walden (1893) was able to transform an optically active compound into its enantiomer. In some cases the product is 100 per cent optically pure, *i.e.*, the *inversion* is quantitative; in other cases the product is a mixture of the (+)- and (–)-forms, but in unequal quantities, *i.e.*, a partial inversion has taken place.

The phenomenon was first discovered by Walden with the following reactions:

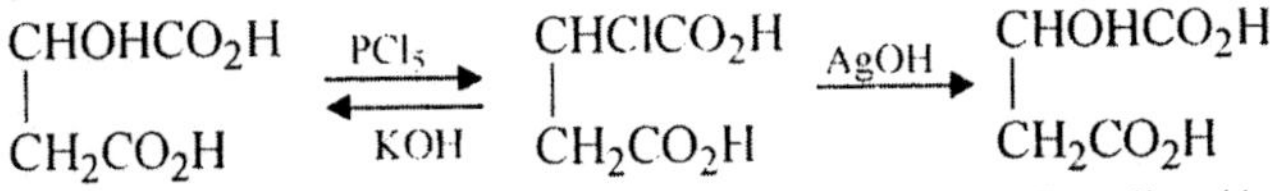

$$\underset{\text{(I)}}{\underset{\text{L-malic acid}}{\begin{matrix}CHOHCO_2H\\|\\CH_2CO_2H\end{matrix}}} \underset{KOH}{\overset{PCl_5}{\rightleftarrows}} \underset{\text{(II)}}{\underset{\text{D-chlorosuccinic acid}}{\begin{matrix}CHClCO_2H\\|\\CH_2CO_2H\end{matrix}}} \xrightarrow{AgOH} \underset{\text{(III)}}{\underset{\text{D-malic acid}}{\begin{matrix}CHOHCO_2H\\|\\CH_2CO_2H\end{matrix}}}$$

In one of the two reactions there must be an interchange of position between two groups; *e.g.*, if the configuration of (I) corresponds with that of (II), the inversion must have taken place between (II) and (III). *The step in which the inversion occurs constitutes a Walden inversion.*

As the above experiment stands, there is no way of telling at which stage inversion has occurred. Change in the sign of rotation does not necessarily indicate an inversion of configuration; *e.g.*, when D(+)-glyceraldehyde is oxidised to glyceric acid, the acid obtained is D(–)-glyceric acid.

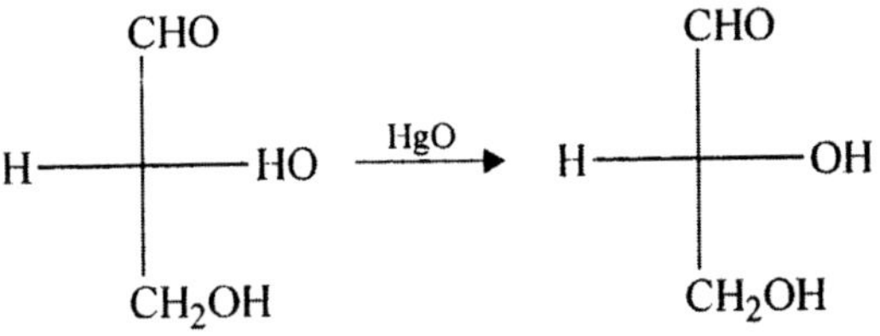

$$\underset{\text{D(+)-glyceraldehyde}}{\begin{matrix} & CHO & \\ & | & \\ H & — & HO \\ & | & \\ & CH_2OH & \end{matrix}} \xrightarrow{HgO} \underset{\text{D(–)-glyceric acid}}{\begin{matrix} & CHO & \\ & | & \\ H & — & OH \\ & | & \\ & CH_2OH & \end{matrix}}$$

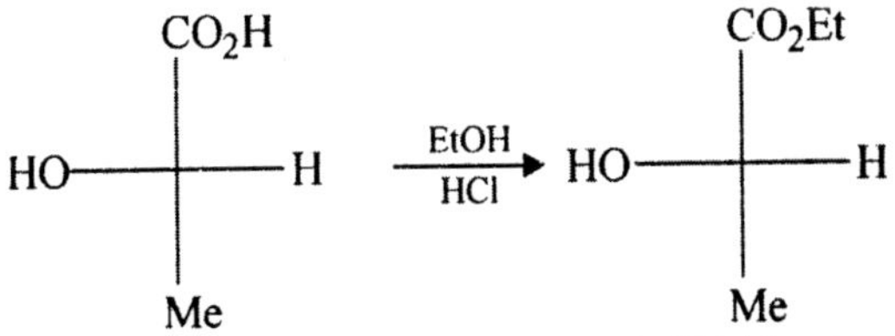

$$\underset{\text{L(+)-latice acid}}{\begin{matrix} & CO_2H & \\ & | & \\ HO & — & H \\ & | & \\ & Me & \end{matrix}} \xrightarrow[HCl]{EtOH} \underset{\text{ethyl L(–)-lactate}}{\begin{matrix} & CO_2Et & \\ & | & \\ HO & — & H \\ & | & \\ & Me & \end{matrix}}$$

In each case, there is no change in configuration, but the signs of rotation are opposite. Thus it is necessary to have methods to determine relative configurations. The above examples are easy to solve: starting material and product both have the same relative configurations, since the asymmetric carbon atom is not affected in the reaction, and so inversion is not possible. At the same time, because the asymmetric carbon atom is not involved, starting with optically pure compounds will lead to optically pure products. In this way, it is easy to measure the *optical purity* of a sample of such products. Optical purity is expressed as a percentage, *e.g.*, if the (maximum) specific rotation of compound A is +5° and an impure sample has a rotation +3°, this sample is *60 per cent optically pure.*

When the asymmetric carbon atom is involved in .various reactions, the problem is far more difficult. Kenyon et at (1925) established a basis for the determination of relative configurations as follows. These authors carried out a series of reactions on optically active hydroxy compounds. Now it has been established that in the esterification of a monocarboxylic acid by an alcohol under ordinary conditions, acyl-oxygen fission occurs; *i.e.*,

$$R^1CO \overset{1}{—} OH + R^2O \overset{1}{—} \longrightarrow R^1COOR^2 + H_2O$$

Kenyon assumed that in all reactions of this type the R^2—bond remained intact, and consequently no inversion of the alcohol is possible. The following chart shows a series of reactions carried out on ethyl (+)-lactate; Ts = p-Mosyl group, *i.e.*, *p*-toluenesulphonyl group, $CH_3C_6H_4SO_2$— Ac = CH_3CO, and the symbo! $\underset{8}{\longrightarrow}$ is used to represent inversion in that step.

Me, CO_2Et, H, OH on C — (+)⁻ (IV) $\xrightarrow{TsCl}$ Me, CO_2Et, H, OTs on C — (+)⁻ (V)

(IV) $\xrightarrow{Ac_2O}$ 2 Me, CO_2Et, H, OAc on C — (–)⁻ (VI)

(V) $\xrightarrow{AcO^-K^+}$ Me, OAc, H, CO_2Et on C — (+)⁻ (VII)

(IV) and (VI) have the same relative configurations although the sign of rotation has changed. Similarly, (IV) and (V) have the same relative configurations. Reaction of (V) with AcO^-K^+, however, produces (VII), the enantiomer of (VI). It should be noted that if inversion is going to occur, the *complete group* attached to the asymmetric carbon atom must be removed. The converse, however is not necessarily true, *i.e.*, removal of a complete group does not invariably result in inversion.

The above set of reactions has been used as a standard, and all closely analogous reactions are assumed to proceed in the same manner, *e.g.*, treatment of (V) with lithium chloride gives (VIII) with inversion.

```
Me     CO2Et              Me     Cl
  \   /                     \   /
    C         LiCl            C
  /   \       ---->         /   \
H      OTs                H      CO2Et
   (V)                      (VIII)
```

Also, since (+)-(IV) is obtained from the esterification of (–)-lactic acid, and (–)-lactic acid has been correlated with D(+)-glyceraldehyde the stereochemical relationships to D(+)-glyceraldehyde are therefore established.

Participation of neighbouring groups in nucleophilic substitution. So far, we have discussed polar and steric effects on rates and mechanisms of reactions (for solvent effects). We have already seen that the phenomenon of *neighbouring group participation* may operate in various rearrangements. In the same way, this effect may also operate in nucleophilic substitution reactions. Here we have a group attached to the carbon atom *adjacent* to the carbon atom where nucleophilic substitution occurs and, during the course of the reaction, becomes bonded or partially bonded to the reaction centre. Neighbouring group participation, however, may also involve a group further removed than the carbon atom adjacent to the reactive centre.

Whether a group Z can enter into neighbouring group participation depends on the nature of Z, and for a given Z it usually depends on the size of the ring that can be formed (by n.g.p.).

When the reaction is accelerated by neighbouring group participation, that reaction is said to be *anchimerically assisted.* As we have seen, various atoms exhibit this phenomenon of neighbouring group participation, *e.g.*, halogen. Brominium (bromonium) ions were first proposed by Roberts *et al.* (1937) as intermediates in the addition of bromine to alkenes. The existence of this cyclic brominium ion (bridged ion) has been demonstrated by Winstein et ill (1939), who found that the action of fuming hydrobromic acid on (–)-threo-3-bromobutan-2-ol gave(±)-2.3-dibromobutane.

OH, H, Me, C_1, C_2, H, OTs, Br (–)-form $\xrightarrow{H^+}$ $\overset{+}{O}H_2$, H, Me, C_1, C_2, H, Me, Br $\xrightarrow{-H_2O}$ H, Me, C_1, Br^+, C_2, H, Me

$\xrightarrow{Br^-}$ Br, H, Me, C_1, C_2, H, Me, Br (–)-form + Br, H, Me, C_1, C_2, H, Me, Br (+)-form

Had no neighbouring group participation occurred, then if the reaction were S_NZ, complete inversion would have occurred only at C_1. If the reaction were S_N1, C_1 would have formed a classical carbonium ion and so racemisation would have occurred at C_1 only.

Since retention or inversion of *both* C_1 and C_2 occurs, the results are explained as shown above. If the Br^- ion attacks C_1 from the back the (–)-form is produced: if it attacks C_2 from the back, the (+)-form is produced.

Further evidence or the existence of the brominium ion has been obtained by Olah *et al.* (1967), who examined the N.M.R. spectrum of the solution of the product formed from the addition of bromine to tetramethylethylene in SbF_5—SO_2 at –60'C.

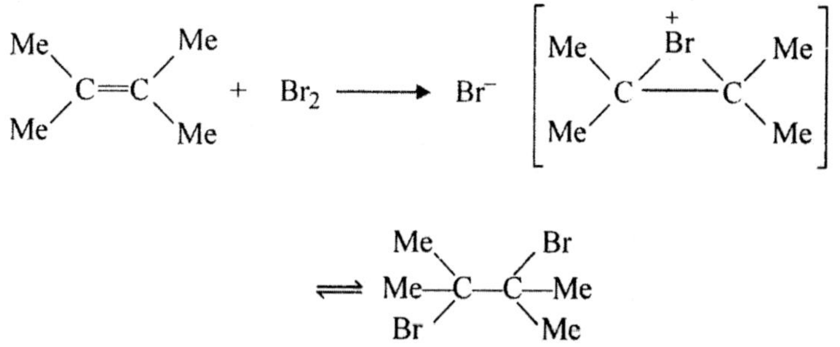

$$Me_2C{=}CMe_2 + Br_2 \longrightarrow Br^- \left[Me_2\overset{}{C}\underset{\overset{+}{Br}}{—}CMe_2 \right]$$

$$\rightleftharpoons Me_2CBr—CBrMe_2$$

The authors have interpreted the results as being consistent with the formation of the brominium bromide salt and also showed that chlorine and iodine formed bridged ions under these conditions, but not fluorine. On the other hand. Wynberg el al. (1969) believe that they have actually isolated a reasonably stable brominium ion as its bromide. $RBr^+ \; Br_3^-$.

The problem of addition reactions of alkenes is more complicated than that given in the account above. In some cases, addition may be stereospecifically trans. *e.g.*, the addition of chlorine to trans-but-2-ene at –90 C in the *absence* of solvent (Poutsma. 1965). On the other hand addition of *e.g.*. DBr to cis-phenylpropene (1) is stereoselectively cis (88 per cent) at 0°C in methylene dichloride (Dewar et. al., 1963). If addition through bridged ions were the only route, and the bridged ions were stable, cis addition would be stereospecific. However, if the classical carbonium ion becomes increasingly stable, (*i.e.*, addition will become increasingly favoured. One factor that increases the stability of the bridged ion is the nature of the addendum, *e.g.*, the order for the halogens (as shown experimentally) is I > Br > Cl > F. Hence stereospecific trans addition operates for iodine. On the other hand when the addendum is not an iodine atom (in the first step) and the classical ion is stabilised. *e.g.*, by resonance, then the rate-determining step is the formation of a classical carbonium ion as an ion-pair (II).

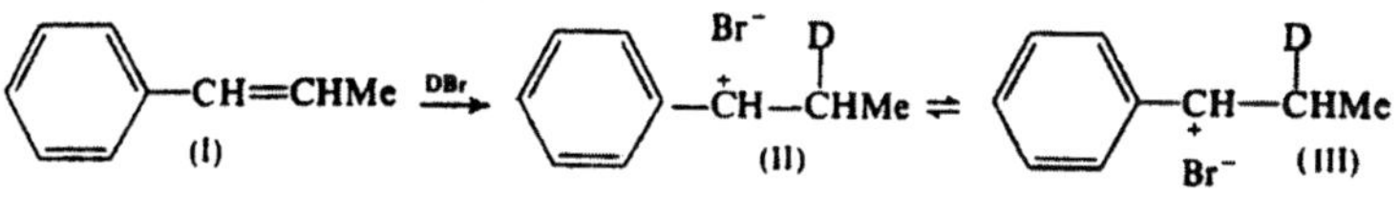

$$\underset{(I)}{C_6H_5—CH{=}CHMe} \xrightarrow{DBr} \underset{(II)}{C_6H_5—\overset{+}{C}H—CHDMe \;\; Br^-} \rightleftharpoons \underset{(III)}{C_6H_5—CHBr—CHDMe}$$

Fig. 3.13

where the halide ion is held on the *same side* of the original double bond as the entering deuteron. Collapse of this ion-pair thus gives the cwadduct but, at the same time, rearrangement of (II) to the isomeric ion-pair (III), gives the irans-nddwl. It has also been found that the nature of the solvent affects the relative amounts of cis and trans products.

Elimination reactions can also be stereoselectively or stereospecifically cis. An example of the former is (Cristol ct al., 1960):

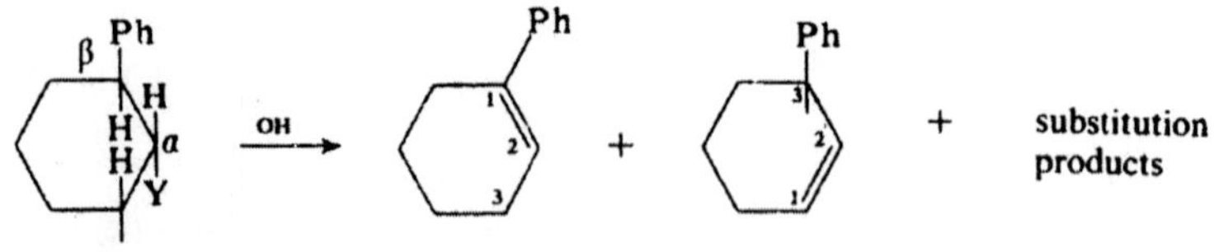

	OH →	Ph (1,2,3)	+	Ph (3,2,1)	+	substitution products
$Y = SMe_2^+$		22%		2%		61%
$Y = NMe_3^+$		64%		2%		8%

Fig. 3.14

The reaction was shown to be E2. and hence I-phenylcyclohexene is formed by cis-elimination. This has been explained by Ingold (1963) as follows. In cis-elimination (E2 mechanism), bond changes do not occur simultaneously (as in trans-elimination). The more difficult it is to detach group Y the further ahead will the proton on C-β be transferred to the hydroxide ion (as compared with the breaking of the C—Y bond). Thus charge builds up on C-β and the electron-pair becomes increasingly available when C-α is "free". In this way phenyl conjugation controls the orientation, and so *cis*-elimination becomes increasingly favoured the more difficult it is to remove Y. Since it has been shown (from other experiments) that it is more difficult to detach the NMe_3^+ group than the SMe_2^+, the former leads to more cis-elimination than the latter.

ASYMMETRIC SYNTHESIS

In ordinary laboratory preparations of optically active compounds, the racemic modification is always obtained. By special means, however, it is possible to prepare optically active compounds from symmetrical compounds (*i.e.*, not optically active) without the necessity of resolution. The method involves the use of optically active compounds, and is known *partial asymmetric synthesis*. The first partial asymmetric synthesis was carried out by Marckwald (1904), who prepared an active (–)-valeric acid (laevorotatory to the extent of about 10 percent of the pure compound) by heating the halfbrucine salt of ethylmethyhnalonic acid at 170°C.

CH_3 CO_2H \ C / C_2H_5 CO_2H $\xrightarrow{(-)\text{-brucine}}$ CH_3 CO_2H [(–)-brucine] \ C / C_2H_5 CO_2H
(I)

CH_3 CO_2H \ C / C_2H_5 CO_2H [(–)-brucine]
(II)

$\xrightarrow{170°C}$ CH_3 CO_2H[(–)-brucine] \ C / C_2H_5 CO_2H
(III)

\+ CH_3 H \ C / C_2H_5 CO_2H [(–)-brucine]
(IV)

$\xrightarrow{HCl}$ CH_3 CO_2H \ C / C_2H_5 H
(V)

\+ CH_3 H \ C / C_2H_5 CO_2H
(VI)

(I) and (II) are diastereoisomers; so are (III) and (IV). (V) and (VI) are enantiomers, and since the mixture is optically active, they must be present in unequal amounts. This was believed to be due to the different rates of decomposition of diastereoisomers (I) and (II).

This reaction was reinvestigated by Eisenlohr and Meier (1938), and they believed that the half-brucine salts were not present in *equal* amounts in the solid form (as thought by Marckwald).

These authors suggested that as the *less* soluble salt crystallispd out (during evaporation of the solution), some of the *more* soluble salt spontaneously changed into the *less* soluble salt to restore the equilibrium between the two; thus the final result was a mixture of the half-brucine salt containing a larger proportion of the *less soluble* diastereoisomer.

If this be the case, then we are dealing with an example of asymmetric transformation. Kenyon and Ross (1952) have also reinvestigated this reaction, and their work appears to show that the above reaction is a true asymmetric synthesis. According to these authors, when the half-brucine salt-is heated, decarboxylation occurs with the intermediate formation of a carbanion. (1) and (II) both produce the same carbanion:

$$(I) \longrightarrow \begin{matrix} CH_3 \\ \quad \diagdown \bar{C}CO_2H[(-)\text{-brucine}] \\ \diagup \\ C_2H_5 \end{matrix} \longleftarrow (II)$$

Combination of this carbanion with a proton would produce diastereoisomers (III) and (IV) in different amounts, since, in general, diastereoisomers are formed at different rates.

McKenzie (1905) reduced with aluminium amalgam pyruvic esters in which the alcohol was optically active, *e.g.*, (–)-amyl alcohol, (–)-menthol, etc. When the product, lactic acid ester was hydrolysed, the resulting lactic acid was found to be slightive laevorotatory.

$$CH_3COCO_2R \xrightarrow{Al/Hg} CH_3CHOHCO_2R \xrightarrow{H^+} CH_3CHOHCO_2H$$

Prelog et. al. (1953) have studied, by conformational analysis, the steric course of the addition of Grignard reagents to benzoylformic esters of asymmetric alcohols, and have found that the configuration of the

$$C_6H_5COCO_2R^1 \xrightarrow[\text{(ii) hydrolysis}]{\text{(i) } R^2MgX} C_6H_5CHOHCO_2H$$

asymmetric carbon atom in the stereoisomer that predominated in this reaction could be correlated with the asymmetric carbon atom in R^1, *e.g.*, (–)-menthol and (–)-borneol are both configurationally related to (–)-glyceraldehyde, and both lead to a predominance of the (–)-hydroxyacid. This method has been used to correlate configurations.

The other type of asymmetric synthesis is absolute asymmetric synthesis. This is the preparation of an optically active compound without the intermediate use of optically active reagents. The first conclusive evidence for an absolute asymmetric synthesis was obtained by Kuhn and Knopf (1930), who irradiated (±)-α-azidopropionic dimethylamide, $CH_3CH(N_3)CON(CH_3)_2$, with right circularly polarised light and obtained a product that was slightly dextrorotatory. When the amide was irradiated with left circularly polarised light, the product was slightly laevorotatory.

HYDROXY-DIBASIC AND POLYBASIC ACIDS

Tartronic acid (hydroxymalonic acid) may be prepared by heating bromomalonic acid with silver oxide suspended in water:

$$CHBr(CO_2H)_2 + \text{'AgOH'} \longrightarrow CHOH(CO_2H)_2 + AgBr$$

It is a crystalline solid which, heated to 160°C, melts with the evolution of carbon dioxide and the formation of polyglycollide, $(C_2H_2O_2)_n$.

Malic acid (*hydroxysuccinic acid, hydroxybulanedioic acid*), m.p. 100°C. occurs in sour apples (Latin: malum, apple), fruits, berries, etc. Malic acid is made synthetically by heating maleic acid with dilute sulphuric acid under pressure:

$$\begin{array}{l} CHCO_2H \\ \| \\ CHCO_2H \end{array} + H_2O \xrightarrow{H_2SO_4} \begin{array}{l} CHOHCO_2H \\ | \\ CH_2CO_2H \end{array} \quad (100\%)$$

It may be conveniently prepared in the laboratory by heating bromosuccinic acid with silver oxide suspended in water:

$$\begin{array}{l} CHBrCO_2H \\ | \\ CH_2CO_2H \end{array} + \text{'AgOH'} \longrightarrow \begin{array}{l} CHOHCO_2H \\ | \\ CH_2CO_2H \end{array} + AgBr$$

Malic acid contains one asymmetric carbon atom, and can therefore exist in the D-. L- and DL-forms. Malic acid from natural sources is L(–); synthetic malic acid is DL ; D(+)-malic acid may be obtained by the careful reduction of (+)-tartaric acid with concentrated hydriodic acid

$$\begin{array}{c} CO_2H \\ | \\ H-C-OH \\ | \\ HO-C-H \\ | \\ CO_2H \end{array} + 2HI \longrightarrow \begin{array}{c} CO_2H \\ | \\ H-C-OH \\ | \\ CH_2 \\ | \\ CO_2H \end{array} + I_2 + H_2O$$

Malic acid behaves both as an alcohol and acid. Inspection of the formula of malic acid shows it to be and α-hydroxyacid with respect to one carboxyl group, and a β-hydroxyacid with respect to the other. Such acids, when heated, undergo the reaction characteristic of the β-acid, *i.e.*, they eliminate a molecule of water to form an unsaturated acid (not the lactide). Thus, on heating, malic acid forms maleic anhydride and fumaric acid (q.v.):

$$\begin{array}{l} CHCO \\ \| \quad\;\; \diagdown \\ \| \qquad\; O \\ \| \quad\;\; \diagup \\ CHCO \end{array} \longleftarrow \begin{array}{l} CHOHCO_2H \\ | \\ CH_2CO_2H \end{array} \longrightarrow \begin{array}{c} H \qquad CO_2H \\ \diagdown \;\; \diagup \\ C \\ \| \\ C \\ \diagup \;\; \diagdown \\ HO_2C \qquad H \end{array}$$

Tartaric acid (α, α′ *dihydroxysuccinic acid*, 2.,3-*dihydroxy-butanedioic acid*) contains two structurally identical carbon atoms, and can therefore exist in the D-, L-, DL- and *meso*-forms, all of these are known (Fig. 17.8).

The configurations of the tartaric acids are a troublesome problem. Fischer wrote the configuration of the natural dextrorotatory acid (*i.e.*, the (+)-acid as shown. It is possible to synthesise (–)-tartaric acid from D(+)-glyceraldehyde, and on this basis the (+)-acid would be L(+)-tartaric acid. This is in agreement with Rosanoff's scheme of building up from D(+)- glyceraldehyde. It is also possible, however, to degrade (+)-tartaric acid into D(-)-glyceric acid, and hence (+)-tartaric acid write be D(+)-tartaric acid. The designation L(+)-tartaric acid is used by those chemists who regard this acid as a carbohydrate derivative. This difficulty may be overcome by specifying (+)-tartaric acid as (RR)-tartaric acid.

When fumaric acid is treated with dilute alkaline permanganate, DL-tartaric acid is formed. Maleic acid under the same conditions, forms mese-tartaric acid.

$$HO_2CCH = CHCO_2H + [O] \xrightarrow{KMnO_4} HO_2CCHOHCHOHCO_2H$$

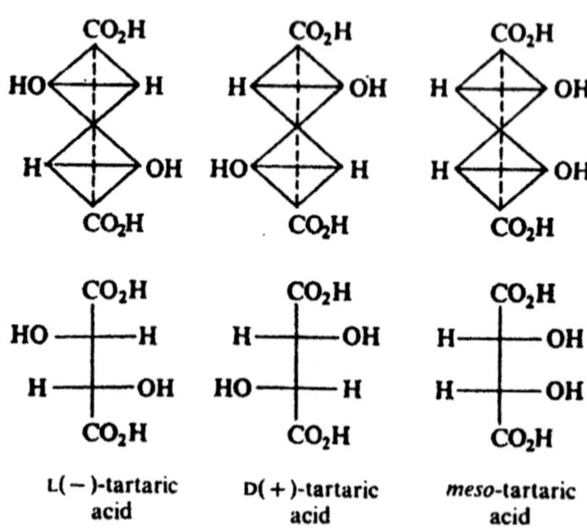

Fig. 3.15

Both DL- and *meso*-tartaric acid are formed when to, α-α′-dibromosucdnic acid is boiled with silver oxide suspended in water:

$$\begin{array}{l} CHBrCO_2H \\ | \\ CHBrCO_2H \end{array} + 2\text{'AgOH'} \longrightarrow \begin{array}{l} CHOHCO_2H \\ | \\ CHOHCO_2H \end{array} + 2AgBr$$

DL- and *meso*-Tartaric acids are also formed by the hydrolysis of glyoxal cyanohydrin:

$$\begin{array}{l} CHO \\ | \\ CHO \end{array} + 2HCN \longrightarrow \begin{array}{l} CHOHCN \\ | \\ CHOHCN \end{array} \xrightarrow{H_2O} \begin{array}{l} CHOHCO_2H \\ | \\ CHOHCO_2H \end{array}$$

All the foregoing synthetic preparations clearly show the structure of tartaric acid.

D(+)-**Tartaric acid**, occurs in the free state and as potassium hydrogen tartrate in juice of grapes. During the fermentation of grapes, the acid potassium salt separates as a reddish-brown crystalline mass which is known as argol. When recrystallised, argol is converted into the purer substance (white), which is known as cream of tartar, from which D(+)-tartaric acid is obtained by dissolving it in water, adding calcium hydroxide, then calcium chloride to the filtrate, and decomposing the collected batches of precipitated calcium tartrate with the calculated quantity of dilute sulphuric acid:

$$2KHC_4H_4O_6 \xrightarrow{Ca(OH)_2} CaC_4H_4O_6 + 2H_2O + K_2C_4H_4O_6$$

$$\xrightarrow{CaCl_2} 2KCl + CaC_4H_4O_6 \xrightarrow{H_2SO_4} H_2C_4O_6 + CaSO_4$$

Anhydrous D(+)-tartaric acid, m.p. 170°C, is obtained from the filtrate by evaporation to crystallisation. When heated, tartaric acid is converted into pyruvic acid, and is reduced by hydi-iodic acid, first to malic acid and then to succinic acid.

$$\begin{array}{l} CHOHCO_2H \\ | \\ CHOHCO_2H \end{array} \longrightarrow CH_3COCO_2H + CO_2 + H_2O$$

Sodium potassium D(+)-tartrate, $NaKC_4H_4O_6.4H_2O$, is known as *Rochelle salt*, and is used in the preparation of Fehling's solution, which is an alkaline solution of cupric ion complexed with two tartrate ions. *Tartar emetic* is potassium antimonyl D(+)-tartrate; its structure is uncertain.

When heated with sodium hydroxide solution (or even with water), D(+)-tartaric acid is converted into DL-tartaric acid (29-33 per cent) and *meso*-tartaric acid (13-17 per cent). D(+)-Tartaric acid is used in the preparation of effervescent drinks. The acid and tartar emetic are both used as mordants in dyeing and printing.

L(–)-**Tartaric acid**, does not occur naturally. It may be obtained by the resolution of DL-tartaric acid. Physically and chemically, it is similar to D(+)-tartaric acid.

DL-**Tartaric acid**, **racemic tartaric acid**, crystallises as the hemihydrate $(C_4H_6O_6)_2.H_2O$, m.p. 206°C. In the solid state DL-tartaric acid exists as the *racemic compound*, but dissociates into the D- and L-forms in solution. It is manufactured by the oxidation of fumaric acid with dilute alkaline permanganate.

***meso*-Tartaric acid** (*i-Tartaric acid*) crystallises as the monohydrate; the melting point of the anhydrous acid is 140°C. *meso*-Tartaric acid is optically inactive by internal compensation, and may be obtained (together with DL- by heating D(+)-tartaric acid with alkali. It may also be prepared by the oxidation of maleic acid with dilute alkaline permanganate.

Citric acid (β-*hydroxytricarballylic acid*, 2-*hydroxypropane*-, 1, 2, 3-*tricarboxylic acid*) occurs in many fruits, especially unripe fruits of the citrus group, *e.g.*, lemon juice contains about 6-10 per cent citric acid. It is manufactured by the fermentation of solutions of glucose, sucrose or purified cane-molasses in the presence of certain inorganic salts, by various moulds or fungi, *e.g.*, *Citromyces pfefferianus*, *Aspergillus wentii*.

Citric acid may be synthesised from glycerol by the following reactions; this synthesis shows its structure:

$$\begin{matrix} CH_2OH \\ | \\ CHOH \\ | \\ CH_2OH \end{matrix} \xrightarrow{HCl} \begin{matrix} CH_2Cl \\ | \\ CHOH \\ | \\ CH_2Cl \end{matrix} \xrightarrow{[O]} \begin{matrix} CH_2Cl \\ | \\ CH \\ | \\ CH_2Cl \end{matrix} \xrightarrow{HCN}$$

$$\begin{matrix} CH_2Cl \\ | \\ C(OH)CN \\ | \\ CH_2Cl \end{matrix} \xrightarrow{KCN} \begin{matrix} CH_2CN \\ | \\ C(OH)CN \\ | \\ CH_2CN \end{matrix} \xrightarrow{H^+} \begin{matrix} CH_2CO_2H \\ | \\ C(OH)CO_2H \\ | \\ CH_2CO_2H \end{matrix}$$

Lawrence (1897) synthesised citric acid by means of the Refonnatsky reaction, starting with ethyl bromoacetate and oxalacetic ester:.

$$\begin{array}{c} CH_2Br \\ | \\ CO_2C_2H_5 \end{array} + \begin{array}{c} COCO_2C_2H_5 \\ | \\ CH_2CO_2C_2H_5 \end{array} + Zn \longrightarrow$$

$$\begin{array}{c} OZnBr \\ | \\ C_2H_5O_2CCH_2CCO_2C_2H_5 \\ | \\ CH_2CO_2C_2H_5 \end{array} \xrightarrow{\text{acid}} \begin{array}{c} OH \\ | \\ HO_2CCH_2CCO_2H \\ | \\ CH_2CO_2H \end{array}$$

The monohydrate of citric acid, obtained by crystallisation from water, loses its water of crystallisation when heated at 130°C, and melts at 153°C.

$$\text{(III)} \begin{array}{c} CHCO_2H \\ | \\ CCO_2H \\ | \\ CH_2CO_2H \end{array} \xleftarrow{-H_2O} \begin{array}{c} CH_2CO_2H \\ | \\ C(OH)CO_2H \\ | \\ CH_2CO_2H \end{array} \xrightarrow{-HCO_2H} \text{(I)} \begin{array}{c} CH_2CO_2H \\ | \\ CO \\ | \\ CH_2CO_2H \end{array} \xrightarrow{-CO_2} \text{(II)} \begin{array}{c} CH_3 \\ | \\ CO \\ | \\ CH_2CO_2H \end{array}$$

(III) $\xrightarrow{-CO_2}$

$$\text{(VI)} \begin{array}{c} CHCO \\ \| \quad \diagdown \\ \| \quad\quad O \\ \| \quad \diagup \\ CCO \\ | \\ CH_3 \end{array} \xleftarrow{-H_2O} \begin{array}{c} CHCO_2H \\ \| \\ CCO_2H \\ | \\ CH_3 \end{array} \text{(III)} \qquad \begin{array}{c} CH_2 \\ \| \\ CCO_2H \\ | \\ C_2CO_2H \end{array} \text{(V)} \xrightarrow{-H_2O} \begin{array}{c} CH_2 \\ \| \\ CCO \diagdown \\ | \quad\quad O \\ | \quad \diagup \\ CH_2CO \end{array} \text{(VII)}$$

(II) $\longrightarrow$

$$\begin{array}{c} CH_3 \\ \| \\ CO \\ | \\ CH_3 \end{array}$$

Citric acid behaves as an alcohol and a tribasic acid, *e.g.*, it forms the acetyl derivative and three series of salts.

Citric acid is used for making beverages and as a mordant in dyeing.

Citric acid is both an α- and β-hydroxy-acid when heated to 150°C, it eliminates a molecule of water to form aconitic acid. On pyrolysis citric acid gives a number of products among which have been isolated aconitic acid, citraconic (mesaconic) and itaconic anhydrides, and acetone (I-VII).

(I) is acetonedicarboxylic acid; (II) acetoacetic acid; (III) aconitic acid; (IV) citraconic and mesaconit acid (cis-trans isomers); (V) itaconic acid; (VI) citraconic (mesaconic) anhydride; (VII) itaconic anhydride.

When citric acid is heated with concentrated sulphuric acid, aconitic acid is obtained (41-44 per cent).

When treated with *fuming* sulphuric acid, citric acid forms acetonedicarboxylic acid (85-90 per cent), a reaction which is characteristic of α-hydroxyacids.

Tricarballylic acid (*propane*-1, 2, 3 *tricarboxylic acid*), m.p. 166°C, occurs in unripe beet-roots. It may be oreoared by the reduction of aconitic acid or from 1, 2, 3-tribromooro Dane as follows:

$$\begin{matrix} CH_2Br \\ | \\ CHBr \\ | \\ CH_2Br \end{matrix} \xrightarrow{KCN} \begin{matrix} CH_2CN \\ | \\ CHCN \\ | \\ CH_2CN \end{matrix} \xrightarrow[\text{(ii) HCl}]{\text{(i) aq. KOH}} \begin{matrix} CH_2CO_2H \\ | \\ CHCO_2H \\ | \\ CH_2CO_2H \end{matrix} \quad (70\%)$$

Tricarballylic acid may also be synthesised by the Michael condensation between diethyl fumarate and malonic ester, and heating the product, ethyl -propane-1, 1, 2, 3-tetracarboxylate, with concentrated hydrochloric acid:

$$C_2H_5O_2CCH{=}CHCO_2C_2H_5 + CH_2(CO_2C_2H_5)_2 \xrightarrow{C_2H_5ONa}$$

$$\begin{matrix} C_2H_5O_2CCHCH_2CO_2C_2H_5 \\ | \\ CH(CO_2C_2H_5)_2 \end{matrix} \xrightarrow{HCl} \begin{matrix} CH_2CO_2H \\ | \\ CHCO_2H \\ | \\ CH_2CO_2H \end{matrix} \quad (88–90\%)$$

Tricarballylic esters have been used as plasticisers.

UNSATURATED DICARBOXYLIC ACIDS

The formula of the simplest unsaturated dicarboxylic acid is $HO_2CCH=CHCO_2H$ this formula actually represents two isomers: maleic acid and fumaric acid.

Maleic acid may be prepared:

(i) By heating malic acid at about 250°C:

$$\begin{matrix} CHOHCO_2H \\ | \\ CH_2CO_2H \end{matrix} \xrightarrow{-H_2O} \begin{matrix} CHCO_2H \\ \| \\ CHCO_2H \end{matrix} \xrightarrow{-H_2O} \begin{matrix} CHCO \\ \| \\ CHCO \end{matrix} \rangle O$$

$$\xrightarrow[\text{(ii) HCl}]{\text{(i) NaOH}} \begin{matrix} CHCO_2H \\ \| \\ CHCO_2H \end{matrix}$$

(ii) By heating bromosuccinic acid with aqueous alkali; some fumaric acid is also obtained:

$$\begin{matrix} CHBrCO_2H \\ | \\ CH_2C_2H \end{matrix} + KOH \longrightarrow \begin{matrix} CHCO_2H \\ \| \\ CHCO_2H \end{matrix} + KBr + H_2O$$

Maleic anhydride is prepared industrially:

(a) By the oxidation of benzene with air in the presence of vanadium pentoxide as catalyst at 410-430°C:

$$2\,C_6H_6 + 90_2 \xrightarrow{V_9O_5} 2 \begin{matrix} CHCO \\ \| \\ CHCO \end{matrix} \rangle O + 4CO_2 + 4H_2O$$

(b) As a by-product in the manufacture of phthalic anhydride from naphthalene:

(c) By the oxidation of but-2-ene (from cracked petroleum) or crotonaldehyde with air in the presence of vanadium pentoxide as catalyst at 450°C, *e.g.*,

$$CH_3CH=CHCH_3 + 3O_2 \xrightarrow{V_2O_5} \begin{matrix} CHCO \\ \| \\ CHCO \end{matrix} \rangle O + 3H_2O$$

Maleic acid is a synthetic compound. It is a crystalline solid, m.p. 130°C, soluble in wate (79 g per 100 ml at 25°C). When heated, some distils unchanged; the rest is converted into maleic anhydride. A much better yield of maleic anhydride is obtained by heating the acid with acetic anhydride.

Maleic acid may be reduced catalytically or electro-lytically to succinic acid. It is oxidised by dilute alkaline permanganate to we-yo-tartaric acid; this may be obtained in excellent yield by replacing the permanganate by potassium chlorate and osmium tetroxide. Prolonged heating of maleic acid at 150°C converts it into fumaric acid. Maleic acid and its anhydride are used in the Diels-Alder synthesis.

Maleic acid inhibits rancidity in milk powders, oils and fats. Maleic anhydride is used of making varnishes and lacquers.

Fumaric acid may be prepared:

(i) By heating maleic acid for some time at 150°C.

(ii) By heating bromosuccinic acid with alkali; maleic acid is also formed.

(iii) By the Knoevenagel reaction; malonic acid is condensed with glyoxylic acid in the presence of pyridine:

$$HO_2CCHO + CH_2(CO_2H)_2 \xrightarrow{\text{pyridine}} HO_2CCH = CHCO_2H + H_2O + CO_2$$

(iv) By the oxidation of furfural with sodium chlorate:

$$\begin{matrix} CH & — & CH & \\ \| & & \| & \\ CH & & CCHO & + 4[O] \\ & \diagdown_{O}\diagup & & \end{matrix} \xrightarrow{NaClO_3}$$

$$\begin{matrix} HO_2CCH & \\ \| & \\ CHCO_2H & + H_2O + CO_2 \quad (50\text{–}88\%) \end{matrix}$$

Fumaric acid is prepared industrially by boiling maleic acid with hydrochloric acid or sodium hydroxide. Another industrial preparation is the fermentation of glucose (and other carbohydration by *e.g.*, *Rhizopus nigricans*.

Fumaric acid occurs in nature in many plants. It is a crystalline solid, m.p. 287°C, slightly soluble in -water (0-7 g per 100 ml at 25°C). It does not form an anhydride of its own, but gives maleic anhydride when heated at 230°C. It may be reduced to succinic acid, and is oxidised alkaline permanganate to DL-tartaric acid; the latter is obtained in excellent yield if potassium chlorate and osmium tetroxide are used as the oxidising agent (cf. maleic acid). Fumaroyl chloride is formed when maleic anhydride is heated with phthaloyl chloride in the presence zinc chloride.

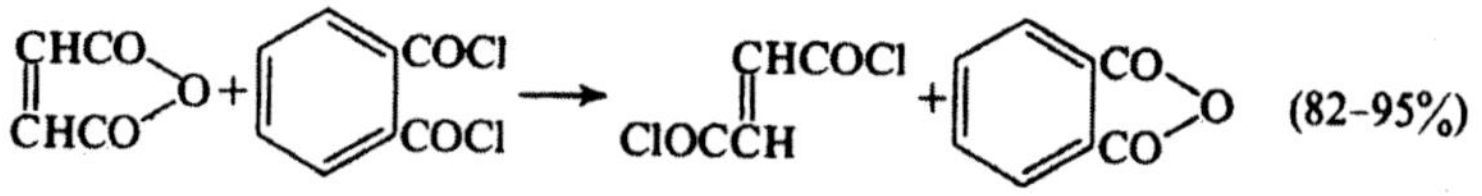

$$\begin{matrix}CHCO\\ \|\\ CHCO\end{matrix}\!>\!O + C_6H_4\begin{matrix}COCl\\ COCl\end{matrix} \longrightarrow \begin{matrix}CHCOCl\\ \|\\ ClOCCH\end{matrix} + C_6H_4\begin{matrix}CO\\ CO\end{matrix}\!>\!O \quad (82\text{–}95\%)$$

Fig. 3.16

Many acid chlorides can be obtained in 95 per cent yield by this method.

	pk_a^1	pk_a^2	K_a^1/K_a^2
Maleic acid	1.92	6.23	20.420
Fumaric acid	3.02	4.38	22.9

At first sight, since the structures are identical, it might have been expected that the ionisation constants would be the same. As can be seen, however, maleic acid is stronger for the first ionisation but weaker for the second. The reason for this is believed to be due to the fact that hydrogen bonding can occur in the maleate anion, whereas this is not possible for the corresponding fumarate anion.

Thus, the former anion (cB) is stabilised with respect to its acid, and so there is a driving force for ionisation which is absent in the case of fumaric acid. Furthermore, since the maleate anion is stabilised by hydrogen bonding, the fumarate anion ionises to the dianion more readily than the former.

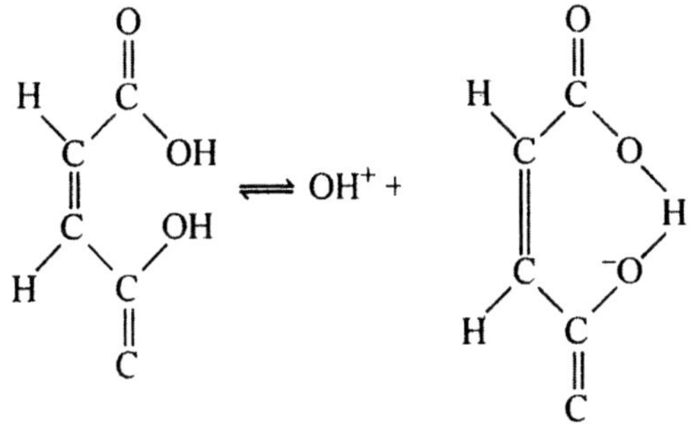

$$\text{HC}(\text{C(=O)OH})\text{=C(H)}\text{–C(OH)=C} \rightleftharpoons OH^+ + \text{HC}(\text{C(=O)–O}\cdots\text{H}\cdots{}^-\text{O})\text{=C(H)–C=C}$$

Maleic acid

$$\text{HO}_2\text{C–CH=CH–CO}_2\text{H} \rightleftharpoons \text{H}^+ + \text{HO}_2\text{C–CH=CH–CO}_2^-$$

Fumaric acid

Not only can configuration affect acid strength but so also can conformation. Table 3.1 gives the pk_a, values of some substituted malonic acids (in water). These differences cannot be explained by the polar effects of the alkyl groups alone. As we have seen, there are the inductive and field effects operating in the dicarboxylic acids. The field effect, since it is a 'distance' effect, will depend very much on the conformation of the molecule. The two carboxyl groups in the substituted malonic acids will be 'crowded' together due to the steric effects of the substituent alkyl groups.

Table 3.1

Acid	pk_a^1	pk_a^2	K_a^1/K_a^2
$CH_2(CO_2)_2$	2.86	5.7	692
$MeCH(CO_2)_2$	3.05	5.76	510
$EtCH(CO_2H)_2$	2.99	5.83	702
$Me_2C(CO_2H)_2$	3.17	6.06	783
$MeEtC(CO2H)_2$	2.86	6.41	3,580
$Et_2C(CO_2H)_2$	2.21	7.29	121,00

$$\text{HO}_2\text{C–CH}_2\text{–CO}_2\text{H} \qquad \text{HO}_2\text{C–CR}_2\text{–CO}_2\text{H}$$

The closer together the two carboxyl groups are, then when one is converted into the carboxylate anion, the more difficult it will be for the other carboxyl group to lose a proton (because of the increased field effect). Thus the K_a^1/K_a^2 ratio will increase, the more so the greater the crowding.

Geometrical Isomerism

The basic principles of geometrical isomerism have already been discussed. The classical example of this phenomenon is the case of maleic and fumaric acids. Both have the same molecular formula $C_4H_4O_4$, but differ in most of their physical and in many of their chemical properties, and neither is optically active. It was originally thought that they were structural isomers, and because of this, different names were assigned to each form (this applies to many other geometrical isomers—see text). It was subsequently shown however, that maleic and fumaric acids were not structural isomers, *e.g.*, both (i) are catalytically reduced to succinic acid; (ii) add on hydrogen bromide to form bromosuccinic acid; (iii) add on water to form malic acid; (iv) are oxidised by alkaline permanganate to tartaric acid (the stereochemical relationships in reactions (ii), (iii) and (iv) have been ignored). Thus both acids have the same structure, viz., $HO_2CCH = CHCO_2H$ and the two possible geometrical isomers are (1) and (II), cis-and trans-butenedioic acid, respectively. Since these are planar molecules, they are superimposable on their mirror images and so are not optically active.

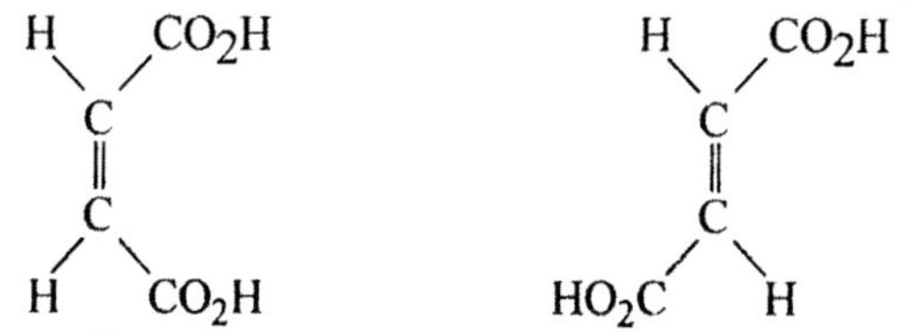

```
H     CO2H            H     CO2H
 \   /                 \   /
   C                     C
   ||                    ||
   C                     C
 /   \                 /   \
H     CO2H         HO2C     H
   (I)                  (II)
```

Geometrical isomers are now also classified as diastereoisomers. According to the new definition, *diastereoisomers* are *any* stereoisomers which are not enantiomers of each other, *i.e.*, the restriction that diastereoisomers are *optical* isomers has been dropped. Thus, this type of isomerism (*geometrical* or *cis-trans*) is a sub-class of the general phenomenon of diastereoisomerism.

The problem now is to decide which formula represents maleic acid, and which fumaric acid. There is no general method for determining the configuration of geometrical isomers; the method used depends on the nature of the compound in question. The methods of stylisation, dipole-moment measurements and spectroscopy may be used to determine the configurations of maleic and fumaric acids.

Method of cyclisation. Wislicenus was the first to suggest the principle that *intramolecular* reactions are more likely to occur the closer together the reacting groups are in the molecule This principle is generally true, but has led to incorrect results (as shown by other work) of certain cases, *e.g.*, the aromatic oximes.

Of the two acids, only maleic acid forms the anhydride when heated. Fumaric acid does not form an anhydride of its own, but when strongly heated, gives maleic anhydride. If we accept the principle of cyclisation, then (I) is maleic acid, and (II) fumaric acid.

In fumaric acid, the two carboxyl groups are too far apart to react with each other.

H CO_2H
C
‖
C
H CO_2H

⟶

H CO
C
‖ O + H_2O
C
H CO

Method of dipole-moment measurements. The use of dipole moments to assign configurations to geometrical isomers must be used with caution. Since dipole moments are vector quantities the sum of two equal and opposite group moments will be zero only if the two vectors are collinear or parallel. When the group moment is directed along the axis of the bond formed by the 'key' atom of the group and the carbon atom to which it is joined, then that group is said to have a linear moment. Such groups are H, halogen. Me, CN, NO_2, etc. On the other hand, groups which have non-linear moments are OH, OR, CO_2H, NH_2, etc. Thus for cis- and trans-olefinic compounds, the dipole moments of the latter will be zero only if the groups attached have linear moments, *e.g.*, $CH_2 = CH_2$, CHCl = CHCl, CHMe = CHMe. When the groups have non-linear moments, their vector sum is not zero, and the difference between the dipole moments of the cis- and trans-forms may be too small to assign configuration with any confidence, *e.g.*, the dipole moment of diethyl maleate is 2-54D and that of diethyl fumarate is 2-38 D.

Spectroscopic methods. Geometrical isomers may often be differentiated by examination of their infrared and NMR spectra, and by their ultra-violet spectra.

Geometrical isomerism is also possible in *cyclic* compounds, the ring structure being comparable to the double bond in olefinic compounds (in giving rise to a more or less rigid structure), *e.g.*, hexahydroterephthalic acids:

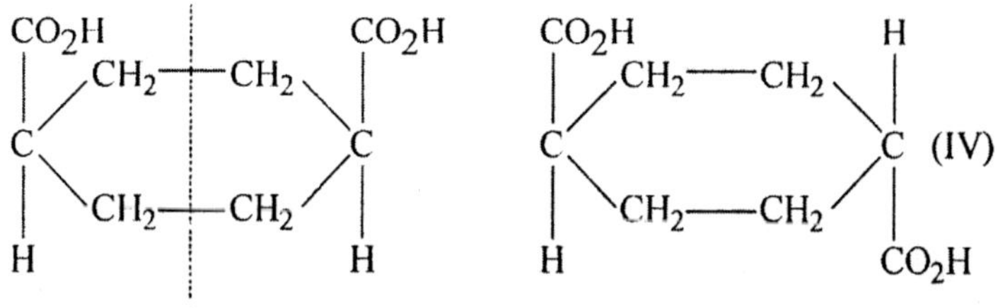

```
CO2H                  CO2H      CO2H                    H
 |    CH2 -|- CH2      |         |    CH2 ----- CH2      |
 |  /      |      \    |         |  /              \     |
 C         |          C          C                    C  (IV)
 |  \      |      /    |         |  \              /     |
 |    CH2 -|- CH2      |         |    CH2 ----- CH2      |
 H         |           H         H                      CO2H
          cis                           trans
```

Both hexahydroterephthalic acids are optically inactive; (III) has a plane, and (IV) a centre of symmetry.

Compounds with a triple bond cannot exhibit geometrical isomerism, *e.g.*, acetylenedi-carboxylic acid, HO, CC ≡ CCO_2H. Using either sp^3 (tetrahedral) hybridisation (Fig. 3.17) or sp (digonal) hybridisation, the molecule is linear and consequently cannot exhibit geometrical (diastereoisomerism) (nor optical) isomerism.

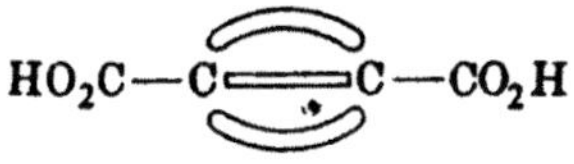

$$HO_2C—C \equiv C—CO_2H$$

Fig. 3.17

Properties of Cis-trans Isomers

Comparison of the properties of cis- and trans-isomers of known configurations shows certain regularities, *e.g.*, the melting point and stability of the cis- are lower than those of the trans-isomer; the density, refractive index, solubility, dipole moment, heat of combustion and the dissociation constant (if an acid) of the *cis* are *greater* than those of the *trans-isomer*. It can be seen from these properties that the cis-isomer is usually the labile form. Based on certain of these generalisations is the *Auwers-Skita rule* (1915, 1920), viz., in a pair of geometrical isomers, the cis has the higher boiling point, density and refractive index. This rule has been used to elucidate configurations, but there are exceptions to the rule.

It is possible, by suitable means, to convert the labile *cis-isomer* into the stable *trans-isomer*, *e.g.*, maleic acid may be converted to fumaric acid by heating the solid, or a solution of the solid in water or benzene, to a temperature above its melting point (130°C). It is more difficult to convert the *trans-isomer* into the *cis-*. Usually the best method is to irradiate the [*trans*-isomer- with ultra-violet light in the presence of a trace of bromine; the product is generally an equilibrium mixture of both isomers.

The Stereochemistry of Carbon Compounds not Containing an Asymmetric Carbon Atom

As pointed out previously, the presence of an asymmetric carbon atom is not essential for optical activity; the essential requirement is *the asymmetry of the molecule as a whole.* Allenes are compounds whose structures are asymmetric, and so should be resolvable. Several compounds of this type have been obtained in optically active forms, *e.g.*, where $R^1 = C_6H_5$ (phenyl) and R^2 = I-$C_{10}H_7$ (l-naphthyl) [Mills and Maitland, 1936].

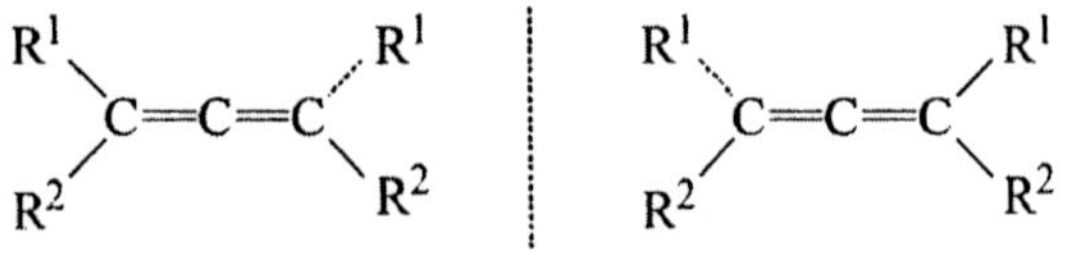

R^1 R^2 C=C=C R^1 R^2 | R^1 R^2 C=C=C R^1 R^2

If both double bonds of allene are replaced by rings, *spirans* are obtained in which the rings are at right angles to each other. Hence by suitable substitution, it should be possible to obtain optically active *spiro*-compounds, *e.g.*. Backer (1928) resolved the following *spiro*heptane derivative.

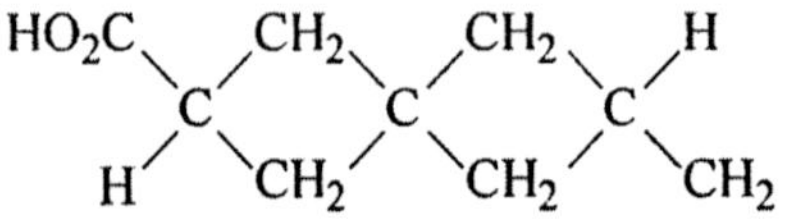

HO_2C CH_2 CH_2 H
C C C
H CH_2 CH_2 CH_2

Suitably *ortho*-substituted biphenyls are also compounds whose structures are asymmetric, the asymmetry arising from restricted rotation about the single bond joining the two benzene rings.

The Optical Isomerism of Elements Other than Carbon

Many quadricovalent elements whose valencies are distributed tetrahedrally have also been obtained in optically active forms, *e.g.*, silicon and tin.

Nitrogen can be tercovalent or quadricovalent unielectrovalent. In the latter compounds, if the charge on the nitrogen atom is ignored, the molecule then closely resembles carbon compounds. Thus, for example, the following compounds have been obtained in optically active forms:

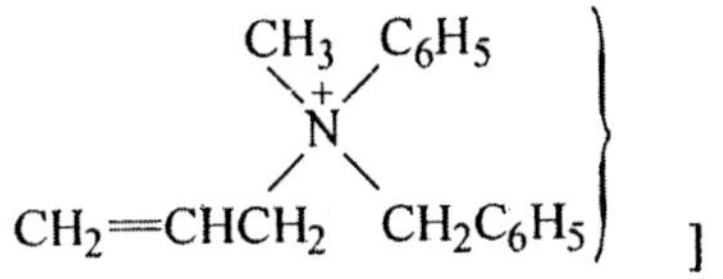

$$\left\{ \overset{+}{N}(CH_3)(C_6H_5)(CH_2{=}CHCH_2)(CH_2C_6H_5) \right\} I^-$$

allylbenzylmethylphenylammonium iodide
(Pope and Peachey, 1899)

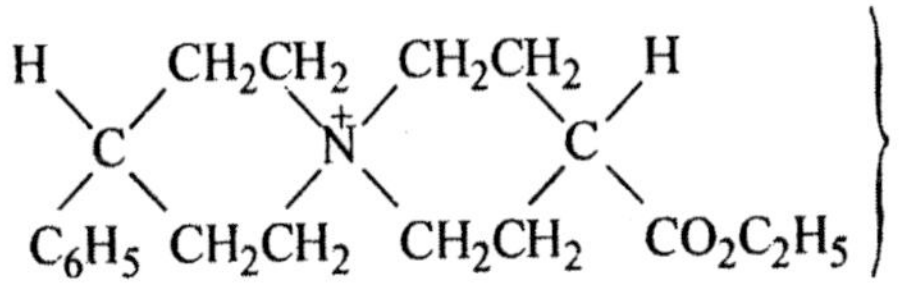

$$\left\{ \begin{matrix} H & & CH_2CH_2 & & CH_2CH_2 & & H \\ & C & & \overset{+}{N} & & C & \\ C_6H_5 & & CH_2CH_2 & & CH_2CH_2 & & CO_2C_2H_5 \end{matrix} \right\} Br^-$$

4-carbethoxy-4'-phenybispiperidinium-1,1'-spiran bromide (Mills and Warren, 1925)

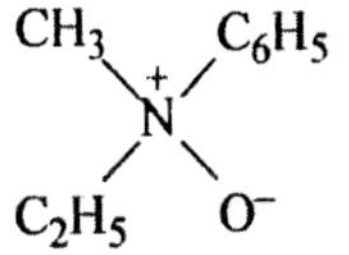

$$\begin{matrix} CH_3 & & C_6H_5 \\ & \overset{+}{N} & \\ C_2H_5 & & O^- \end{matrix}$$

ethylmethylphenylamine oxide (Meisenheimer, 1908)

Recemisation of compounds of the type $\overset{+}{N}abde\ X-$ is effected far more readily than with the carbon compounds, Cabde. The mechanism is beloved to be due to the ready dissociation.

$$\overset{+}{N}abde\,n \} \ X^- \rightleftharpoons Nabd + eX$$

The amine, *Nabd readily* recemises (see blow), and so the quaternary compound will recemise.

Tercovalent nitrogen offers a very interesting problem from the point of view of stereoisomerism. No tertiary amine, Nabd, has yet been resolved. It was therefore suggested that such molecules were planar, but physico-chemical evidence, *e.g.*, dipole moment measurements, absorption spectra, etc, shows that the configuration of ammonia and amines is tetrahedral, the nitrogen atom being at one corner of the tetrahedron with a valency angle of about 109°. This failure to resolve tertiary amines has been explained by rapid oscillation of the nitrogen atom through the plane of the three groups a, b, and d. This oscillation of regarded as an 'umbrella' switch of bonds, *i.e.*, the bond lengths remain

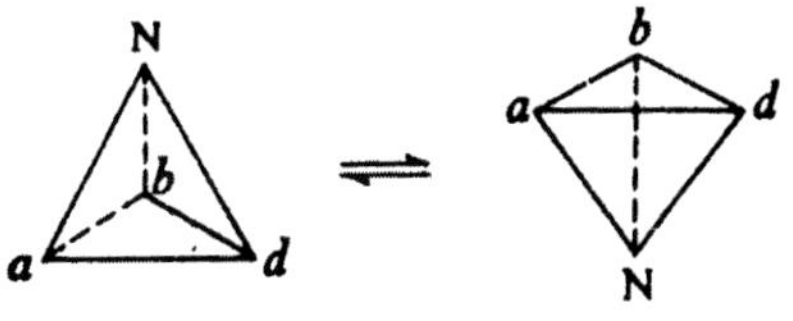

Fig. 3.18

unaltered and only the nitrogen valency angles change. Theoretical considerations have shown that if the nitrogen atom were 'anchored' by forming part of a ring, then the inversion (due to oscillation) would be inhibited. This has been confirmed by the resolution of Troger's base.

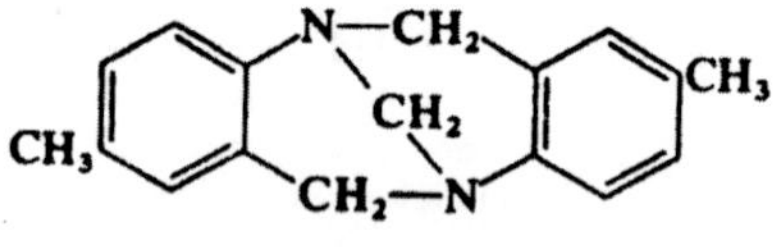

Fig. 3.19

Oximes are also compounds of tercovalent nitrogen. They exhibit geometrical isomerism, and some have also been obtained in optically active forms, *e.g.*, the oxime of cyclohexanone-4-carboxylic acid (Mills and Bain, 1910).

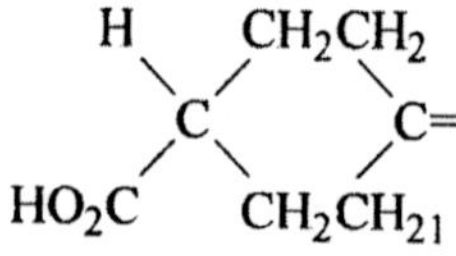

Several tertiary phosphines have been resolved, but many quadricovalent phosphorus compounds have been obtained in optically active forms, *e.g.*,

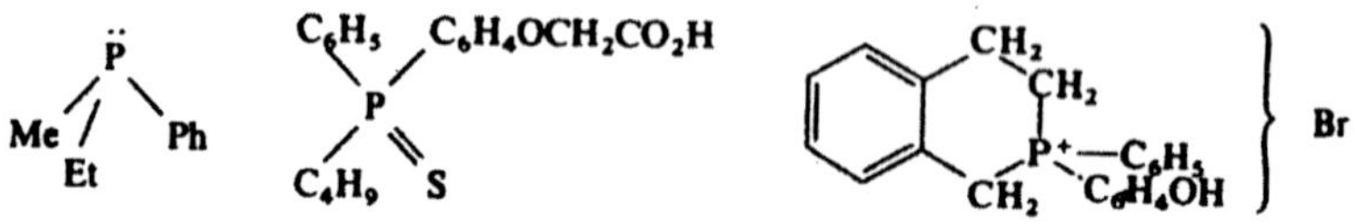

(Horner, 1961) (Davies and Mann, 1944) (Holliman and Mann, 1947)

Fig. 3.20

Tercovalent and quadricovalent arsenic compounds have been resolved, *e.g.*,

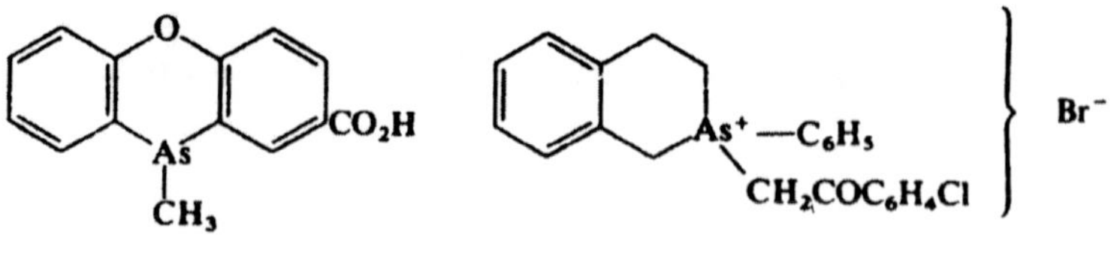

(Lesslie and Turner, 1934) (Holliman and Mann, 1943)

Fig. 3.21

Various sulphur compounds have also been resolved, *e.g.*,

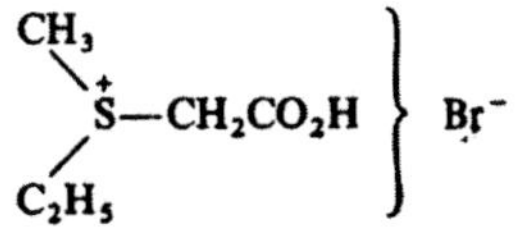

(Pope and Peachey, 1900)

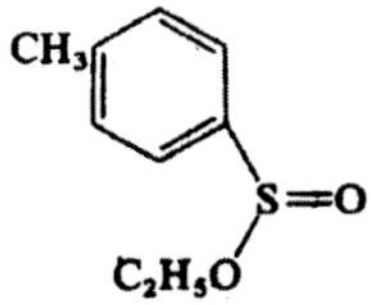

(Phillips, 1925)

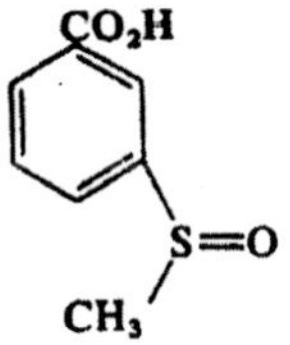

(Phillips, 1926)

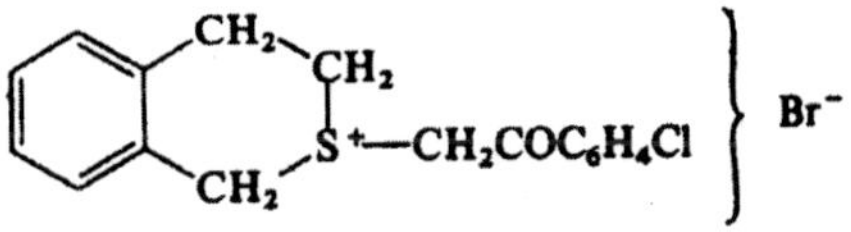

(Mann and Holliman, 1946)

Fig. 3.22

EXERCISES

1. Draw all possible stereoisomers for the following and state which are optically active, inactive, and which are *meso* forms.

   (i) MeCHOHCHOHMe,

   (ii) MeCHOHCHOHCHO,

   (iii) MeCHOHCOCHOHMe, (iv) $MeCHClCO_2H$,

   (v) $HO_2CCHBrCH_2CO_2Et$, (vi) $CH_2 = CHMe$,

   (vii) MeCH = CHEt, (viii) $MeCHBrCH = CH_2$,

   (ix) MeCHOHCH = CHMe, (x) $CMe_2 = C = CH_2$, (xi) MeEtPrN,

   (xii) $MeEtPrPr^iN^+Br^-$.

   (xiii) $PhMeEtN^+ - \bar{O}$, (xiv) PhMeSO, (xv) CMeCl = C = CHMe.

2. Complete the following equations and state the relations between the configurations of the reactant and product. Give our reasons.

   (i) D-MeCHOHEt + $Ac_2O \longrightarrow$?

   (ii) L-MeCH(OAc)Et + NaOH $\longrightarrow$?

   (iii) L-MeCHOHEt + HBr $\longrightarrow$?

   (iv) D-$MeCHClCO_2H$ + 2NaOH $\longrightarrow$?

   (v) L-MeEtPrCOMe + HI $\longrightarrow$?

3. Complete the equations of the reactions of malaeic and fumaric acids with the following reagents and comment.

   (i) bromine, (ii) dilute alkaline permanganate solution,

   (iii) perbenzoic acid, (iv) catalytic hydrogenation.

4. Label the following compounds with R or S and give your reasons.

$$\begin{array}{c} CO_2H \\ H_2N - \!\!+\!\! - H \\ Me \end{array} \qquad \begin{array}{c} CO_2H \\ H_2N - \!\!+\!\! - H \\ CH_2SH \end{array} \qquad \begin{array}{c} CO_2H \\ H - \!\!+\!\! - OH \\ H - \!\!+\!\! - OH \\ CH_2OH \end{array}$$

(i) (ii) (iii)

$$\begin{array}{c} Me \\ OH - \!\!+\!\! - H \\ Cl - \!\!+\!\! - H \\ Me \end{array} \qquad \begin{array}{c} CH_2OH \\ HO - \!\!+\!\! - H \\ Me \end{array} \qquad \begin{array}{c} CH_2OAc \\ H - \!\!+\!\! - OAc \\ CH_2OMe \end{array} \qquad \begin{array}{c} CH_2OH \\ H - \!\!+\!\! - OH \\ CH_2OMe \end{array}$$

(iv) (v) (vi) (vii)

5. Draw the Fischer projection formula for.

   (i) (R)-3-methylpentan-1-ol, (ii) -2,3, dimethylhexane,

   (iii) (R)-3-methoxy-2-methylpropan-l-ol,

   (iv) (2S, 3 R) – 2,3-dichloropentane.

6. Write out the structures and I.U.P.A.C. names of the isomeric hydroxyacids with the molecular formula $C_4H_8O_3$. How would each acid behave towards:

   (i) heat; (ii) dilute sulphuric acid on warming?

7. Complete the equations, and assuming all but one of the steps are S_N2, label each product with D or L.

   Comment.

$$? \longleftarrow \begin{array}{c} CO_2H \\ H - \!\!+\!\! - Br \\ Me \end{array} \xrightarrow{NaN_3} ? \xrightarrow[Ni]{H_2} ?$$

8. Complete and comment:

(i) $\overset{\ulcorner\text{—O—}\urcorner}{CH_2CH_2CO} + NaCl \xrightarrow{H_2O} ?$

(ii) $(HO_2CCH_2)_2C(OH)CO_2H \xrightarrow[H_2SO_4]{\text{fumng}} ? \xrightarrow[HCl]{EtOH} ?$

(iii) $EtO_2CC \equiv CCO_2Et + CH_2(CO_2Et)_2 \xrightarrow[EtOH]{EtONa}$

(iv) $\overset{\ulcorner\text{———O———}\urcorner}{CH_2(CH_2)_2CH_2CO} \xrightarrow{LAH} ?$

9. Discuss the following data:

	$DH_f^{\ominus}$ (kJ)	$S_f^{\ominus}$ (JK^{-1})
Maleic acid	–787.85	159.4
Fumaric acid	–810.9	166.1

10. At which steps do inversions occur? Explain.

$PhCH_2CHMeOH \xrightarrow[(i)]{TsCl} PhCH_2CHMeOTs$

$\xrightarrow[(ii)]{AcONa} PhCH_2CHMeOAc \xrightarrow[(iii)]{OH^-} PhCH_2CHMeOH$

11. *meso*-2,3-Dibromobutane reacts with iodide ion more rapidly than does the (±)-compound.

    What are the respective alkenes formed? Explain.

12. Complete the equations:

(i) $n\text{-}C_5H_{11}\overset{\ulcorner\text{——O——}\urcorner}{CH(CH_2)_2CO} + 2MeMgBr \xrightarrow[\text{(ii) aq. } NH_4Cl]{\text{(i) } Et_2O} ?$

(ii) $\overset{\ulcorner\text{——O——}\urcorner}{CH_2(CH_2)_2CO} + EtOH + HBr(g) \longrightarrow ?$

(iii) $CH_3CHOHCOH \xrightarrow{Na} ?$

13. Mustard gas, $(ClCH_2CH_2\text{—})_2S$, is hydrolysed by water to $ClCH_2CH_2SCH_2CH_2OH$ *much faster* than expected for a primary alkyl halide. Offer an explanation.

14. An aliphatic ketone, M.W. 86, when converted into its oxime which is then reduced, gives an amine that can be resolved. What is the ketone?

If more than one is possible, how could they be distinguished by a physical method?

15. What are A-D in the following reactions?

$$
\begin{array}{c}
\text{EtCHCO} \\
| \quad \diagdown \\
\quad\quad \text{O} \\
| \quad \diagup \\
\text{CH}_2\text{CH} \\
| \\
\text{MeC}{=}\text{CH}_2
\end{array}
\xrightarrow{\text{A}}
\begin{array}{c}
\text{EtCHCO} \\
| \quad \diagdown \\
\quad\quad \text{O} \\
| \quad \diagup \\
\text{CH}_2\text{CH} \\
| \\
\text{MeCO}
\end{array}
\xrightarrow{\text{B}}
\begin{array}{c}
\text{EtCHCH}_2\text{OH} \\
| \\
\text{CH}_2\text{CHOH} \\
| \\
\text{MeCHOH}
\end{array}
$$

$$
\begin{array}{c}
\text{EtCHCH}_2\text{OH} \\
| \\
\text{CH}_2\text{CHOH} \\
| \\
\text{MeCHOH}
\end{array}
\xrightarrow{\text{HIO}_4} \textbf{C} + \textbf{D}
$$

4

SATURATED DICARBOXYLIC AND MONOCARBOXYLIC ACIDS

INTRODUCTION

The general formula of the saturated dicarboxylic acids is $C_nH_{2n}(CO_2H)_2$ (n = 0 for oxalic acid), and the best-known examples are those which have the two carboxyl groups at the opposite ends of the carbon chain.

NOMENCLATURE

The dicarboxylic acids are commonly known by names which indicate their source, *e.g.*, HO_2CCO_2H oxalic acid; this occurs in plants of the *oxalis* group (for further examples, see the individual acids).

In this trivial system of nomenclature, the positions of side-chains or substituents are indicated by Greek letters, *e.g.*,

$HO_2CCH(CH_3)CH_2CH_2CHClCO_2H$ α-chloro-α′-methyladipic acid

According to the I.U.P.A.C. system of nomenclature, the class suffix is *-dioic*, *e.g.*,

HO_2CCH,CO,H propanedioic acid

$HO_2CCH(CH_3)CH(CH_3)CH_2CO_2H$ 2,3-dimethylpentanedioic acid

When this method leads to cumbrous names, the alternative' scheme is. to regard the carboxyl group as a substituent, and the name of the acid is then obtained by adding the suffix carboxylic acid, *e.g.*,

$HO_2CCH_2CH_2CO_2H$ 1,2-ethanedicarboxylic acid or

ethane-1,2-dicarboxylic acid

General Methods of Preparation

1. The *cyanide synthesis* of dicarboxylic acids is a very useful method: the starting material may be either a halogen-acid or a polymethylene dibromide:

(i) $ClCH_2CO_2H \xrightarrow{KCN} CH_2(CN)CO_2H \xrightarrow{H_3O^+} CH_2(CO_2H)_2$

(ii) $Br(CH_2)_2Br \xrightarrow{KCN} NC(CH_2)_2CN \xrightarrow{H_3O^+} HO_2C(CH_2)_2CO_2H$

2. The **Crum-Brown** and **Walker electrolytic method** (1891, 1893). This is the electrolysis of an aqueous solution of the potassium alkyl esters of the dicarboxylic acids (cf. Kolbe's method):

$$\begin{matrix} K—O_2C—(CH_2)_n—CO_2C_2H_5 \\ K—O_2C—(CH_2)_n—CO_2C_2H_5 \end{matrix} + 2H_2O \longrightarrow$$

$$\begin{matrix} (CH_2)_nCO_2C_2H_5 \\ | \\ (CH_2)_nCO_2C_2H_5 \end{matrix} + 2CO_2 + 2KOH + H_2 \quad (f.g.–g.)$$

It is obvious that this method can be used to prepare only *even* homologues.

Dicarboxylic acids may also be prepared by the electrochemical hydrodimerisation of suitable α, β-unsaturated compounds.

3. Dicarboxylic acids may be prepared by the acetoacetic ester synthesis and better, by the malonic ester synthesis.

4. Cyclic ketones may be oxidised to dicarboxylic acids, *e.g.*, cyclohexanone gives adipic acid:

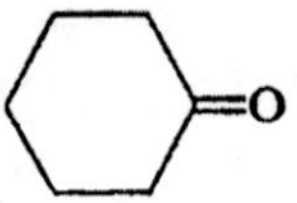

$$C_6H_{10}{=}O \xrightarrow{HNO_3} HO_2C(CH_2)_4CO_2H$$

Fig. 4.1

Alternatively, lactones can be converted into dicarboxylic acids.

5. The preparation of higher homologues from lower homologues may be carried out in several ways; which method is used

depends on the homologue desired. Many *even* higher homologues can be prepared by the Crum-Brown and Walker method (method 2). Any dicarboxylic acid can be stepped up by two carbon atoms as follows:

$$(CH_2)_n(CO_2H)_2 \xrightarrow{\text{LAH}} (CH_2)_n(CH_2OH)_2 \xrightarrow{\text{HBr}}$$

$$(CH_2)_n(CH_2Br)_2 \xrightarrow{\text{KCN}} (CH_2)_n(CH_2CN)_2$$

$$\xrightarrow{H_3O^+} (CH_2)_n(CH_2CO_2H)_2$$

By using malonic ester instead of potassium cyanide, the acid may be stepped up by four carbon atoms:

$$(CH_2)_n(CH_2Br)_n + 2[CH(CO_2C_2H_2)_2]^- Na^+ \longrightarrow$$

$$2NaBr + (CH_2)_n [CH_2CH(CO_2C_2H_5)_2]_2 \xrightarrow[\text{(ii) HCl}]{\text{(i) KOH}}$$

$$(CH_2)_n[CH_2CH(CO_2H)_2]_2 \xrightarrow{150-200°C} (CH_2)_n(CH_2CH_2CO_2H)_2 + 2CO_2$$

Dicarboxylic acids can be stepped up by ten carbon atoms, by starting with a di-acyl chloride and an examine.

Dicarboxylic acids may also be prepared from oc-bromo-esters as follows.

$$\text{cyclohexane-1,3-dione} \xrightarrow[\text{(ii) BrCH}_2\text{CO}_2\text{Et}]{\text{(i) KOH - MeOH}} \text{2-}(CH_2CO_2Et)\text{-cyclohexane-1,3-dione} \xrightarrow[\text{(ii) HCl}]{\text{(i) } N_2H_4,\ OH^-,\ \text{glycol}} HO_2C(CH_2)_6CO_2H$$

Fig. 4.2

General Properties

All the dicarboxylic acids are crystalline solids, the lower members being soluble in water, the solubility decreasing with increase in molecular weight: the odd acids are more soluble than the even. None is steam volatile, and except for oxalic acid, the dicarboxylic acids are stable towards oxidising agents. Their melting points follow the oscillation rule, the even acids having higher melting points than the odd. This alternation is due to the relative positions of the carboxyl groups in the crystal structures. These acids dissociate in two steps, the dissociation constant of the first being much greater than that of the second.

Furthermore, the acid strength of the dicarboxylic acids decreases as the series is ascended, Table 4.1.

Table 4.1

Acid	pK_a^1	pK_a^2	K_a^1/K_a^2
Oxalic	1.271	4.266	989
Malonic	2.86	5.70	692
Succinic	4.21	5.64	26.9
Glutaric	4.34	5.27	8.51
Adipic	4.41	5.28	7.41

Since these acids possess two carboxyl groups, ionisation pair occur in two steps:

$$HO_2C(CH_2)_nCO_2H + H_2O \rightleftharpoons HO_2C(CH_2)_n \quad ...(1)$$

$$HO_2C(CH_2)_nCO_2^- + H_2O \rightleftharpoons {}^-O_2C(CH_2)_nCO_2^- + H_3O^+ \quad ...(2)$$

In the-first-stage ionisation (1), the acid can lose a proton from two positions, but the 'half-ion' (conjugate base, cB_1) can add a proton to only one position. Hence, if no other factors operated, it would be anticipated that the acid in equilibrium (1) is twice as strong as a monocarboxylic acid, RCO_2H, *i.e.*, $K_a^1 = 2 \times K_a$, of RCO_2H. In the second-stage ionisation (2), the half-ion can lose a proton from only one position, but the dianion (cB_2) can add a proton to two positions. Therefore for equilibrium (2) $K_a^2 = 1/2\ K_a$, of RCO_2H. Thus, equilibrium (1) is statistically four times as favoured as equilibrium (2). Hence, on the assumption above, *i.e.*, that no other factors operate, the value of K_a^1 would be expected to be four times K_a^2. In practice, the two constants differ from this factor of 4 Table 4.1.

An explanation for the experimental results may be as follows. Bjerrum (1923) assumed that the electrostatic effect of the negatively charged carboxylate ion, cB_1, would require additional energy to separate the proton of the carboxyl group, *i.e.*, there is a field effect through the solvent and consequently increases the ratio K_a^1/K_a^2. At the same time, the carboxylate anion exerts a + I effect with resultant weakening of the acid. It would therefore follow that as the chain increases in length, *i.e.*, as n increases, both the field and + I effects would decrease, and so the ratio would approach 4. This is borne out in practice, Table 4.1: see also conformational effects.

The reactions of the dicarboxylic acids depend, to a large extent, on the length of the carbon chain. The dicarboxylic acids, therefore, will be described individually.

Oxalic acid (*ethanedioic acid*) is one of the most important dicarboxylic acids. It occurs in rhubarb, in sorrel and other plants of the *oxalis* group (hence its name). Oxalic acid is one of the final products of oxidation of many organic compounds, *e.g.*, sugars, starch, etc., give oxalic acid when oxidised with concentrated nitric acid.

Preparation

Oxalic acid is prepared industrially by heating sodium formate rapidly to 360°C.

$$2HOC_2Na \longrightarrow (CO_2Na)_2 + H_2$$

The free acid is obtained from its sodium salt by adding calcium hydroxide solution, then treating the precipitated calcium oxalate with the calculated amount of dilute sulphuric acid, removing the precipitated calcium sulphate, and finally evaporating the filtrate to crystallisation.

The usual laboratory method for preparing oxalic acid is to oxidise sucrose or molasses with concentrated nitric acid in the presence of vanadium pentoxide as catalyst:

$$C1_2H_{22}O_{11} + 18\,[O] \xrightarrow[V_2O_5]{HNO_3} 6(CO_2H)_2 + 5H_2O\ (25\%)$$

Properties and Reactions

Oxalic acid crystallises from water as colourless crystals with two molecules of water of crystallisation; the melting point of the hydrate is 101.5°C; that of the anhydrous acid is 189.5°C. Oxalic acid is poisonous, soluble in water and ethanol but almost insoluble in ether. The dihydrate loses water when heated at 100-105°C, and when heated at about 200°C, oxalic acid decomposes into carbon dioxide, carbon monoxide, formic acid and water:

$$(CO_2H)_2 \longrightarrow CO_2 + HCO_2H \longrightarrow CO + H_2O$$

The anhydrous acid is conveniently obtained by heating the hydrate with carbon tetrachloride.

When heated with concentrated sulphuric acid at 90°C, oxalic acid is decomposed:

$$(CO_2H)_2 \xrightarrow{H_2SO_4} CO + CO_2 + H_2O$$

It is oxidised by permanganate to carbon dioxide:

$$(CO_2H)_2 + [O] \longrightarrow 2CO_2 + H_2O$$

It is only very slowly oxidised by concentrated nitric acid. When fused with potassium hydroxide, it evolves hydrogen:

$$(CO_2H)_2 + 2C_2H_5OH \longrightarrow (CO_2C_2H_5)_2 + 2H_2O \ (80–90\%)$$

The anhydride of oxalic acid is unknown. When anhydrous oxalic acid is refluxed with ethanol, ethyl oxalate is formed:

$$(CO_2)_2 + 2C_2H_5OH \longrightarrow (CO_2C_2H_5)_2 + 2H_2O \ (80–90\%)$$

A more general method for preparing di-esters is to heat a mixture of the dicarboxylic acid, alcohol, toluene and concentrated sulphuric acid. Sulphuric acid is the catalyst, and the toluene removes the water by forming a ternary azeotrope of alcohol, water and toluene; the yield of di-ester is 94-98 per cent from oxalic to sebacic acid.

The esterification may be carried out in the absence of toluene: a larger amount of sulphuric acid is required and the yields are lower (70-90 per cent).

The ethyl and propyl esters of oxalic acid are liquid; the methyl ester is solid, and hence has been used to prepare pure methanol by hydrolysis with sodium hydroxide solution. Ethyl oxalate undergoes the Claisen condensation with esters containing two α-hydrogen atoms to form keto-esters, *e.g.*, with ethyl acetate, oxalacetic ester is formed:

$$C_2H_5O_2CCO_2C_2H_5 + CH_3CO_2C_2H_5 \xrightarrow[\text{ethanol}]{C_2H_5ONa\text{ in}}$$

$$\begin{array}{l} CHCO_2C_2H_5 \\ \| \\ C(ONa)CO_2C_2H_5 \end{array} \xrightarrow{CH_3CO_2H} \begin{array}{l} CH_2CO_2C_2H_5 \\ | \\ COCO_2C_2H_5 \end{array} \quad (60–70\%)$$

When oxalic acid is heated with ethylene glycol, the cyclic compound, ethylene oxalate, is obtained:

$$HOCH_2CH_2OH + HO_2CCO_2H \longrightarrow \begin{array}{c} CH_2—CH_2 \\ O \qquad\quad O \\ CO—CO \end{array} + 2H_2O$$

This reaction is characteristic of oxalic acid; the other dicarboxylic acids usually react with glycol to form polyesters. When oxalic acid is heated with glycerol, formic acid or allyl alcohol is obtained according to the conditions.

Oxalic acid forms the diamide, *oxamide*. This may be prepared by shaking ethyl oxalate with concentrated ammonium hydroxide solution:

$$C_2H_5O_2CCO_2C_2H_5 + 2NH_3 \longrightarrow NH_2COCONH_2 + 2C_2H_5OH$$

It may also be prepared by passing cyanogen into *cold* concentrated hydrochloric acid:

$$(CN)_2 + 2H_2O \xrightarrow{HCl} (CO_2NH_4)_2$$

Oxamide (decomp. 350°C). in aqueous solution slowly changes into ammonium oxalate, a change which is brought about rapidly by hydrochloric acid (on warming):

$$(CONH_2)_2 \xrightarrow{H_3O^+} (CO_2NH_4)_2$$

When heated with phosphorus pentoxide, oxamide is dehydrated to cyanogen:

$$(CONH_2)_2 \xrightarrow{P_2O_5} (CN)_2 + 2H_2O$$

The monoamide of oxalic acid is also known. It is called oxamic acid, the suffix of the names of all the monoamides of the dicarboxylic acids is -amide, Oxamic acid, m.p. 210°C, may be prepared by the action of concentrated ammonium hydroxide solution on ethyl hydrogen oxalate, or by heating ammonium hydrogen oxalate.

When reduced with zinc and sulphuric acid, oxalic acid forms *glycollic acid*, $HOCH_2CO_2H$. Electrolytic reduction with a lead cathode gives glycollic and glyoxylic acids, the latter being obtained in better yield by reducing oxalic acid with magnesium and sulphuric acid.

When oxalic acid is treated with *excess* of phosphorus pentachloride, *oxalyl chloride*, b.p. 64°C, is formed:

$$(CO_2H)_2 \xrightarrow{PCl_5} (COCl)_2 \qquad \text{(f.–g)}$$

If an excess of phosphorus pentachloride is not used, oxalic acid is decomposed, possibly via the intermediate formation of the half-acid chloride.

$$(CO_2H)_2 \xrightarrow{PCl_5} [HO_2CCOCl] \longrightarrow CO_2 + CO + HCl$$

Alkanes can be carboxylated by oxalyl chloride under the influence of light (Kharasch and Brown, 1942):

$$RH + (COCl)_2 \longrightarrow RCOCl + CO + HCl$$

Oxalic acid is used for the manufacture of ink and for bleaching straw. Its antimony salts are used as mordants in printing and dyeing.

Malonic acid (*propanedioic acid*), m.p. 136°C, was first obtained by the oxidation of malic acid (hence its name):

$$HO_2CCHOHCH_2CO_2H + [O] \xrightarrow{K_2Cr_2O_7/H_2SO_4} \underset{\text{oxalacetic acid}}{HO_2CCOCH_2CO_2H}$$

$$\xrightarrow{[O]} HO_2CCH_2CO_2H + CO_2$$

Malonic acid may be prepared by heating potassium chloroacetate with aqueous potassium cyanide and hydrolysing the product, potassium cyanoacetate, with hydrochloric acid:

$$ClCH_2CO_2K \xrightarrow{KCN} CH_2(CN)CO_2K \xrightarrow{H_3O^+} CH_2(CO_2H)_2\ (84\%)$$

When heated to 140-150°C, or when refluxed in sulphuric acid solution, it eliminates carbon dioxide.

$$HO_2CCH_2CO_2H \longrightarrow C_3CO_2 + CO_2$$

All dicarboxylic acids which have both carboxyl groups attached to the *same* carbon atom are decomposed in a similar manner. When malonic acid is heated with phosphorus pentoxide, a small amount of *carbon suboxide* is obtained:

$$HO_2CCH_2CO_2H \xrightarrow{P_2O_5} O = C = C = C = O + 2H_2O$$

Carbon suboxide, b.p. 7°C, may be regarded as a diketen; it combines with water to form malonic acid. Malonic acid does not form a cyclic anhydride, but dimethylmalonic acid does.

When malonic acid is treated with nitrous acid and the product hydrolysed, **mesoxalic acid** *(keto-malonic acid)* is obtained:

$$(HO_2C)_2CH_2 \xrightarrow{HNO_2} (HO_2C)_2C = NOH \xrightarrow{H_3O^+}$$

$$HO_2CCOCO_2H$$

Mesoxalic acid crystallises from water with one molecule of water which is held firmly, and hence is believed to be water of constitution:

$(HO)_2C(CO_2H)_2$ (cf. glyoxylic acid). Ethyl malonate may be oxidised directly to ethyl mesoxalate by selenium dioxide (yield: 23 per cent), or with dinitrogen trioxide (HNO_3 + AS_2O_3; yield: 74-76 per cent).

Ethyl malonate is far more important than the acid because of its synthetic uses; the acid contains an active methylene group, the reactivity of which is more pronounced in the ester. Both the acid and ester may be readily brominated, *e.g.*, monobromomalonic acid, CHBr(CO,H), is formed when a suspension of malonic acid in ether is treated with bromine; owing to the high reactivity of the methylene group, no red phosphorus is required as catalyst.

Ethyl malonate does not form the diamide, *malonamide,* when shaken with concentrated ammonium hydroxide solution; the dimethyl ester, however, gives a good yield of malonamide. Ethyl malonate forms *barbituric acid* when heated with urea in the presence of sodium ethoxide.

Succinic acid *(butanedioic acid),* m.p. 185°C, was originally obtained by the distillation of amber (Latin: *succinum,* amber). It may be synthesised by the following methods:

(i) From ethylene dibromide:

$$\begin{matrix} CH_2Br \\ | \\ CH_2Br \end{matrix} \xrightarrow{KCN} \begin{matrix} CH_2CN \\ | \\ CH_2CN \end{matrix} \xrightarrow{H_3O^+} \begin{matrix} CH_2CO_2H \\ | \\ CH_2CO_2H \end{matrix} \quad (80\%)$$

(ii) By the reaction between malonic ester (I molecule) and ethyl chloroacetate, or between malonic ester (2 molecules) and iodine.

Alkyl-substituted succinic acids may be prepared by using monoalkylmalonic ester and α-halogeno-acid esters or iodine.

Alternatively, they may be prepared by the addition of hydrogen cyanide to α, β-unsaturated esters and hydrolysing the Bcyanacomplex produced.

(iii) By heating malic acid, in a sealed tube, with constant boiling hydriodic acid and red phosphorus:

$$\begin{matrix} CHOHCO_2H \\ | \\ CH_2CO_2H \end{matrix} + 2HI \longrightarrow \begin{matrix} CH_2CO_2H \\ | \\ CH_2CO_2H \end{matrix} + I_2 + H_2O \quad (60\%)$$

Succinic acid is prepared industrially by the catalytic (or by the electrolytic) reduction of maleic acid:

$$\begin{array}{l} \text{CHCO}_2\text{H} \\ \| \\ \text{CHCO}_2\text{H} \end{array} + \text{H}_2 \xrightarrow{\text{Ni}} \begin{array}{l} \text{CH}_2\text{CO}_2\text{H} \\ | \\ \text{CH}_2\text{CO}_2\text{H} \end{array}$$

When heated, a large amount sublimes, the rest being converted into the *cyclic* anhydride, succinic anhydride:

$$\begin{array}{l} \text{CH}_2\text{CO}_2\text{H} \\ | \\ \text{CH}_2\text{CO}_2\text{H} \end{array} \longrightarrow \begin{array}{l} \text{CH}_2\text{CO}_2\text{H} \\ | \\ \text{CH}_2\text{CO}_2\text{H} \end{array} \text{O} + \text{H}_2\text{O}$$

When heated with excess of glycol, succinic acid forms high-polymer esters (*polyesters*, which belong to the group known as the *alkyl resins*). These esters are acidic, the end groups being succinic acid residues:

$$\text{HO}_2\text{C(CH}_2)_2\text{CO}_2\text{H} + \text{HOCH}_2\text{CH}_2\text{OH} + \text{HO}_2\text{C(CH}_2)_2\text{CO}_2\text{H}$$
$$+ \text{HOCH}_2\text{CH}_2\text{OH} + \ldots \longrightarrow \text{HO}_2\text{C(CH}_2)_2\text{CO}$$
$$\text{—[—O(CH}_2)_2\text{OCO(CH}_2)_2\text{CO—]—OH}$$

Dicarboxylic acids from malonic to adipic acid form these polyesters with ethylene glycol. These esters are linear polymers, but if glycerol is used instead of glycol, three dimensional polymeric esters are obtained.

Succinic anhydride (*butanedioic anhydride*) is obtained in excellent yield by distilling succinic acid with acetic anhydride, acetyl chloride, or phosphoryl chloride.

The anhydride, and not the acid chloride, is obtained when succinic acid is heated with thinly chloride.

Succinic anhydride, m.p. 119°C, when boiled with water or alkalis, is converted into succinic acid:

$$\begin{array}{l} \text{CH}_2\text{CO} \\ | \\ \text{CH}_2\text{CO} \end{array} \text{O} + \text{H}_2\text{O} \longrightarrow \begin{array}{l} \text{CH}_2\text{CO}_2\text{H} \\ | \\ \text{CH}_2\text{CO}_2\text{H} \end{array}$$

When reduced with sodium and ethanol, succinic anhydride is converted first into γ-butyro- lactone, and finally into tetramethylene glycol:

$$\begin{array}{l} \text{CH}_2\text{CO} \\ | \\ \text{CH}_2\text{CO} \end{array} \text{O} \xrightarrow{\text{e; H}^+} \begin{array}{l} \text{CH}_2\text{CH}_2 \\ | \\ \text{CH}_2\text{CO} \end{array} \xrightarrow{\text{e; H}^+} \begin{array}{l} \text{CH}_2\text{CH}_2\text{OH} \\ | \\ \text{CH}_2\text{CH}_2\text{OH} \end{array}$$

If the reduction is carried out with sodium amalgam in *acid* solution, the lactone is obtained in good yield; some butyric acid is also formed.

Succinic anhydride is very useful for preparing the 'half-derivatives' of succinic acid, *e.g.*, with alcohols it forms the acid-ester:

$$\begin{array}{l} CH_2CO \\ CH_2CO \end{array}\!\!>O + ROH \longrightarrow \begin{array}{l} CH_2CO_2R \\ | \\ CH_2CO_2H \end{array}$$

With Grignard reagents, succinic anhydride forms γ-keto-acids:

$$\begin{array}{l} CH_2CO \\ CH_2CO \end{array}\!\!>O + RMgX \longrightarrow RCOCH_2CH_2CO_2H$$

The carbonyl groups (C = O stretch) in *five*-membered ring anhydrides absorb in the regions 1870-1830 (s) and 1800-1760 cm^{-1}(s). Thus these anhydrides can be distinguished from acyclic anhydrides. The molecular formula should also help here for all cyclic anhydrides.

$$\begin{array}{l} CH_2CO \\ CH_2CO \end{array}\!\!>O + NH_3 \longrightarrow \begin{array}{l} CH_2CO \\ CH_2CO \end{array}\!\!>NH + H_2O \quad (v.g.)$$

Fig. 4.3 is the infrared spectrum of succinic anhydride (as a Nujol mull).

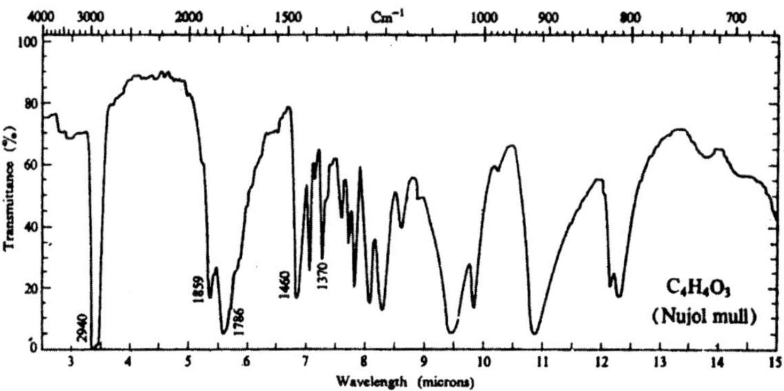

Fig. 4.3

2940, 1460 and 1370cm^{-1} : Bands for Nujol. 1859 and 1786cm^{-1}: C = O (str.) in 5-ring anhydrides

Succinimide (*butanimide*) is formed when succinic anhydride is heated in a current of dry ammonia:

Succinamic acid (m.p. 157°C), (I), and succinamide (m.p. 243°C), (II), are both readily converted into succinimide when heated:

$$\begin{array}{l} CH_2CONH_2 \\ | \\ CH_2CO_2H \end{array} \xrightarrow{-H_2O} \begin{array}{l} CH_2CO \\ | \qquad\quad \rangle NH \\ CH_2CO \end{array} \xleftarrow{-NH_3} \begin{array}{l} CH_2CONH_2 \\ | \\ CH_2CONH_2 \end{array}$$

Succinimide, m.p. 125°C, when boiled with water or alkalis is converted into succinic acid:

$$\begin{array}{l} CH_2CO \\ | \qquad\quad \rangle NH \\ CH_2CO \end{array} + 2H_2O \longrightarrow \begin{array}{l} CH_2CONH_2 \\ | \\ CH_2CO_2H \end{array} + NH_3$$

When succinimide is distilled with zinc dust, *pyrrole*, (III), is obtained. When reduced will sodium and ethanol, succinimide forms *pyrrolidine*, (IV), and when reduced electrolytical it forms *pyrrolidone*, (V):

$$\begin{array}{c} CH — CH \\ \| \qquad \| \\ CH \qquad CH \\ \diagdown NH \diagup \\ (III) \end{array} \qquad \begin{array}{c} CH_2 — CH_2 \\ | \qquad | \\ CH_2 \qquad CH_2 \\ \diagdown NH \diagup \\ (IV) \end{array} \qquad \begin{array}{c} CH_2 — CH_2 \\ | \qquad | \\ CH_2 \qquad CO \\ \diagdown NH \diagup \\ (V) \end{array}$$

Succinimide is acidic, *e.g.*, it reacts with potassium hydroxide to form potassium succinimide:

$$\begin{array}{l} CH_2CO \\ | \qquad\quad \rangle N^- \\ CH_2CO \end{array} \longleftrightarrow \begin{array}{l} CH_2C{=}O^- \\ | \qquad\quad \rangle N^- \\ CH_2CO \end{array} \longleftrightarrow \begin{array}{l} CH_2CO \\ | \qquad\quad \rangle N^- \\ CH_2CO{—}O^- \end{array}$$

Succinimide, however, is only very weakly acidic; the potassium salt is decomposed by carbon dioxide, the imide being regenerated.

A very important derivative of succinimide is N-*bromosuccinimide*, which may be prepared by the action of bromine on succinimide at 0°C in the presence of sodium hydroxide:

$$\begin{array}{l} CH_2CO \\ | \qquad\quad \rangle NH \\ CH_2CO \end{array} + Br_2 \xrightarrow{NaOH} \begin{array}{l} CH_2CO \\ | \qquad\quad \rangle NBr \\ CH_2CO \end{array} + HBr$$

N-bromosuccinimide (NBS) is a valuable reagent or brominating olefinic compounds in the allyl position (Ziegler. 1942).

$$-CH_2-CH{=}CH_2 + \begin{array}{l} CH_2CO \\ | \\ CH_2CO \end{array}\!\!\!\!\!\!\!\!\!\!\!\!\!\!\!\!\begin{array}{l} \\ \diagdown \\ \;\; NBr \\ \diagup \end{array} \longrightarrow -CHBrCH{=}CH_2 + \begin{array}{l} CH_2CO \\ | \\ CH_2CO \end{array}\begin{array}{l} \diagdown \\ NH \\ \diagup \end{array}$$

This reaction offers a means of splitting off three carbon atoms from a compound by oxidative degradation, provided the compound contains one double bond or can have one produced, *e.g.*,

$$RCH_2CH_2CH_2OH \xrightarrow[350°C]{Al_2O_3} RCH_2CH_2CH = CH_2$$

$$\xrightarrow{\text{N-bromosuccinimide}} RCH_2CHBrCH = CH_2$$

$$\xrightarrow[\text{etanol}]{KOH} RCH = CHCH = CH_2 \xrightarrow{\text{ozonolysis}} RCHO$$

This method has found great use in steroid chemistry.

Normally, *N*-bromosuccinimide *substitutes* alkenes in the ally position, and this reaction is believed to take place by a free-radical mechanism; it is catalysed by peroxides and is promoted by light (both are free-radical producing agents). If there are two allylic positions then two monobromo-derivatives may be obtained. Since a secondary hydrogen is substitute more readily than a primary hydrogen atom, the former gives rise to the predominan product.

The details of the mechanism are uncertain. According to Goldfinger *et al.* (1953, 1956), the function of the *N*-bromosuccinimide is to supply a *low concentration of molecular bromine.*

$$>NBr + HBr \longrightarrow >NH + Br_2$$

$$Br_2 \rightleftharpoons Br\cdot$$

$$Br\cdot + >CHCH{=}CH_2 \longrightarrow >\dot{C}CH{=}CH_2 + HBr$$

$$Br_2 + >\dot{C}CH{=}CH_2 \longrightarrow CBrCH{=}CH_2 + Br\cdot$$

N-bromosuccinimide also brominates unsaturated esters, *e.g.*,

$$CH_3CH{=}CHCO_2CH_3 \xrightarrow{\rangle N-Br} BrCH_2CH{=}CH_2 + Br\cdot$$

In addition to substitution, Af-bromosuccinimide may also produce addition compounds, but these are usually formed only in small amount. Braude *et al.* (1952), however, have shown that the addition reaction is catalysed by tetraethylammonium salts, *e.g.*, *N*-bromosuccinimide and cyclohexene give 3-bromocyclohexene (Ziegler *et al.*, 1942), but in the presence of, *e.g.*, tetraethylammonium bromide, 1,2-dibromocvclohexane is the main product.

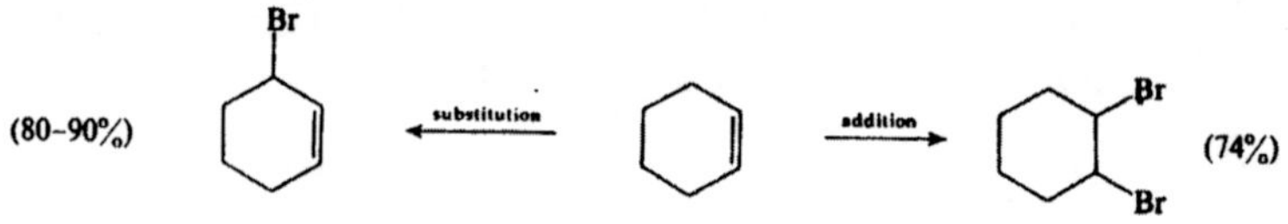

(80–90%) Br ← substitution — ⟶ addition → Br Br (74%)

Fig. 4.4

The addition reaction probably involves heterolytic fission

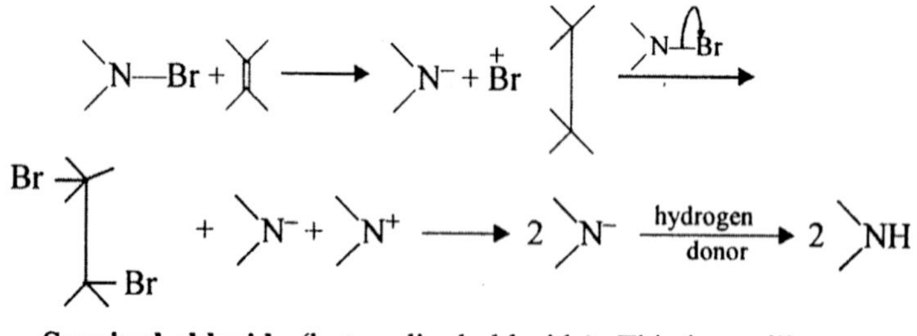

$$\rangle N-Br + \rangle C{=}C\langle \longrightarrow \rangle N^- + \overset{+}{Br} \quad \xrightarrow{\rangle N-Br}$$

$$Br-C-C-Br + \rangle N^- + \rangle N^+ \longrightarrow 2\rangle N^- \xrightarrow[\text{donor}]{\text{hydrogen}} 2\rangle NH$$

Succinyl chloride (butanedioyl chloride). This is readily prepared by the action of phosphorus pentachloride on succinic acid. It was formerly believed that succinyl chloride existed in two forms, the chain isomer, (I), and the ring isomer, (II). However, all recent work indicates structure (1), *e.g.*, Schmid (1965, 1966), from NMR studies, showed that the molecule is symmetrical, and from inferred studies that a y-lactone oxo group (in II) is absent.

$(CH_2COCl)_2$ (I) $\qquad$ CH_2CO—O—CH_2CO (ring) (II)

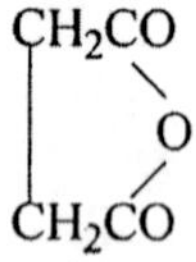

Isomeric with is succinic acid is *methylamine acid* or *isosuccinic acid*, m.p. 130°C. It may be readily prepared from sodiomalonic ester and methyl iodide, or as follows:

$$CH_3C_2CO_2H \xrightarrow{Br_2/P} CH_3CHBrCO_2H$$

$$\xrightarrow{KCN} CH_3CH(CN)CO_2H \xrightarrow{H^+} CH_3CH(CO_2H)_2$$

Oxalacetic ester is the diethyl ester of ketosuccinic acid, and may be prepared by the Claisen condensation between ethyl oxalate and ethyl acetate. It is a colourless liquid which can be distilled under reduced pressure (b.p. 132°C/24 mm), but at atmospheric pressure it eliminates a molecule of carbon monoxide to form malonic ester:

$$C_2H_5O_2CCOH_2CO_2C_2H_5 \longrightarrow CH_2(CO_2C_2H_5)_2 + CO$$

Hydrogen of the methylene group adjacent to the carbonyl group may be replaced by alkyl groups by reactions similar to those used for acetoacetic ester, *e.g.*,

$$C_2H_5O_2CCOCH_2CO_2C_2H_5 \xrightarrow{C_2H_5ONa}$$

$$[C_2H_5O_2CCOCHCO_2C_2H_5]^-Na^+ \xrightarrow{CH_3I}$$

$$C_2H_5O_2CCOCH(CH_3)CO_2C_2H_5 \xrightarrow[\text{(ii) } C_2H_5I]{\text{(i) } C_2H_5ONa}$$

$$C_2H_5O_2CCOC(CH_3)(C_2H_5)CO_2C_2H_5$$

Oxalacetic ester and its alkyl derivatives undergo 'acid hydrolysis' when boiled with alkalis, and 'ketonic hydrolysis' when boiled with dilute sulphuric acid.

'*Acid hydrolysis*'

$$C_2H5O_2CCOCHRCO_2C_2H_5 \xrightarrow[\text{(ii) } H^+]{\text{(i) NaOH}}$$

$$(CO_2H)_2 + RCH_2CO_2 + 2C_2H_5OH$$

'*Ketonic hydrolysis*'

$$C_2H_5O_2CCOCH_2CO_2C\,_2H_5 + 2H_2O \xrightarrow{HCl} [HO_2CCOCHRCO_2H]$$

$$\longrightarrow CO_2 + RCH_2COCO_2H$$

Hydrolysis of oxalacetic ester with concentrated hydrochloric acid in the cold gives oxalacetic acid (*ketosuccinic acid*):

$$C_2H_5O_2CCOCH_2CO_2H_5 + 2H_2O \xrightarrow{HCl} HO_2CCOCH_2CO_2H$$

$$+ 2C_2H_5OH$$

A simpler preparation of oxalacetic acid is to oxidise maleic acid with Fenton's reagent.

Oxalacetic acid is a fairly stable substance, soluble in water, the aqueous solution giving a red colour with ferric chloride, thus indicating the existence of an enolic form. Oxalacetic acid exists in two forms, one with m.p. 155°C and the other 184°C. These two forms are actually hydroxymaleic acid, (I), and *hydroxyfumaric acid*, (II), respectively:

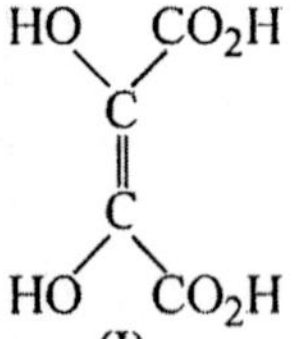

```
HO   CO2H          HO2C   H
  \ /                  \ /
   C                    C
   ‖                    ‖
   C                    C
  / \                  / \
HO   CO2H             H   CO2H
  (I)                  (II)
```

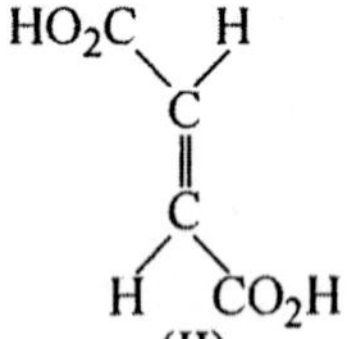

(I) is convertible into (II) by 30 per cent sulphuric acid, and the salts of (II) give (I) on treatment with dilute acid.

According to Kumler *et al.* (1962), who have examined the NMR spectrum of oxalacetic acid in deuterium oxide, at equilibrium there is 50 per cent of keto-form and 8 per cent of enol, the rest being present in some other form.

Glutaric acid (*pentanedioic acid*), m.p. 97°C (relationship to glutamic acid and tartaric acid gave rise to its name), may be prepared by a number of methods, *e.g.*:

(i) By refluxing trimethylene dicyanide with concentrated hydrochloric acid:

$$NC(CH_2)_3CN + 4H_2O \xrightarrow{HCl} HO_2C(CH_2)_3CO_2H + 2NH_3$$

(83–53%)

(ii) By the action of methylene di-iodide on sodiomalonic ester:

$$2[CH(CO_2C_2H_5)_2]^- Na^+ + CH_2I_2 \longrightarrow {}_2NaI + CH_2[CH(CO_2C_2H5)_2]_2$$

$$\xrightarrow[(ii)\ HCl]{(i)\ KOH} CH_2[CH(CO_2H_2)_2]_2$$

$$\xrightarrow{150-200°C} HO_2C(CH_2)3CO_2H + {}_2CO_2$$

(iii) By condensing formaldehyde with malonic ester in the presence of diethylamine:

$$2(C_2H_5O_2C)_2CH_2 + CH_2O$$

$$\xrightarrow[\text{reflux}]{(C_2H_5)_2NH} (C_2H5O_2C)_2CHCH_2CH(CO_2C_2H5)_2 \quad (61\%)$$

This tetracarboxylic ester may now be treated as in (ii), or may be converted into glutaric acid directly by refluxing with concentrated hydrochloric acid:

$$(C_2H_5O_2C)_2CHCH_2CH(CO_2H_5)_2$$
$$\xrightarrow{HCl} HO_2C(CH_2)_3CO_2H \qquad (76–80\ \%)$$

When heated with acetic anhydride or thionyl chloride, glutaric acid is converted into glutaric anhydride:

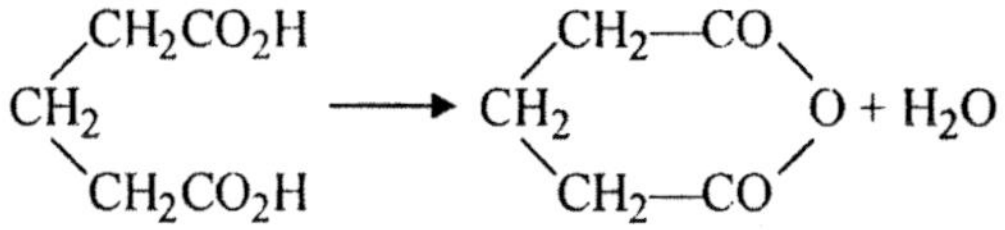

$$CH_2(CH_2CO_2H)_2 \longrightarrow CH_2\begin{matrix}CH_2—CO\\CH_2—CO\end{matrix}\!\!>O + H_2O$$

Succinic and glutaric acids differ from their higher homologues in that they readily form the cyclic anhydride when heated with acetic anhydride.

Adipic acid (*hexanedioic acid*), m.p. 150°C, received its name from .the fact that it was first obtained by the oxidation of fats (Latin: adeps, fat). It may be synthesised from sodiomalonic ester and ethylene dibromide, but is prepared industrially by the oxidation of cyclohexane (from the hydrogenation of benzene) by air in the presence of a catalyst, *e.g.*, cobalt acetate:

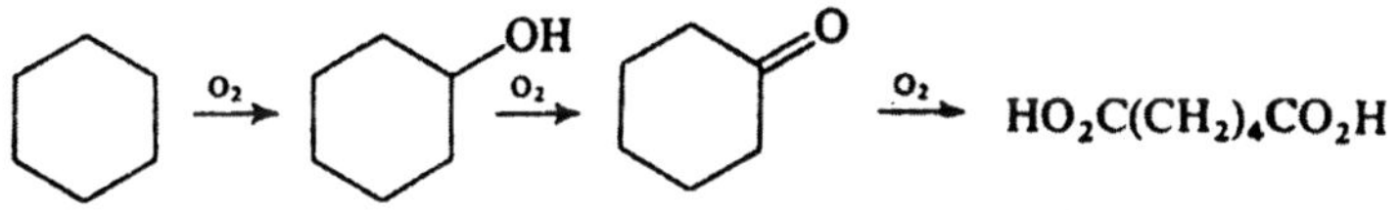

$$\text{cyclohexane} \xrightarrow{O_2} \text{cyclohexanol (OH)} \xrightarrow{O_2} \text{cyclohexanone (=O)} \xrightarrow{O_2} HO_2C(CH_2)_4CO_2H$$

Fig. 4.5

When adipic and pimelic acids are heated with acetic anhydride and the product distilled at 300°C, a cyclic ketone is obtained in each case:

Adipic acid is used for the preparation of resins (poly-esters), and is an intermediate in the manufacture of *nylon.* Nylon is used as the generic name for all synthetic fibre-forming polymeric amides having a protein-like structure. Nylon yarns and fabrics are practically non-inflammable. Nylon '66' is formed from adipic acid and hexa-methylenediamine, and polymerisation is effected by melting the mixture in the presence of various catalysts.

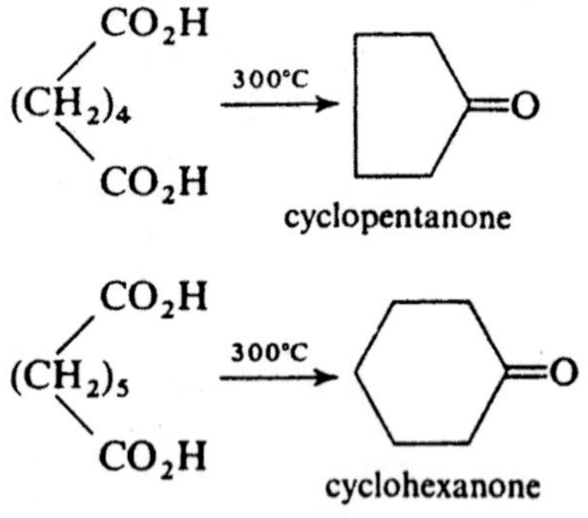

$$(CH_2)_4(CO_2H)_2 \xrightarrow{300^\circ C} \text{cyclopentanone}$$

$$(CH_2)_5(CO_2H)_2 \xrightarrow{300^\circ C} \text{cyclohexanone}$$

Fig. 4.6

$$HO_2C(CH_2)_4CO_2H + H_2N(CH_2)_6NH_2 + HO_2C(CH_2)_4CO_2H + H_2N\text{—} \ldots \longrightarrow$$

$$+ HO_2C(CH_2)_4CO\text{—}[\text{—}NH(CH_2)_6NHCO(CH_2)_4CO\text{—}]_n NH(CH_2)_6NH_2$$

Nylon '6' is made from *caprolactam*, which is manufactured in several ways starting from cyclohexane:

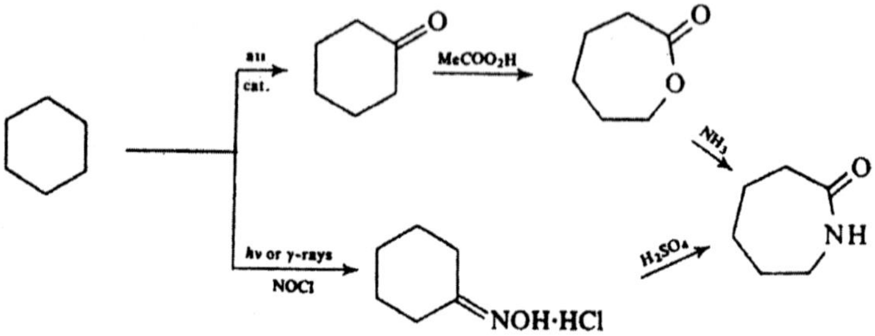

air, cat.; $MeCOO_2H$; NH_3; $h\nu$ or γ-rays, NOCl; H_2SO_4; NOH·HCl; NH

Fig. 4.7

The conversion of the oxime to caprolactam occurs via the Be Loann rearrangement, and the caprolactam is polymerised in the presence of a catalyst:

$$\text{caprolactam} \longrightarrow \text{—}[\text{—}CO(CH_2)_5NH\text{—}]_n\text{—}$$

Pimelic acid (*heptanedioic acid*), m.p. 104°C, received its name from the fact that it was obtained originally from the oxidation of fats (Greek: *pimele*, fat). It may be prepared by the hydrolysis of pentamethylene dicyanide (formed by the action of potassium cyanide on pentamethylene dibromide), or by the malonic ester synthesis, using trimethylene (*cf.* adipic acid). Pimelic acid may also be prepared by the reduction of salicylic acid with sodium and isopentanol, followed by the addition of water and then hydrochloric acid:

$$\text{(OH, } CO_2H\text{ benzene ring)} + H_2O \xrightarrow{e;\, H^+} HO_2C(CH_2)_5CO_2H \quad (43\text{–}50\%)$$

Fig. 4.8

Subcric acid (*octanedioic acid*), $HO_2C(CH_2)_6CO_2H$, m.p. 144°C, may be synthesised by the electrolysis of potassium ethyl glutarate. It is prepared industrially by the oxidation of cork (Latin: *suber*, cork) with concentrated nitric acid.

Azelaic acid (*nonanedioic acid*), m.p. 107°C, may be synthesised from malonic ester and pentamethylene dibromide. It may be obtained .in the laboratory by refluxing castor oil with ethanolic potassium hydroxide, acidifying with sulphuric acid, and oxidising the crude ricinoleic acid so obtained with alkaline permanganate:

$$CH_3(CH_2)_5CHOHCH_2CH = CH(CH_2)_7CO_2H$$
$$\xrightarrow{[O]} HO_2C(CH_2)_7CO_2H \qquad (32\text{-}36\%)$$

Azelaic acid is prepared industrially by the ozonolysis folic acid.

Sebacic acid (*decanedioic acid*), m.p. 133°C, is prepared industrially by heating castor oil with sodium hydroxide.

$$CH_3(CH_2)_5CHOHCH_2CH = CH(CH_2)_7CO_2Na$$
$$\xrightarrow{NaOH} NaO_2C(CH_2)_8CO_2Na + C_6H_{13}CHOHCH_3$$

Many higher dicarboxylic acids are known.

The acid esters of the dicarboxylic acids may be prepared by refluxing a mixture of the acid with half its equivalent of the di-ester in the presence of concentrated hydrochloric acid. This is an example of acidolysis. Alternatively, the acid ester may be obtained by the half saponification of the di-ester. In either case the di-ester is separated from the acid ester by fractional distillation.

Another method of preparing a half-ester is to heat equimolecular amounts of dicarboxylic acid and alcohol: the products are unchanged acid, half-ester and di-ester, which are then separated. If the dicarboxylic acid is substituted at one a-carbon atom, then the other carboxyl group is readily esterified by the Fischer-Speier method; this is probably a question of steric hindrance.

When a cyclic anhydride is available, this is a better way to prepare the half-ester. Here again, if one α-carbon atom is substituted, it is the half-ester of the other carboxyl group that is formed.

Half-esters are useful starting materials in various syntheses, *e.g.*,

(i) $$\begin{matrix} CH_2CH_2CO_2H \\ | \\ CH_2CH_2CO_2Et \end{matrix} \xrightarrow[\text{(ii) } Br_2]{\text{(i) } SOCl_2} \begin{matrix} CH_2CHBrCOCl \\ | \\ CH_2CH_2CO_2Et \end{matrix}$$

$$\xrightarrow{ROH} \begin{matrix} CH_2CHBrCO_2R \\ | \\ CH_2CH_2CO_2Et \end{matrix}$$

(ii) $HO_2C(CH_2)_8CO_2Et \xrightarrow[Br_2/CCl_4]{HgO} Br(CH_2)_nCO_2Et$

(iii) $HO_2C(CH_2)_nCO_2Et \xrightarrow{Na-EtOH} HO_2C\ (CH_2)_nCH_2OH$

When the calcium, barium, or better still, the thorium salts of the carboxylic acids from adipic to nonadecanedioic are distilled, varying yields of cyclic ketones are obtained; *e.g.*, adipic acid gives cyclopentanone:

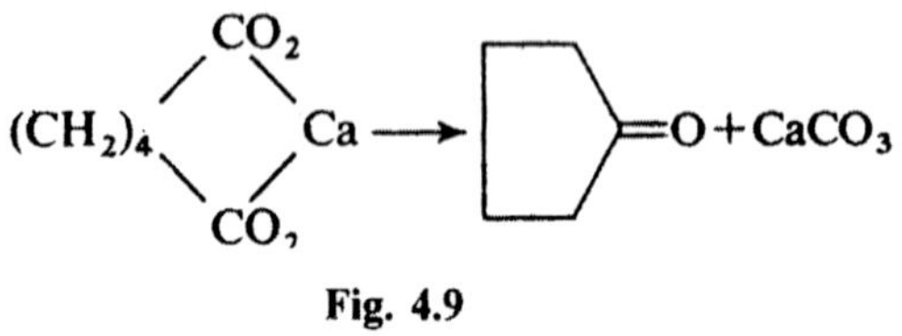

$$(CH_2)_4\begin{matrix} CO_2 \\ \quad \\ CO_2 \end{matrix}Ca \longrightarrow \text{cyclopentanone}{=}O + CaCO_3$$

Fig. 4.9

Based on the experimental results of the ease of formation of cyclic anhydrides and cyclic ketones, is **Blanc's rule** (1905). Blanc found that dicarboxylic acids, on heating with acetic anhydride, and then distilling at 300°C (or distilling directly at 300°C), gave cyclic anhydrides or cyclic ketones according to the relative positions of the two carboxyl groups. 1,4- and 1,5-dicarboxylic acids gave cyclic anhydrides. 1,6- and 1,7- gave cyclic ketones: this is Blanc's rule. By using Blanc's rule it is possible to determine the size of rings. A double bond is introduced into the ring, and the ring opened by oxidation to the corresponding dicarboxylic acid. This is then heated with acetic anhydride and distilled. If a cyclic anhydride is obtained, the acid is either 1,4- or 1,5-; if a cyclic ketone, either 1,6- or 1,7-; if there is no change, the acid is 1,8-or more. In naturally occurring compounds the rings are usually five- or six-membered. Hence the formation of the anhydride is taken to mean a five-membered ring, and of the ketone a six-membered ring. Blanc's rule, however, is only satisfactory for simple cyclic compounds. For substituted rings, the results may be misleading, since substituents in the chain bring about ring closure much more easily; *e.g.*, substituted adipic acids may give rise to the anhydride and not to the cyclic ketone.

CARBONIC ACID AND ITS DERIVATIVES

Orthocarbonic acid, C(OH), is unknown in the free state, but its esters have been prepared. They may be obtained by the reaction between sodium alkoxide and nitrochloroform:

$$4RONa + CCl_3NO_2 \longrightarrow C(OR)_4 + 3NaCl + NaNO_2 \quad (f.g.)$$

The alkyl orthocarbonates are ethereal smelling liquids.

Carbonic acid (*metacarbonic acid*), $0 = C(OH)_2$, is unknown in the free state, but its salts and esters have been prepared. Alkyl carbonates may be readily formed by heating alcohols with carbonyl chloride in the presence of pyridine:

They may also be prepared by heating silver carbonate with alkyl iodide:

$$Ag_2CO_3 + 2RI \longrightarrow CO(OR)_2 + 2AgI$$

The alkyl carbonates are ethereal smelling liquids, readily soluble in water. In recent years they have found various uses in organic synthesis, *e.g.*,

(i) They may be used to introduce the carbalkoxy group, CO_2R, into ketones to produce β-ketoesters. The reaction is carried out by heating a ketone with sodium or sodium alkoxide in the presence of a large excess of alkyl carbonate:

$$R^2COCH_3 + (R^1O)_2CO + NaOR^1 \longrightarrow [R^2COCHOR^1]^- Na^+ + 2R^1OH$$

$$\xrightarrow{H^+} R_2COCH_2CO_2R^1$$

(ii) Alkyl carbonates react with primary alkyl cyanides in the presence of sodium alkoxide to form α-cyanoesters:

$$[R^2C(CO_2C_2H_5)_2]^- Na^+ + (R^1O)_2 CO \longrightarrow R^1R^2C(CO_2C_2H_5)_2 + R^1OCOONa$$

(iii) Alkyl carbonates may be used to alkylate sodiomalonic esters:

$$[R^2C(CO_2C_2H_5)_2]^- Na^+ + (R^1O)_2CO \longrightarrow R^1R^2C(CO_2C_2H_5)_2 + R^1OCOONa$$

The acid chloride of carbonic acid is *carbonyl chloride* (*phosgene*). It may be obtained by the action of chlorine on carbon monoxide under the influence of sunlight, or in the presence of heated charcoal (200°C) as catalyst; the latter is the commercial method.

Carbonyl chloride is a colourless liquid, b.p. 8°C, and has been used as a toxic gas in warfare. It is used in various organic syntheses, behaving as an acid chloride. It is very slowly decomposed by water:

$$COCl_2 + H_2O \longrightarrow CO_2 + 2HCl$$

When treated with slight excess of one equivalent of alcohol in the cold, carbonyl chloride forms chloroformic esters:

$$COCl_2 + ROH \longrightarrow ClCO_2R + HCl$$

When treated with an excess of alcohol in the presence of pyridine, alkyl carbonates are formed (see above). Carbonyl chloride reacts with ammonia to form urea:

$$COCl_2 + 2NH_3 \longrightarrow CO(NH_2)_2 + 2HCl$$

With primary or secondary amines, substituted ureas are formed, *e.g.*,

$$COCl_2 + 2RNH_2 \longrightarrow CO(NHR)_2 + 2HCl$$

Chloroformic acid (*chlorocarbonic acid*), $ClCO_2H$, is not known in the free state, but its esters have been prepared. These may be readily obtained by treating carbonyl chloride with slight excess of one equivalent of alcohol in the cold (see above). Chloroformic esters are acid-chloride esters, and the chlorine atom reacts with compounds containing active hydrogen; thus ethyl chloroformate, b.p. 94°C, reacts with water, alcohols, ammonia, primary and secondary amines; *e.g.*, are than is formed with ammonia:

$$ClCO_2C_2H_4 + NH_3 \longrightarrow NH_2CO_2C_2H_5 + HCl$$

Ethyl chloroformate is useful for introducing the carbethoxy group on a nitrogen atom; *e.g.*, ethyl *N*-tricarboxylate is formed when ethyl chloroformate is added to a mixture of urethan, ether and sodium, this mixture having been previously heated:

$$NH_2C\ O_2C_2H5 + 2ClCO_2H_2H_5 + 2Na \longrightarrow N(CO_2C_2H_5)_3 + 2NaCl + H_2 \quad (51\text{–}57\%)$$

Ethyl chloroformate may also be used to prepare ethyl esters by reaction with Grignard reagents.

Amides of carbonic acid. Since carbonic acid is dibasic, two amides are possible: the mono- and diamide. These are known, respectively, as *carbamic acid*, (1), and *urea*, (II):

$$O{=}C\begin{matrix} \diagup NH_2 \\ \diagdown OH \end{matrix} \qquad O{=}C\begin{matrix} \diagup NH_2 \\ \diagdown NH_2 \end{matrix}$$

(I) (II)

Carbamic acid is not known in the free state, but its salts and esters have been prepared. Ammonium carbamate is formed when dry ammonia reacts with dry carbon dioxide:

$$2NH_3 + CO_2 \longrightarrow NH_2COONH_4$$

It is a white crystalline solid, very soluble in water. When its aqueous solution is warmed to 60° C, ammonium carbamate is hydrolysed to ammonium carbonate:

$$NH_2COONH_4 + H_2O \longrightarrow CO(ONH_4)_2$$

Esters of carbamic acid are known as **urethans**. These may be prepared:

(i) By treating a chloroformic ester with ammonia, *e.g.*, ethyl chloroformate produces ethyl carbamate:

$$ClCO_2C_2H_5 + NH_3 \longrightarrow NH_2CO_2C_2H_5 + HCl$$

(ii) By refluxing urea in an alcohol, *e.g.*, n-butanol gives n-butyl carbamate when refluxed for 30 hours:

$$CO(NH_2)_2 \longrightarrow NH_3 + HNCO$$

$$HNCO + CH_3CH_2CH_2CHOH \longrightarrow NH_2CO_2CH_2CH_2CH_2CH_3 \quad (75\text{–}76\%)$$

(iii) N-substituted urethans may be prepared by the Curtius reaction. The acid azide is refluxed in benzene solution and then an alcohol is added:

$$R^1CON_3 \longrightarrow N_2 + R^1NCO \xrightarrow{R^2OH} R^1NHCO_2R^2$$

N-phenylurethans may be prepared by reaction between phenyl isocyanate and alcohol:

$$C_6H_5NCO + ROH \longrightarrow C_6H_5NHCO_2R$$

These are crystalline solids and may be used to characterise the alcohols.

Readily dehydrated alcohols, particularly t-alcohols, do not form urethans; alkenes and diphenyl-urea are produced:

$$R_2C(OH)CH_2CH_3 \longrightarrow R_2C = CHCH_3 + H_2O$$

$$PhNCO + H_2O \longrightarrow CO_2 + PhNH_2 \xrightarrow{PhNCO} (PhNH)_2CO$$

Ethyl carbamate, usually known as *urethan*, is a crystalline solid, m.p. 50°C. It reacts with ammonia to form urea:,

$$NH_2CO_2C_2H_5 + NH \longrightarrow CO(N\ H_2)_2 + C_2H_5OH$$

It is decomposed by aqueous sodium hydroxide on warming:

$$NH_2CO_2C_2H_5 + {}_2NaOH \longrightarrow Na_2CO_3 + NH_3 + C_2H_5OH$$

Ethyl carbamate has hypnotic properties; many urethans are valuable hypnotics, *e.g.*, aponal (t-amyl carbamate, t-amylurethan), $NH_2COOC(CH_3)_2(C_2H_5)$.

Urea (*carbamide*) is very important physiologically. It is the chief nitrogenous product of protein metabolism; adults excrete about 30 g per day in the urine, from which it can be extracted by evaporating the urine to small bulk and adding nitric acid, whereupon the slightly soluble urea nitrate, $CO(NH_2)_2 \cdot HNO_3$, is precipitated. Urea is historically very important because Wohler (1828) synthesised it by evaporating a solution containing potassium isocyanate and ammonium sulphate, ammonium isocyanate, which is formed first, undergoes molecular rearrangement:

$$NH_4NCO \rightleftharpoons CO(NH_2)_2$$

The reaction is reversible; the solution contains about 5 per cent ammonium isocyanate.

The mechanism of this rearrangement is uncertain. One possibility is as follows (via dissociation):

$$NH_4NCO \rightleftharpoons NH_3 + HNCO$$

$$H-\ddot{N}=C=O + \ddot{N}H_3 \rightleftharpoons H-\bar{N}-\underset{\underset{{}^{+}NH_3}{|}}{C}=O \rightleftharpoons H_2N-\underset{\underset{NH_2}{|}}{C}=O$$

Urea may be prepared in the laboratory by the action of ammonia on carbonyl chloride, alkyl carbonates, chloroformates or urethans, *e.g.*,

$$COCl_2 + 2NH_3 \longrightarrow CO(NH_2)_2 + {}_2HCl \quad (f)$$

$$(C_2H5O)_2CO + {}_2NH_3 \longrightarrow CO(NH_2)_2 + {}_2C_2H_5OH \quad (f)$$

Industrially, urea is prepared by allowing liquid carbon dioxide and liquid ammonia to interact, and heating the ammonium carbamate so formed to 130-150°C under about 35atm pressure:

$$2NH_3 + CO_2 \longrightarrow NH_2COONH_4 \longrightarrow CO(\ NH_2)_2 + H_2O$$

Structure of urea. Although urea appears to be a simple molecule, it is only recently that its structure has been ascertained with any degree of certainty. The diamide structure, $CO(NH_2)_2$, seemed to be indicated by its synthesis from carbonyl chloride and ammonia:

$$OC\begin{matrix} \diagup Cl \\ \diagdown Cl \end{matrix} + 2NH_3 \longrightarrow OC\begin{matrix} \diagup NH_2 \\ \diagdown NH_2 \end{matrix} + 2HCl$$

This structure, however, did not appear to explain all the reactions of urea and so gave rise to a great deal of controversy. Physical methods of structure determination have now shown that the diamide structure is correct, at least in the solid state. Crystal structure studies have shown that in solid urea both nitrogen atoms are identical. Bond-length measurements in urea give the C—N distance as 1-37Å. In aliphatic amines the C—N bond length is 1-47Å. This indicate; that the C—N bond in urea has some double bond character (about 28 per cent): this can be explained by resonance:

$$\begin{matrix} H_2N & & NH_2 \\ & \diagdown\ \diagup & \\ & C & \\ & \| & \\ & O & \end{matrix} \longleftrightarrow \begin{matrix} H_2\overset{+}{N} & & NH_2 \\ & \diagdown\!\!\diagdown\ \diagup & \\ & C & \\ & \| & \\ & O^- & \end{matrix} \longleftrightarrow \begin{matrix} H_2N & & \overset{+}{N}H_2 \\ & \diagdown\ \diagup\!\!\diagup & \\ & C & \\ & \| & \\ & O^- & \end{matrix}$$

Both nitrogen atoms are identical in the hybrid molecule. Furthermore, the negatively charge oxygen atom is capable of co-ordinating with one proton (therefore urea will be a 'monoacidic' base), and thus the salt may be formulated as a resonance hybrid:

$$H_2\overset{+}{N}{=}C(NH_2)OH \longleftrightarrow H_2N{-}C(OH){=}\overset{+}{N}H_2 \longleftrightarrow H_2N{-}C(NH_2){=}\overset{+}{O}H$$

NMR spectra of crystalline salts of urea (and thiourea) are consistent with protonation at the oxygen atom (or sulphur atom; Redpath *et al.*, 1962).

Properties and Reactions of Urea

Urea is a white crystalline solid, m.p. 132°C, soluble in water and ethanol, but insoluble in ether. It is used for preparing formaldehyde-urea plastics, barbiturates, as a fertiliser, etc. A recent use is for the manufacture of hydrazine; urea is treated with alkaline sodium hypochlorite (in effect, the Hofmann preparation of primary amines is applied to urea):

$$NH_2CONH_2 + NaOCl + 2NaOH \longrightarrow N_2H_4 + NaCl + Na_2CO_3 + H_2O$$

(i) Urea behaves as a 'monoacidic' base; the nitrate and oxalate are the most important, since neither is very soluble in water.

When urea nitrate is added to cold concentrated sulphuric acid, nitrourea is formed:

(ii) Urea is hydrolysed by boiling with acids or alkalis:

$$CO(NH_2)_2 + H_2O \longrightarrow CO_2 + 2NH_3$$

The enzyme urease (which occurs in soyabeans) produces the same change.

(iii) When gently heated, urea loses ammonia to form biuret:

$$(NH_2)_2\,CO \rightleftharpoons NH_3 + HNCO \xrightarrow{(NH_2)_2CO} NH_2CONHCONH_2$$

When an aqueous biuret solution is treated with sodium hydroxide solution and a drop of copper sulphate solution, a violet colour is produced. This is known as the *biuret reaction*, which is characteristic of all compounds containing the grouping —CONH—, *e.g.*, proteins.

When heated rapidly, urea evolves ammonia and forms cyanic acid which rapidly polymerises to cyanuric acid. When refluxed with alcohols, urea forms urethans (see above).

(iv) Nitrous acid reacts with urea with the liberation of nitrogen which, however, is not evolved quantitatively:

$$CO(NH_2)_2 + 2HNO_2 \longrightarrow CO_2 + 3H_2O + 2N_2$$

Nitrogen is also evolved, again not quantitatively, when urea is treated with excess of alkaline hypobromite:

(v) Acid chlorides and acid anhydrides react with urea to form weides, *e.g.*, acetyl chloride forms acetylurea:

$$CH_3COCl + NH_2CONH_2 \longrightarrow CH_3CONHCONH_2 + CHI$$

Many of these ureides are useful drugs, particularly when the acid group has a branched chain, *e.g.*, *bromural* (α-bromoisovalerylurea), $(CH_3)_2CHCHBrCONHCONH_2$.

Dicarboxylic acids react with urea in the presence of phosphoryl chloride to form cyclic ureides; *e.g.*, oxalic acid forms parabanic acid (oxalylurea):

$$\begin{matrix} CO_2H \\ | \\ CO_2H \end{matrix} + CO(NH_2)_2 \xrightarrow{POCl_3} + 2H_2O + \begin{matrix} CO\text{—}NH \\ | \\ CO\text{—}NH \end{matrix}\! \begin{matrix} \diagdown \\ \quad CO \\ \diagup \end{matrix}$$

Cyclic ureides may also be prepared by refluxing a di-ester with urea in ethanolic solution containing sodium ethoxide; *e.g.*, malonic ester forms barbituric acid (malonylurea):

$$CH_2\begin{matrix} \diagup CO_2C_2H_5 \\ \\ \diagdown CO_2C_2H_5 \end{matrix} + \begin{matrix} H_2N\diagdown \\ \\ H_2N\diagup \end{matrix} CO \xrightarrow{C_2H_5ONa} \begin{matrix} & O & \\ & \| & \\ HN^{1} & {}^{6} & {}^{5} \\ {}^{2} & {}^{3} & {}^{4} \\ O\!\!=\; & N & \;\!\!=O \\ & H & \end{matrix} + 2C_2H_5OH \quad (72\text{–}78\%)$$

Fig. 4.10

Barbituric acid and its 5- or 5,5-derivatives are used in medicine as hypnotics and sedatives, *e.g.*, barbitone (5,5-diethylbarbituric acid), and *phenobarbitone* (5-phenyl-5-ethylbarbituric acid).

Inclusion Complexes

Several types of inclusion complexes are known, and in all of them molecules of one component, the 'guest' molecules, are physically imprisoned in the cavities of the crystalline structure of the second component, the 'host' molecules. Two important types of inclusion complexes are the *channel* (*canal*) *complexes* and the *clathrate* (*cage*) *complexes* (for a discussion of the latter).

Channel complexes are those cases where the host crystallises in a form with parallel, approximately cylindrical channels in which molecules of the guest are enclosed lengthwise.

Urea normally forms a crystal structure which is closely packed, but in the presence of various *straight-chain* molecules, *e.g.*, n-alkanes, n-alcohols, n-acids, n-esters, etc., urea crystallises in a more open structure which contains long channels enclosing the guest molecules. These channels contain a number of guest molecules and, in general, the number is inversely proportional to the length of the guest molecule. Branched-chain and cyclic structures cannot 'fit' into these channels, and so this property affords a means of separating straight-chain from branched-chain compounds (but see below). The complexes are decomposed by melting or by dissolving away the urea with water. The formula of the channel complexes is usually A_nB (A is urea) where n, usually not a whole number, has values of 4 or more and increases as the length of B increases. Furthermore, since channel complexes are characterised by the fact that guest molecules bear some structural resemblance to each other, it is not possible to separate homologues by channel complexes. For each homologous series there is a certain minimum chain-length for channel complex formation, *e.g.*, six carbons for alkanes, seven for alcohols, five for acids, etc.

Channel complex formation has been used to resolve racemic modifications, and also to separate geometrical isomers. It is also interesting to note that if a molecule is sufficiently long, then 'small' side-chains will not prevent such a molecule behaving as a guest, *e.g.*, Schiessler *et al.* (1957) have shown that 2-methyldecane forms a channel complex with urea.

X-ray analysis of these channel complexes has shown that the diameter of the channel is about 5 Å (Smith, 1952; see also thiourea).

Substituted ureas may be prepared by a reaction similar to Wohler's synthesis of urea; the hydrochloride or sulphate of a primary or secondary amine is heated with potassium isocyanate, *e.g.*, methylamine hydrochloride (I molecule) forms methylurea:

$$CH_3NH_2{\cdot}HCl + KNCO \longrightarrow KCl + CH_3NH_2{\cdot}HNCO \longrightarrow CH_3NHCONH_2$$

If excess of amine salt is used, N, N′-disubstituted ureas are obtained, *e.g.*, excess of aniline hydrochloride forms N, N′-diphenylurea::

$$C_6H_5NH_2{\cdot}HCl + KNCO \longrightarrow KCl + C_6H_5NH_2{\cdot}HNCO \longrightarrow$$

$$C_6H_5NHCONH_2 \xrightarrow{C_6H_5NH_2{\cdot}HCl} C_6H_5NHCONHC_6H_5 + NH_4Cl$$

Alternatively, N,N′-diphenylurea may be obtained by refluxing an aqueous solution of aniline hydrochloride with urea; the urea produces ammonia and isocyanic acid, and the reaction proceeds as shown above. Both phenylurea (38-40 per cent) and N,N′-diphenylurea (52-55 per cent) may be isolated.

N,N′-Substituted ureas may be obtained by the action of carbonyl chloride on a primary or secondary amine:

$$COCl_2 + 2RNH_2 \longrightarrow CO(NHR)_2 + 2HCl$$

The reaction between phenyl isocyanate and a primary or secondary amine also produces an 5-disubstituted urea, *e.g.*, with ethylamine, N-ethyl-N'-phenylurea is formed:

$$C_6H_5NCO + C_2H_5NH_2 \longrightarrow C_6H_5NHCONHC_2H_5$$

This reaction is used to characterise amines.

A very convenient method of preparing alkylureas is to evaporate an aqueous or ethanolic solution of an amine and nitrourea (see above); the latter decomposes into isocyanic acid and nitramide, NH_2NO_2, which readily decomposes into nitrous oxide:

$$NH_2CONHNO_2 \longrightarrow NHCO + NH_2NO_2$$

$$HNCO + R_2NH \longrightarrow R_2NCONH_2$$

$$NH_2NO_2 \longrightarrow N_2O + H_2O$$

When urea is treated with methyl sulphate in faintly alkaline solution, methylisourea is obtained:

$$NH_2CONH_2 \xrightarrow{Me_2SO_4} NH = C(OMe)NH_2$$

COMPOUNDS RELATED TO UREA

Semicarbazide (*aminourea*), m.p. 96°C, may be prepared by treating hydrazine sulphate with potassium cyanate:

$$H_2NNH_2{\cdot}HNCO \longrightarrow NH_2CONHNH_2$$

It is manufactured by heating urea with hydrazine hydrate.

$$NH_2CONH_2 + N_2H_4 \cdot H_2O \longrightarrow NH_2CONHNH_2 + NH_3 + H_2O$$

A more recent method is the electrolytic reduction of nitrourea in sulphuric acid solution using a lead anode:

$$NH_2CONHNO_2 \xrightarrow{e;H^+} NH_2CONHNH_2 + 2H_2O \ (61\text{–}69\%)$$

Semicarbazide is an important reagent for the identification of carbonyl compounds, with which it forms *semicarbazones*; it is also used in the Wolff-Kishner reduction.

Guanidine (*ammomethanamidine*), m.p. 50° C, is found in beet juice, and is one of the degradation products of the purines. It may be prepared by heating cyanamide with ammonium chloride:

$$NH_2CN + NH_4Cl \longrightarrow (NH_2)_2C = NH \cdot HCl$$

Guanidine is a *strong* '*monoacid*' base, even forming a carbonate. Its strength as a base may be explained by resonance. There are unlike charges in the neutral molecule, whereas the ion has only one positive charge which, is spread, and at the same time all the resonating structures are equivalent (cf. urea).

Neutral molecule

$$(H_2N)(H_2N)C{=}NH \longleftrightarrow (H_2N^+)(H_2N)C{=}NH^- \longleftrightarrow (H_2N)(H_2N^+)C{=}NH^-$$

Ion

$$(H_2N)(H_2N)C{=}NH_2^+ \longleftrightarrow (H_2N^+)(H_2N)C{=}NH_2 \longleftrightarrow (H_2N)(H_2N^+)C{=}NH_2$$

X-ray analysis of guanidinium iodide shows that the three nitrogen atoms are symmetrically placed round the carbon atom. Furthermore, the C—N distance has been found to be 1.38Å (Theilacker, 1935; cf. urea). These facts are in keeping with the assumption that the guanidinium ion is a resonance hybrid.

Careful hydrolysis with barium hydroxide solution converts guaridine into urea:

$$(NH_2)_2C = NH + H_2O \longrightarrow (NH_2)_2C = O + NH_3$$

Guanidine nitrate may be prepared by heating dicyandiamide with ammonium nitrate:

$$(NH_2)_2C = NCN + 2NH_4NO_3 \longrightarrow 2(NH_2)_2C = NH{\cdot}HNO_3\ (85\%)$$

When treated with concentrated sulphuric acid, guanidine nitrate is converted into nitroguanidine, $(NH_2)_2C = NNO_2$ (cf. urea nitrate). Nitroguanidine is used for making flashless powders.

Thiourea (*thiocarbamide*), m.p. 180°C, may be prepared by heating ammonium thiocyanate at 170°C for some time:

The structure of thiourea is similar to that of urea; it behaves as a 'monoacidic' base and, like urea, forms channel complexes. However, in this case, the diameter of the channel is about 7Å, and so can accommodate larger guest molecules (than can urea).

Thus, thiourea readily encloses branched-chain Carbohydrates and Proteins and even cycloalkanes. Apparently the cross-section of straight-chain alkanes is too small to support the channels in thiourea and consequently these compounds do not form channel complexes. When heated with alkalis, thiourea is hydrolysed:

$$S = C(NH_2)_2 + 2H_2O \xrightarrow{NaOH} CO_2 + H_2S + 2NH_3$$

Oxidation with *alkaline* permanganate or alkaline hydrogen peroxide converts thiourea into urea:

$$S = C(NH_2)_2 + [O] \longrightarrow O = C\ (NH_2)_2 + S$$

Oxides of lead, silver or mercury remove a molecule of hydrogen sulphide from thiourea into room temperature to form cyanamide:

$$(NH_2)_2CS + HGO \longrightarrow NH_2CN + HgS + H_2O$$

When treated with alkyl halide, thiourea forms S-*alkyl*-ψ-thiouronium salts (S-*alkyliso- thiouronium salts*):

$$CH_3I + S{=}C(NH_2)_2 \longrightarrow \left[CH_3SC\begin{matrix}\nearrow\!\!\!\!\nearrow NH_2 \\ \searrow NH_2\end{matrix}\right] I^-$$

These compounds are used to characterise sulphonic acids, with which they form insoluble salts. They may also be used to prepare thioalcohols, and also to prepare sulphonyl chlorides by oxidation with chlorine-water.

$$RSC(=\overset{+}{N}H_2Cl^-)NH_2 \xrightarrow[H_2O]{Cl_2} RSO_2Cl$$

s-*Diphenyllhiourea* or *diphenylthiocarbanilide*, $(C_6H_5NH)_2CS$, which is used as a rubber accelerator, may be prepared by heating aniline with carbon disulphide in an ethanolic solution containing potassium hydroxide.

Dithiocarbamic acid, $S = C(NH_2)SH$, although unstable in the free state, gives rise to stable salts. These may be prepared by heating a primary or secondary mine with carbon disulphide. This reaction is complicated and the mechanism is uncertain.

A possibility is via the formation of the unstable N-alkyldithioca basic acid which forms the stable dithiocarbamic acid salt with another molecule of amine:

$$R\ddot{N}H_2 \quad S{=}C{=}S \longrightarrow \left[R{-}\overset{H}{\underset{H}{N^+}}{-}C(S^-){=}S\right]$$

$$\xrightarrow{RNH_2} \left[RNH{-}C(S^-){=}S\right] RNH_3^+$$

The Hofmam mustard oil reaction is believed to take place via the formation of a dithio-carbamic acid salt, which is then decomposed by the mercuric chloride, possibly as follows:

$$\left[RNH{-}C(S^-){=}S \quad RNH_3^+\right]$$

$$\xrightarrow{HgCl_2} RNH_3^+Cl^- + R{-}\overset{H}{N}{-}C(S{-}Hg{-}Cl){=}S$$

$$\longrightarrow RN{=}C{=}S + HgS + HCl$$

SATURATED MONOCARBOXYLIC ACIDS AND THEIR DERIVATIVES

Introduction : The saturated monocarboxylic acids are also referred to as the fatty acids, a name which arose from the fact that some of the higher members, particularly palmitic and stearic acids, occur in natural fats. Their general formula is $C_nH_{2n}O_2$ but, because their functional group is the carboxyl group, $—CO_2H$, they are more conveniently expressed as $C_nH_{2n+1}CO_2H$, since these show the nature of the functional group. The structure of the carboxyl group is written as $—C\begin{matrix}{}^{\nearrow O}\\{}_{\searrow OH}\end{matrix}$ (or —COOH), but, as we shall later, it does not represent accurately the behaviour of this group. Also, since only the carboxyl group hydrogen atom is replaceable by a metal, the monocarboxylic acids are *monobasic.*

NOMENCLATURE

The monocarboxylic acids are commonly known by the trivial names, which have been derived from the source of the particular acid; *e.g.*, formic acid, HCO_2H, was so named because it was first obtained by the distillation of ants; the Latin word for ant is *formica.* Acetic acid, CH_3CO_2H, is the chief constituent of vinegar, the Latin word for which is *acetum*; etc.

Another system of nomenclature considers the acids, except formic acid, as derivatives of acetic acid, *e.g.*,

$CH_3CH_2CO_2H$	methylacetic acid
$(CH_3)_3CCO_2H$	trimethylacetic acid
$(CH_3)_2CHCH_2CO_2H$	isopropylacetic acid

In the above two systems the positions of substituents in the chain are indicated by Greek letters, the α-carbon atom being the one joined to the caboxyl group, *e.g.*,

$CH_3CH(OH)CH_2CH_2CO_2H$	γ-hydroxyvaleric acid
$CH_3CHClCH(CH_3)CO_2H$	β-chloro α-methylbutyric acid

According to the I.U.P.A.C. system of nomenclature, the suffix of the monocarboxylic acids is *-oic*, which is added to the name of the alkane corresponding to the longest carbon chain containing the carboxyl group, *e.g.*,

HCO_2H methanoic acid

$CH_3CH_2CO_2H$ propanoic acid

The positions of side-chains (or substituents) are indicated by numbers, *the carboxyl group always being given number* 1:

$CH_3CH(CH_3)CH(CH_3)CH_2CO_2H$ 3,4-dimethylpentanoic acid

Alternatively, the carboxyl group is regarded as a substituent, and is denoted by the suffix carboxylic acid, *e.g.*,

$\overset{4}{C}H_3\overset{3}{C}H_2\overset{2}{C}H(CH_3)\overset{1}{C}H_2CO_2H$ 2-methylbutane-1-carboxylic acid

General Methods of Preparation

1. Oxidation of alcohols, aldehydes or ketones with acid dichromate yields acids:

$$RCH_2OH \xrightarrow{[O]} RCHO \xrightarrow{[O]} RCO_2H \ (g.\text{–}v.g.)$$

$$R_2CHOH \xrightarrow{[O]} R_2CO \xrightarrow{[O]} R^1CO^2H + R^2CO_2H \ (g.)$$

Many *primary* alcohols, on oxidation with acid dichromate, form esters in addition to acids, *e.g.*, n-butanol gives n-butyric acid and n-butyl n-butyrate (41–47 per cent). Several mechanisms have been proposed, but the generally accepted one involves hemiacetal formation.

$$RCH_2OH \xrightarrow{[O]} RCHO \xrightarrow{RCH_2OH} RCH(OH)OCH_2R \xrightarrow{[O]} RCO_2CH_2R$$

2. A very good synthetic method is the hydrolysis of cyanides with or alkali:

$$RC\equiv N \xrightarrow{H_2O} \left[R\text{—}C\begin{matrix}{\nearrow OH} \\ {\searrow NH}\end{matrix} \right] \longrightarrow R\text{—}C\begin{matrix}{\nearrow O} \\ {\searrow NH_2}\end{matrix} \xrightarrow{H_2O} RCO_2H + NH_3 \quad (g.\text{-}r.g.)$$

The amide, $RCONH_2$, may be isolated if suitable precautions are taken.

3. By the reaction between a Grignard reagent and carbon dioxide:

$$RMgX + CO_2 \longrightarrow RCO_2MgX \xrightarrow{H_2O} RCO_2H$$

4. Many monocarboxylic acids may be conveniently synthesised from alkyl halides and ethyl malonate or acetoacetic ester. These syntheses will be discussed in detail later.

5. A number of higher acids may be obtained by the hydrolysis of natural fats or may be prehalf-ester of a dicarboxylic acid (Linstead *et al.*, 1950; *cf.*).

$$RCO_2H + HO_2C(CH_2)_nCO_2CH_3 \longrightarrow H_2 + 2CO_2 + R(CH_2)_nCO_2CH_3$$

The chain of a monocarboxylic acid can be extended by five or six carbon atoms in a synthesis involving enamines.

6. Monocarboxylic acids may be obtained by heating a sodium alkoxide with carbon monoxide under pressure.

$$R\overset{-}{O}\overset{+}{Na} + CO \longrightarrow RCO_2^-\overset{+}{Na}$$

Acids may also be obtained by heating the alcohol with carbon monoxide under pressure in the presence of a catalyst.

The sodium salts of acetic and propionic acids are formed by the interaction of methylsodium or ethylsodium, respectively, and carbon dioxide, *e.g.*,

$$C_2\overline{H}_5\overset{+}{Na} + CO_2 \longrightarrow C_2H_5CO_2Na$$

Monocarboxylic acids may be prepared by heating an alkene with carbon monoxide and steam under pressure at 300 – 400°C in the presence of a catalyst, *e.g.*, phosphoric acid:

$$CH_2{=}CH_2 + CO + H_2O \longrightarrow CH_3CH_2CO_2H$$

$$CH_3CH{=}CH_2 + CO + H_2O \longrightarrow (CH_2)_2CHCO_2H$$

General Properties of the Monocarboxylic Acids

The first three acids are colourless, pungent-smelling liquids; the acids from butyric, $C_4H_8O_2$, to nonanoic, $C_9H_{18}O_2$, are oils which smell like goats' butter; and those higher than decanoic acid, $C_{10}H_{20}O_2$, are odourless solids. The lower members are far less volatile than is to be expected from their molecular weight. This can be explained by hydrogen bonding, and electron diffraction studies (Pauling, 1934) have shown that an eight-membered ring is present, *i.e.*, the acids exist as cyclic dimers. The lower members exist as dimers in the vapour phase and in aqueous solution, but in the liquid phase they exist as 'polymers'.

```
      O···H—O
     //       \
R—C           C—R
     \        //
      O—H···O
```

The melting point of the n-monocarboyxlic acids show alternation or oscillation from one member to the next, the melting point of an 'even' acid being higher than that of the 'odd' acid immediately below and above it in the series (The physical constants of the individuals). A number of homologous series follow this oscillation or 'saw-tooth' rule.

The first four members are very soluble in water, and the solubility decreases as the molecular weight increases. This solubility is due to the acids being capable of forming hydrogen bonds with water.

The infra-red absorption region of the carboxyl group contains a number of bands. The carbonyl group (stretch) absorbs in the same region as that of acyclic ketones. 1725–1700 cm^{-1}(s), but the (free) hydroxyl group (stretch) absorbs at 3560–3500 cm^{-1}(m), which is in the lower frequency region than for a alcohols (3670–3580 cm^{-1}).

Fig. 4.11 is the infra-red spectrum of stearic acid (as a 1 per cent solution in carbon tetrachloride). It should be noted that the band for the bonded O—H (str.) is hidden by the strong C—H (str.))

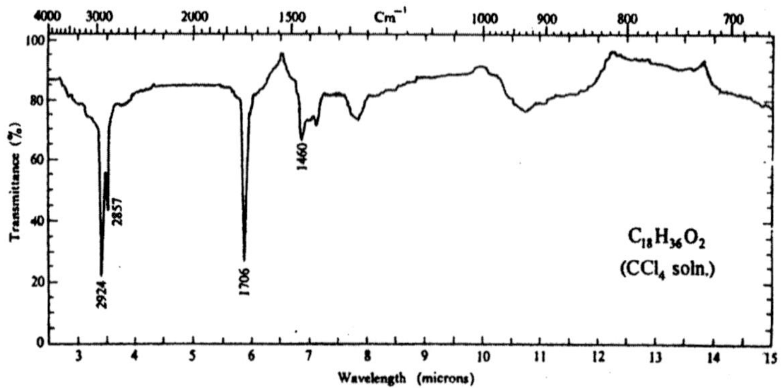

Fig. 4.11

1706 cm^{-1}: C=O (str.) in (saturated ketones and) saturated aliphatic acids. 2942 and 2857 cm^{-1}: C—H (str.) in Me and/or CH_2. 1460 cm^{-1}: C—H (def.) in Me and/or CH_2.

On the other hand Fig. 4.12 is the spectrum of stearic acid as a Nujol mull. In this case, the stearic acid is in the solid (dispersed) state, and it can be seen that there is an extremely large difference in the region

1300–1180 cm^{-1}. In this region there are eight maxima, and this pattern is characteristic of the spectra of long-chain n-alkyl compounds (in the solid state), the number of peaks depending on the chain length Nujol (a petroleum fraction), since it is a hydrocarbon mixture, has the bands characteristic of alkanes. The O—H (str.) is hidden (Fig. 4.11). These two spectra illustrate the influence of the physical state of a compound on its infra-red spectrum.

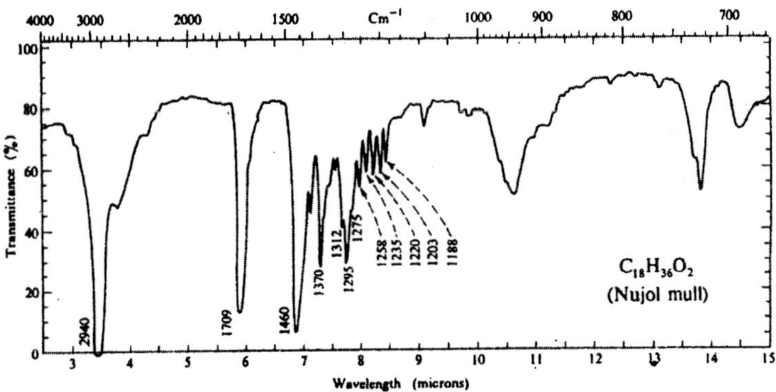

Fig. 4.12

2940, 1460 and 1370 cm^{-1}: Bands for Nujol. 1709 cm^{-1}: C═O (str.) as for Fig. 4.12. 1312, 1295, 1279, 1258, 1235, 1220, 1203 and 1188 cm^{-1}: These bands are characteristic of long-chain n-alkyl compounds (in the solid state).

Because of hydrogen bonding (intermolecular) and exchange reactions, the proton peak of the carboxyl group can be found in many different positions in the NMR spectrum. This applies to hydroxyl protons in general. However, it has been found that in non-polar solvents, the τ-value of the carboxyl proton in aliphatic acids is around – 1.0 p.p.m. (*i.e.*, there is a great shift to lower field due to strong hydrogen bonding).

Mass spectra. Acids, esters, and amides undergo similar fragmentation patterns, *e.g.*, α-cleavage for lower members (Z═OH, OR, NH_2):

$$\mathrm{R\cdot} + \overset{+}{\mathrm{O}}\equiv\mathrm{C}-\ddot{\mathrm{Z}} \longleftrightarrow \mathrm{O}=\mathrm{C}=\overset{+}{\mathrm{Z}} \longleftarrow \left[\mathrm{R}-\overset{\overset{\displaystyle\mathrm{O}}{\|}}{\mathrm{C}}-\mathrm{Z}\right]^{+} \longrightarrow$$

$$m/e = 28 + Z$$

$$[\mathrm{R}-\mathrm{CO}]^{+} + \mathrm{Z}\cdot$$

$$\mathrm{M}-\mathrm{Z}$$

Also, R and Z may carry the positive charge. Hence, peaks are obtained at $M-17$ and *m/e* 45 (Z = OH), $M-31$, M–45, and *m/e* 59, 73 (Z = OMe and OEt, respectively), and $M-16$ and *m/e* 44 (Z = NH_2). Since they are more volatile than their corresponding acids, esters are used preferentially. All can undergo the McLafferty rearrangement if they contain a γ-hydrogen atom:

$$\begin{matrix} & H & \\ CH_2 & & \overset{\cdot +}{O} \\ | & & \| \\ CH_2 & & CZ \\ & CH_2 & \end{matrix} \longrightarrow \begin{matrix} CH_2 \\ \| \\ CH_2 \end{matrix} + \left[\begin{matrix} HO \\ \quad C—Z \\ \| \\ CH_2 \end{matrix}\right]^{\dot{+}}$$

m/e 60 (Z = OH):
74 (Z = OMe);
88 (Z = OEt);
59 (Z = NH_2)

When the alkyl group in O of esters is higher than methyl, esters can also undergo the McLafferty rearrangement to give alkene and acid, *e.g.*,

$$\left[\begin{matrix} & H & \\ CH_3CH & & O \\ | & & \| \\ CH_2 & & CR \\ & O & \end{matrix}\right] \longrightarrow \left[\begin{matrix} CH_3CH \\ \| \\ CH_2 \end{matrix}\right]^{+} + RCO_2H$$

General Reactions of the Monocarboxylic Acids

1. The acids are acted upon by the strongly electropositive metals with the liberation of hydrogen and formation of a salt:

$$RCO_2H + Na \longrightarrow RCO_2^-Na^+ + \frac{1}{2}H_2$$

Salts are also formed by the reaction between an acid and an alkali:

$$RCO_2H + NaOH \longrightarrow RCO_2^-Na^+ + H_2O$$

2. Monocarboxylic acids react with alcohols to form *esters*:

$$R^1CO_2H + R^2OH \rightleftharpoons R^1CO_2R^2 + H_2O$$

3. Phosphorus trichloride, phosphorus pentachloride or thionyl chloride act upon the monocarboxylic acids to from *acid chlorides*:

$$\begin{aligned} 3RCO_2H + PCl_3 &\longrightarrow 3RCOCl + H_3PO_3 \\ RCO_2H + PCl_5 &\longrightarrow RCOCl + HCl + POCl_3 \\ RCO_2H + SOCl_2 &\longrightarrow RCOCl + HCl + SO_3 \end{aligned}$$

The acid series may be 'stepped up' via the acid chloride, but since the process involves many steps, the yield of the higher acid homologue is usually only *f.-f.g.*:

(*a*) $RCO_2H \xrightarrow{PCl_5} RCOCl \xrightarrow[\text{catalyst}]{H_2} RCHO \xrightarrow[\text{catalyst}]{H_2}$

$RCH_2OH \xrightarrow{PBr_3} RCH_2Br \xrightarrow{KCN} RCH_2CN \xrightarrow{H^+} RCH_2CO_2H$

(*b*) $RCO_2H \xrightarrow{PCl_5} RCOCl \xrightarrow{KCN} RCOCN$

$\xrightarrow{H^+} RCOCO_2H \xrightarrow[\text{reduction}]{\text{Clemmensen}} RCH_2CO_2H$

Although method (*b*) involves fewer steps than (*a*), the final yield of acid by route (*a*) is higher than that by (*b*), since the yields in each step of (*b*) are generally low. Furthermore, the yield from (*a*) can be improved by direct reduction of acid to alcohol by means of various metallic hydrides.

4. When the ammonium salts of the monocarboxylic acids are strongly heated, the *acid amide* is formed by the elimination of water:

$$RCO_2NH_4 \longrightarrow RCONH_2 + H_2O$$

5. When the anhydrous sodium salt of a monocarboxylic acid is heated with soda-lime, an alkane and other products are formed.

The mechanism of this decarboxylation is uncertain, but there is much evidence to show that *salts* decompose by an S_E1 reaction :

$$R{-}C(=O)O^- \longrightarrow \bar{R}: + O{=}C{=}O \xrightarrow{H^+} R{-}H$$

The thermal decarboxylation of *free* acids may be:

$$R{-}C(=O){-}O{-}H \longrightarrow \overset{+}{H} + R{-}C(=O)O^- \longrightarrow \overset{+}{H} + \bar{R}: + CO_2 \longrightarrow R{-}H + CO_2$$

Monocarboxylic acids may be 'stepped down' by heating the calcium salt with calcium acetate, and the methyl ketone produced is oxidised with acid dichromate:

$$RCH_2CO_2Ca/_2 + CH_3CO_2Ca/_2 \longrightarrow RCH_2COCH_3$$

$$\xrightarrow{[O]} RCO_2H + CH_3CO_2H$$

It should be noted that the second step is based on the general rule that when an unsymmetrical ketone is oxidised, the carbonyl group remains chiefly with the smaller alkyl group. Since monocarboxylic acids containing an odd number of carbon atoms occur only in small amounts, this method of descending the series affords a useful means of obtaining 'odd' acids from 'even'.

When a mixture of the calcium salt (or any of the others mentioned above) and calcium formate is heated, an aldehyde is obtained. The pyrolysis of iron salts gives very good yields of symmetrical ketones. The pyrolysis of iron slats gives very good yields of symmetrical ketones.

Metallic salts, particularly the silver slats, are converted by bromine into the alkyl bromide. This reaction may also be used to 'step down' the acid series.

6. When a concentrated aqueous solution of the sodium or potassium salt of a monocarboxylic acid is electrolysed, an alkane is obtained.

$$2RCO_2K + 2H_2O \longrightarrow R{-}R + 2CO_2 + 2KOH + H_2$$

7. Monocarboxylic acids react slowly with chlorine or bromine in the cold, but at higher temperatures and in the presence of a small amount of phosphorus, reaction proceeds smoothly to give a-halogeno-acids; *e.g.*,

$$RCH_2CO_2H \xrightarrow{P/Br_2} RCHBrCO_2H \xrightarrow{P/Br_2} RCBr_2CO_2H$$

8. All the acids, except formic acids, are extremely resistant to oxidation, but prolonged heating with oxidising agents ultimately produces carbon dioxide and water.

9. The reduction products of monocarboxylic acids depend on the nature of the reducing agent. Thus, heating with hydrogen iodide-red phosphorous under pressure, or with hydrogen under pressure at elevated temperature in the presence of a nickel catalyst, produces an alkane.

$$RCO_2H + 3H_2 \xrightarrow{Ni} RCH_3 + 2H_2O$$

If the catalyst is Ru—C, the alcohol is formed. Alcohols are also produced by reduction with various metallic hydrides.

10. The acids undergo the *Schmidt reaction* to form a primary amine:

$$RCO_2H + HN_3 \xrightarrow{H_2SO_4} RNH_2 + CO_2 + N_2$$

Schmidt's reaction with acids is a modification of the Curtius reaction. The following mechanism has been proposed :

$$\text{R—C(=O)—OH} \xrightarrow{H^+} \text{R—}\overset{+}{\text{C}}\text{(OH)}_2 \xrightarrow{HN_3} \text{R—C(OH)}_2\text{—NH—}\overset{+}{\text{N}}\equiv\text{N} \xrightarrow{-H_2O} \text{R—C(OH)—NH—}\overset{+}{\text{N}}\equiv\text{N} \xrightarrow{-N_2}$$

$$\text{O=C(R)—}\overset{+}{\text{N}}\text{—H} \xrightarrow{-H^+} \text{O=C=NR} \xrightarrow{H_2O} RNH_2 + CO_2$$

The rearrangement (1,2-shift) has been shown to be intramolecular, *e.g.*, Keynon *et al.* (1939) showed retention of optical activity when α-phenylpropionic acid underwent the Schmidt reaction:

$$C_6H_5CH(CH_3)CO_2H \longrightarrow C_6H_5CH(CH_3)NH_2$$

Formic acid (*methanoic acid*) is prepared industrially by heating sodium hydroxide with carbon monoxide at 210°C, and at a pressure of 6–10 atmospheres:

$$NaOH + CO \longrightarrow HCO_2Na$$

Anhydrous formic acid is obtained from the aqueous solution (70–77 per cent) by the addition of butyl formate followed by distillation. The first fraction is an azeotrope of ester and water, and then the excess of ester is removed from the formic acid by fractionation.

The most convenient laboratory preparation of formic acid is to heat glycerol with oxalic acid at 100–100 C. Glyceryl monoxalate is produced, and decomposes into glyceryl monoformate (monoformin) and carbon dioxide. When the evolution of carbon dioxide ceases, more oxalic acid is added, whereupon formic acid is produced:

$$\begin{array}{l}CH_2OH\\ |\\ CHOH\\ |\\ CH_2OH\end{array} + \begin{array}{l}COOH\\ |\\ COOH\end{array} \longrightarrow \begin{array}{l}CH_2OCOCOOH\\ |\\ CHOH\\ |\\ CH_2OH\end{array} + H_2O \xrightarrow{100\text{-}110°C}$$

$$CO_2 + \begin{array}{l}CH_2OCHO\\ |\\ CHOH\\ |\\ CH_2OH\end{array} \xrightarrow{(COOH)_2} \begin{array}{l}CH_2OCOCOOH\\ |\\ CHOH\\ |\\ CH_2OH\end{array} + HCO_2H \xrightarrow{100\text{-}110°C}$$

The distillate contains formic acid and water. The aqueous formic acid solution cannot be fractionated to give anhydrous formic acid because the boiling point of the acid is 100.5°C. The procedure adopted is to neutralise the aqueous acid solution with lead carbonate, and concentrate the solution until lead formate crystallises out. The precipitate is then recrystallised, dried, and heated at 100°C in a current of hydrogen sulphide:

$$(HCO_2)_2Pb + H_2S \longrightarrow 2HCO_2H + PbS$$

The anhydrous formic acid which distils over contains a small amount of hydrogen sulphide, and may be freed from the latter by adding some dry lead formate and redistilling. This procedure for obtaining the anhydrous acid from its aqueous solution can only be used for volatile acids.

Formic acid is a pungent corrosive liquid, m.p. 8.4°C, b.p. 100.5°C, miscible in all proportions with water, ethanol and ether. It forms salts which, except for the lead and silver salts, are readily soluble in water. Formic acid is a stronger acid than any of its homologues.

Formic acid is dehydrated to carbon monoxide by concentrated sulphuric acid and when heated under pressure at 160°C, is decomposed into carbon dioxide and hydrogen:

$$CO + H_2O \xleftarrow{H_2SO_4} HCO_2H \xrightarrow{\text{heat}} CO_2 + H_2$$

The same decomposition takes place at room temperature in the presence of a catalyst such as iridium, rhodium, etc.

When metallic formates are heated with an alkali, hydrogen is evolve.

$$HCO_2Na + NaOH \longrightarrow H_2 + Na_2CO_3$$

When calcium or zinc formate is strongly heated, formaldehyde is produced:

$$(HCO_2)_2Ca \longrightarrow HCHO + CaCO_3$$

When sodium or potassium formate is rapidly heated to [illegible] hydrogen is evolved and the oxalate is formed.

$$2HCO_2Na \longrightarrow (COONa)_2 + H_2$$

Formic acid forms esters, but since it is a relatively strong acid it is not necessary to use a catalyst: refluxing 90 per cent formic acid with the alcohol is usually sufficient.

Formic acid differs from the rest of the members of the monocarboxylic acid series in being a powerful reducing agent; it reduces ammoniacal silver nitrate and the salts of many of the heavy metals, *e.g.*, it converts mercuric chloride into mercurous chloride.

Acetic acid (*ethanoic acid*) is prepared industrially by various methods, *e.g.*, (*i*) the air oxidation of acetaldehyde in acetic acid in the presence of manganous ion as catalyst: (*ii*) the air oxidation of n-butane in the liquid phase under pressure and at 130–230°C in the presence of a suitable catalyst, *e.g.*, manganous ion; (*iii*) by reaction between carbon monoxide and methanol under pressure in the presence of cobalt octacarbonyl at about 210°C:

$$CO + CH_3OH \longrightarrow CH_3CO_2H.$$

One of the earliest methods for preparing acetic acid was by the destructive distillation of wood to give pyroligneous acid. This contains about 10 per cent acetic acid, and was originally treated by neutralising with lime and then distilling off the volatile compounds (these are mainly methanol and acetone). On distillation with dilute sulphuric acid, the residue gives dilute acetic acid. More recently, the acetic acid is extracted by means of solvents, *e.g.*, isopropyl ether.

Vinegar, which is 6–10 per cent aqueous solution of acetic acid, may be made in several ways. Malt vinegar is prepared by the oxidation of wort by means of the bacteria *Mycoderma aceti*:

$$CH_3CH_2OH + O_2 \longrightarrow CH_3CO_2H + H_2O$$

In the 'quick' vinegar process', beech shavings, contained in barrels, are moistened with strong vinegar containing the bacteria. A 10 per cent aqueous solution of ethanol containing phosphates and inorganic salts (which are necessary for the fermentation) is then poured through the barrels, and the ethanol is thereby oxidised to acetic acid. A plentiful supply of air is necessary, otherwise the oxidation is incomplete and acetaldehyde is produced.

Acetic acid is a pungent corrosive liquid, m.p. 16.6°C, b.p. 118°C, miscible in all proportions with water, ethanol ad ether It is stable towards oxidising agents, and so is a useful solvent for chromium trioxide oxidations. Acetic acid is commonly used as a solvent, and in the preparation of acetates, acetone, acetic anhydride, etc.

Strengths of the monocarboxylic acids. If the structure of the carboxyl group were simply (I), then, because of the – I effect of the carbonyl group (as shown in (II), proton release will be facilitated as compared with alcohols.

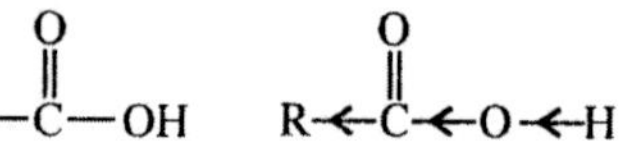

$$\underset{\text{(I)}}{-\overset{\overset{\displaystyle O}{\|}}{C}-OH} \qquad \underset{\text{(II)}}{R \leftarrow \overset{\overset{\displaystyle O}{\|}}{C} \leftarrow O \leftarrow H}$$

Thus carboxylic acids will be stronger acids than alcohols. However, the inductive effect alone cannot account for the very large difference in acid strength. One explanation is that the carboxyl group is a resonance hybrid and because of the positive charge on the oxygen atom of the OH group, proton release is facilitated; this positive charge is absent in alcohols since they are not resonance hybrids.

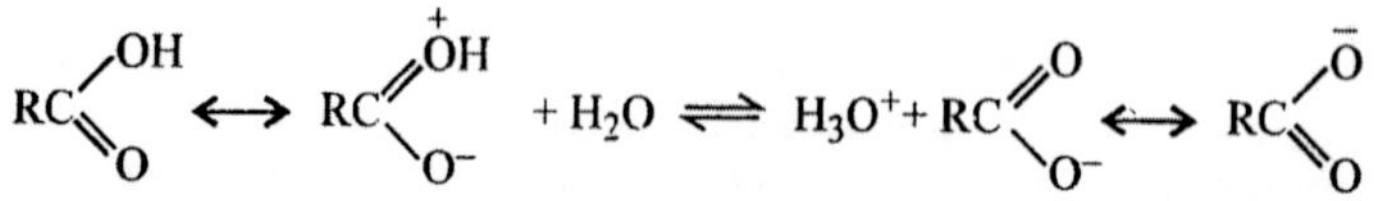

$$RC\overset{OH}{\underset{O}{\diagdown\!\!\!\diagup}} \longleftrightarrow RC\overset{\overset{+}{O}H}{\underset{O^-}{}} + H_2O \rightleftharpoons H_3O^+ + RC\overset{O}{\underset{O^-}{}} \longleftrightarrow RC\overset{\bar{O}}{\underset{O}{}}$$

This argument can be extended as follows. When the acid ionises, *i.e.*, donates its proton to water (which is the solvent), the conjugate base of the acid. *i.e.*, the carboxylate ion, is a resonance hybrid of two *equivalent* resonating structures. At the same time, since the negative charge is spread, the resonance energy of this anion will be greater than that of the un-ionised acid, which is a resonance hybrid of two different resonating structures, one with separation of *unlike* charges. Therefore the internal energy of the anion is lower than that of the un-ionised acid. In the case of alcohols, thee is no resonance stabilisation in the alkoxide ion. It therefore follows that $\Delta G^\ominus$ for the ionisation of monocarboyxlic acids will be more negative than that of alcohols. Hence, the equilibrium constant of the former will be greater than that of the latter, *i.e.*, the pK_a values of the acids will be lower than those of the alcohols. Thus, alcohols are weaker acids than the monocaboxylic acids.

X-ray and electron-diffraction measurements of the carboxylate ion have shown that the two C—O bonds are of equal length. This is in keeping with the ion being a resonance hybrid.

The treatment of the carboxyl group and the carboxylate ion from the M.O. point of view also leads to the same results as those obtained by the V.B. method, the increased stability now being due to *delocalisation*. Let us first consider the *undissociated* carboxylic acid (Fig. 4.13*a*). The carbon atom is linked to the hydroxyl group by a σ-bond, and to the oxygen of the carbonyl group by one σ- and one π-bond. This is shown in (Fig. 4.13*b*). If the oxygen atom of the hydroxyl group has sp^2 hybridisation, this would leave a p orbital perpendicular to the plane of the carboxyl group, and it would contain a pair of electrons (Fig. 4.13*c*).

This orbital could now overlap with both the *p* orbital of the carbon atom and that of the oxygen atom of the carbonyl group. The combination of these three orbitals would give rise to M.O.s embracing *three* nuclei (Fig. 4.13*d*). Since *four* electrons are involved, they must be accommodated, in the ground state, in the lowest two M.O.s. The M.O. with the lowest energy level has *one* node (Fig. 4.13*e*), and the next energy level has *two* nodes (Fig. 4.13*f*).

Table 4.2

Acid	pK_a	*Acid*	pK_a
Formic	3.75	Chloroacetic	2.85
Acetic	4.76	Dichloroacetic	1.25
Propionic	4.87	Trichloroacetic	0.66
n-Butyric	4.82	Trifluoroacetic	0.23
Isobutyric	4.86	Bromoacetic	2.9
n-Valeric	4.86	Iodoacetic	3.16
Isovaleric	4.78	α-Chloropropionic	2.80
Trimethylacetic	5.05	β-Chloropropionic	4.08
Hexanoic	4.88	Nitroacetic	1.68
Heptanoic	4.89	Hydroxyacetic	3.83
Octanoic	4.90	Hydrochloric	– 7.00

Thus (Fig. 4.13*g*) represents the ground state of the undissociated carboxyl group. Because of delocalisation, this group has become more stable. Since a hydroxyl group is more electron-attracting than an oxygen atom, the delocalisation of the pair of electrons in the *p* orbital is not very large, *i.e.*, although this lone pair embraces three nuclei, this pair is far more likely to be found in the region of the donor atom than anywhere else. Since the oxygen atom of the hydroxyl group has lost 'full-control' of this lone pair, it will acquire a small positive charge, and since the oxygen atom of the CO group has acquired a small share of this lone pair, this oxygen atom acquires a small negative charge (actually this oxygen atom already has a small negative charge due to the inductive effect).

The greater electronegativity of oxygen in a hydroxyl group over that of an oxygen atom alone is due to the hydrogen atom sharing one pair of electrons in forming the bond and thereby decreasing, to some extent, the electron density on that oxygen atom.

Now let us consider the *carboxylate ion* (Fig. 4.14*a*). The oxygen atom of the original hydroxyl group retains the σ-electrons when the hydrogen atom is removed as a proton. As in the case of the *carboxy group*, one lone pair enters into conjugation with the π-bond of the C=O group (Fig. 4.14*b*). In the latter case, however, delocalisation is complete, *i.e.*, the lone pair is now likely to be found equally well or either oxygen atom (each atom will have a charge of – 1/2). Since delocalisation is complete, the resonance (delocalisation) energy is much greater than in the case of the undissoicated carboxyl group. Because the carbon atom is sp^2 hybridised in both the carboxyl group and carboxylate ion, the O—C—O angle in both would be expected to be about 120°. This has been observed in practice.

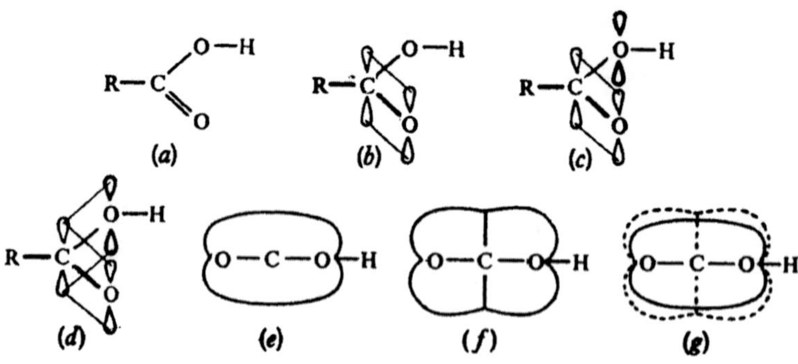

Fig. 4.13

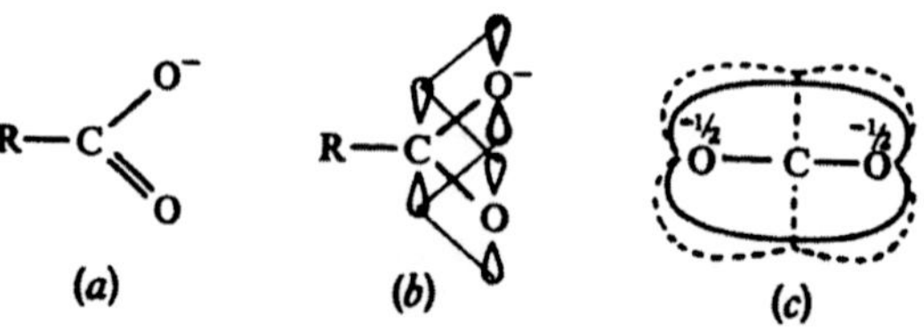

Fig. 4.14

In Figs. 4.13 and 4.14, the orbitals drawn in thicker type indicate that the atoms concerned contain *two* electrons in these orbitals. It should also be noted that lobes and not spheres (with + and – signs) have been used in these diagrams, and that overlapping orbitals have been drawn in an alternative way.

Peracetic (peroxyacetic) acid, $CH_3C\overset{\displaystyle /\!\!/O}{\underset{\displaystyle \diagdown O-OH}{}}$ may be prepared by treating acetic anhydride with concentrated hydrogen peroxide, and then distilling under reduced pressure.

Swern *et al.* (1962) have prepared per-acids (85–97 per cent yield) by treating a solution or slurry of an aliphatic (or aromatic) acid in methanesulphonic acid with 90–95 per cent hydrogen peroxide. This appears to be the best method so far.

Peracetic acid is an unpleasant-smelling liquid, f.p. + 0.1°C, soluble in water, ethanol and ether. It explodes violently when heated above 110°C. It is a powerful oxidising agent; it oxidises the olefinic bond to the oxide.

$$>C=C< + CH_3COO_2H \longrightarrow >C\overset{O}{—}C< + CH_3CO_2H$$

It also oxidises primary aromatic amines to nitroso-compounds, *e.g.*, aniline is converted into nitrosobenzene:

$$C_6H_5NH_2 + 2CH_3COO_2H \longrightarrow C_6H_5NO + 2CH_2CO_2H + H_2O$$

Infra-red absorption spectra measurements of per-acids show that these acids exist in solution very largely in the monomeric, intramolecularly hydrogen-bonded form (*inter alia*, Minkoff, 1954).

$$CH_3C\overset{\displaystyle /\!\!/O\cdots H}{\underset{\displaystyle \diagdown O-O}{}}$$

Propionic acid (*propanoic acid*) is prepared industrially by the oxidation of n-propanol:

$$CH_3CH_2CH_2OH \xrightarrow{[O]} CH_3CH_2CO_2H$$

It is a colourless liquid with an acrid odour, m.p. – 22°C, b.p. 141°C, miscible with water, ethanol and ether in all proportions.

Butyric acids. There are two isomers possible, and both are known.

n-Butyric acid occurs as the glyceryl ester in butter, and as the free acid in perspiration. It is prepared industrially by the oxidation of n-butanol.

n-Butyric acid is a viscous unpleasant-smelling liquid, m.p. – 4.7°C, b.p. 162°C, miscible with water, ethanol and ether. It is the liberation of free n-butyric acid that gives stale butter its rancid odour.

Isobutyric acid occurs in the free state and as its esters in many plants. It is prepared industrially by the oxidation of isobutanol and may be prepared synthetically as follows:

$$(CH_3)_2CHOH \xrightarrow{PBr_3} (CH_3)_2CHBr \xrightarrow{KCN}$$

$$(CH_3)_2CHCN \xrightarrow[acid]{H_2O} (CH_3)_2CHO_2H$$

Isobutyric acid is a liquid, m.p. – 47°C, b.p. 154°C. Its calcium salt is more soluble in hot water than in cold, whereas calcium butyrate is more soluble in cold water than in hot.

Valeric acids. There are four isomers possible, and all are known:

n-Valeric acid, $CH_3CH_2CH_2CH_2CO_2H$, m.p. – 34.5°C, b.p. 187°C.

Isovaleric acid, $(CH_3)_2CHCH_2CO_2H$, m.p. – 51°, b.p. 175°C.

Ethylmethylacetic acid or *active valeric acid*, CH_3CH_2CH-$(CH_3)CO_2H$, p.b. 175°C.

Trimethylacetic acid or *pivalic acid*, $(CH_3)_3CCO_2H$, m.p. 35.5°, p.b. 164°C.

The higher monocarboxylic acids which occur in nature usually have straight chains, and usually contain an *even* number of carbon atoms. **Caproic** (*hexanoic*), $C_6H_{12}O_2$ (m.p. – 9.5°, b.p. 205°C), **caprylic** (*octanoic*), $C_8H_{16}O_2$ (m.p. 16°, p.b. 237°C) and *capric* (*decanoic*) acid, $C_{10}H_{20}O_2$ (m.p. 31.5° p.b. 270°C), are present as glyceryl esters in goats' butter. *Lauric* (*dodecanoic*), $C_{12}H_{24}O_2$ (m.p. 44°C) and **myristic** (*tetradecanoic*) acid, $C_{14}H_{28}O_2$ (m.p. 58°C), occur as their glyceryl esters in certain vegetable oils. The most important higher acids are *palmitic* (*hexadecanoic*), $C_{16}H_{32}O_2$ (m.p. 64°C), and **stearic** (*octadecanoic*), $C_{18}H_{36}O_2$ (m.p. 72°C), which are very widely distributed as their glyceryl esters (together with oleic acid) in most animal ad vegetable oils and fats. The sodium and potassium salts of palmitic and stearic acids are the constituents of ordinary soaps.

Some still higher acids are found in waxes: **arachidic** (*eicosanoic*), $C_{20}H_{40}O_2$ (m.p. 77°C), **behenic** (*docosanoic*), $C_{22}H_{44}O_2$ (m.p. 82°C), **lignoceric** (*tetracosanoic*), $C_{24}H_{48}O_2$ (m.p. 83.5°C), **cerotic** (*hexacosanoic*), $C_{26}H_{52}O_2$ (m.p. 87.7°C), and **melissic** acid (*triacontanoic*), $C_{30}H_{60}O_2$ (m.p. 90°C).

The odd fatty acids may be obtained by degrading an even acid. Two odd acids which may be prepared readily are **n-heptanoic** or

oenanthylic acid, $C_7H_{14}O_2$ (m.p. – 10°, p.b. 223.5°C), and **nonanoic** or **pelargonic** acid $C_9H_{18}O_2$ (m.p. 12°, p.b. 254°C). n-Heptanoic acid is prepared by the oxidation of n-heptaldehyde, which is obtained by destructively distilling castor oil which contains ricinoleic acid (*q.v.*):

$$CH_3(CH_2)_5CHO + [O] \xrightarrow[H_2SO_4]{KMnO_4} CH_3(CH_2)_5CO_2H \ (68\text{–}70\%)$$

Nonanoic acid is obtained by the oxidation of oleic acid (*q.v.*).

Margaric acid (*heptadecanoic acid*), $C_{16}H_{33}CO_2H$, m.p. 61°C, is prepared by heating a mixture of the calcium salts of stearic and acetic acids, and then oxidising the heptadecyl methyl ketone so produced.

$$C_{17}H_{35}CO_2ca + CH_3CO_2ca \longrightarrow CaCO_3 + C_{17}H_{35}COCH_3 \xrightarrow{[O]} C_{16}H_{33}CO_2H$$

Odd-numbered monocarboxylic acids occur naturally in very small amounts up to $n = 23$, and their main source is butter fat. Branched-chain acids have also been isolated from butter fat.

General Reactions of the Monocarboxylic Acids and Their Derivatives

Acids and their derivatives may be represented as follows, where Z = OH, OR, X, NH_2, etc.

$$RC(=O)\ddot{Z} \longleftrightarrow R\overset{+}{C}(=O^-)\ddot{Z} \longleftrightarrow RC(O^-)=Z^+$$

Thus, all are resonance hybrids and the behaviour of the 'carbonyl' group depends on the nature of Z. Z may have a – I and a + R effect and consequently the actual weights of the canonical forms will depend on the relative contributions of these two electronic effects, *e.g.*, for Z = NH_2 (acid amide), + R > – I, whereas for Z = Cl (acid chloride), – I > + R. Hence, the rates of reaction of the carbonyl group in these compounds can be expected to be different.

Many of the reactions of the acids and their derivatives may be generalised as:

$$\text{(i)}\quad R-\underset{Z}{\overset{O}{\overset{\|}{\underset{|}{C}}}} + Y: \xrightleftharpoons{r/d} R-\underset{Z}{\overset{O^-}{\overset{\|}{\underset{|}{C}}}} + Y \longrightarrow R-\overset{O}{\overset{\|}{C}}-Y + Z:$$

$$\text{(ii)}\quad R-\overset{\overset{\displaystyle O}{\|}}{\underset{\underset{\displaystyle Z}{|}}{C}}+H^+ \rightleftharpoons R-\overset{\overset{\displaystyle \overset{+}{O}H}{\|}}{\underset{\underset{\displaystyle Z}{|}}{C}} \longrightarrow R-\overset{\overset{\displaystyle OH}{|}}{\underset{\underset{\displaystyle Z}{|}}{C^+}} \xrightarrow[r/d]{Y:} R-\overset{\overset{\displaystyle OH}{|}}{\underset{\underset{\displaystyle Z}{|}}{C}}-Y$$

$$\longrightarrow R-\overset{\overset{\displaystyle O}{\|}}{C}-Y+HZ$$

These reactions are typical of the behaviour of the carbonyl group (in aldehydes and ketones), and the greater the positive charge on the carbonyl atom, the more easily is this carbon atom attacked by a nucleophile (in the r/d step). Since the + R effect tends to neutralise the positive charge, the greater the + R effect of Z, the slower will be the reaction. Experimental work has shown the rates of hydrolysis are in the order $RCONH_2 < RCO_2R < RCOCl$, and this is the reverse order of the + R effects of the groups, *i.e.*, $NH_2 > OR > Cl$.

ESTERS

Esters are compounds which are formed when the hydroxylic hydrogen atom in oxygen acids is replaced by an alkyl group; the acid may be organic or inorganic. The most important esters are derived from the caboxylic acids. The general formula of the carboxylic esters is $C_2H_{2n}O_2$, which is the same as that of the carboxylic acids, and they are named as the alkyl salts of the acid. *e.g.*,

$CH_3COOC_2H_5$ ethyl acetate

$(CH_3)_2CHOOCH(CH_3)_2$ isopropyl isobutyrate

Carboxylic esters are formed by the action of the acid on an alcohol: acid + alcohol:

acid + alcohol $\rightleftharpoons$ ester + water

The reaction is reversible, the *forward* reaction being known as *esterification*, and the *backward* reaction as *hydrolysis* (for mechanisms).

General Methods of Preparation of the Carboxylic Esters

1. The usual method is *esterification*. The reaction is always slow, but is speeded up by the presence of small amounts of inorganic acids as catalysts, *e.g.*, when the acid is refluxed with the alcohol in the presence of 5–10 per cent concentrated sulphuric acid:

$$R^1COOH + R^2OH \xrightarrow{H_2SO_4} R^1COOR^2 + H_2O \ (v.g.)$$

Alternatively, hydrogen chloride is passed into the mixture of alcohol and acid until there is a 3 per cent increase in weight, and the mixture is refluxed (the yields are very good). This is known as the *Fische-Speier method* (1895), and is more satisfactory for secondary and tertiary alcohols than the sulphuric acid method, which tends to dehydrate the alcohol to alkene.

Esterification without the use of catalysts, and starting with one mole of acid and one mole of alcohol, gives rise to about 2/3 mole of ester. The yield of ester may be increased by using excess of acid or alcohol, the cheaper usually being the one in excess. Increased yields may also be effected by dehydrating agents, *e.g.*, concentrated sulphuric acid behaves both as a catalyst and a dehydrating agent. The same effect may be obtained by removing the water or ester from the reaction mixture by distillation, which is particularly useful for high-boiling acids and alcohols. On the other hand, the water may be removed from the reaction mixture by the addition of benzene or carbon tetrachloride, each of which forms a binary mixture with water (and may form a ternary mixture with water and the alcohol), the azeotropic mixtures boiling at a lower temperature than any of the components.

2. Acid chlorides or anhydrides react rapidly with alochols to form esters:

$$R^1COCl + R_2OH \longrightarrow R^1CO_2R^2 + HCl \ (v.g.\text{–}ex.)$$

$$(R^1CO)_2O + R^2OH \longrightarrow R^1CO_2R^2 + R^1CO_2H \ (v.g.\text{–}ex.)$$

The reaction with tertiary alcohols is very show and is often accompanied by side reactions (alkene or alkyl halide formation), but by using the appropriate conditions, esters are produced. Esters of tertiary alochols may also be conveniently prepared by means of a Grignard reagent.

3. Esters may be prepared by refluxing the silver salt of an acid with an alkyl halide in ethanolic solution:

$$R^1CO_2Ag + R^2Br \longrightarrow R^1CO_2R^2 + AgBr \ (v.g. \text{–}ex.)$$

This method is useful where direct esterification is difficult.

4. *Methyl* esters are very conveniently prepared by treating an acid with an ethereal solution of diazomethane (*q.v.*):

$$RCO_2H + CH_2N_2 \longrightarrow RCO_2CH_3 + N_2 \ (ex.)$$

According to Morrison *et al.* (1961), the BF_3—MeOH reagent is better than diazomethane for the methylation of carboxylic acids.

5. The interaction between an ether and carbon monoxide at 125–180°C, under 500 atmospheres pressure, in the presence of boron trifluoride plus a little water, produces esters:

$$R_2O + CO \longrightarrow RCO_2R$$

Esters may also be prepared by the Tischenko reaction (with aldehydes); ethyl acetate is made commercially this way.

General Properties of Esters

The carboxylic esters are pleasant-smelling liquids or solids. The boiling points of the straight chain isomers are higher than those of the branched-chain isomers. The boiling points of the methyl and ethyl esters are lower than those of the corresponding acid, and this is due to the fact that the esters are not associated since they cannot form intermolecular hydrogen bonds. The esters of low molecular weight are fairly soluble in water–hydrogen bonding between ester and water is possible–and the solubility decreases as the series is ascended; all esters are soluble in most organic solvents.

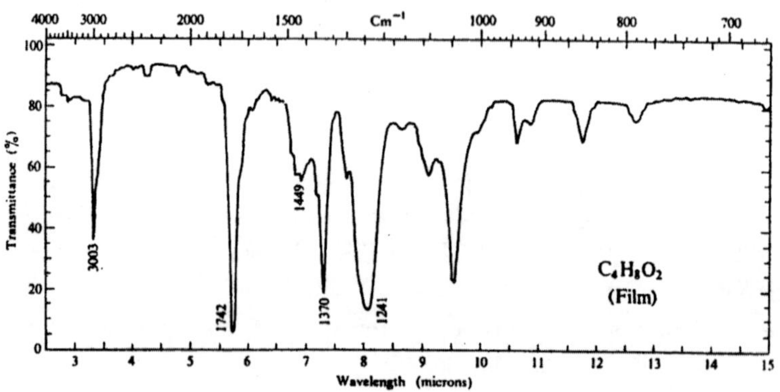

Fig. 4.15

1742 cm^{-1}; C═O (str.) in saturated aliphatic esters. 1241 cm^{-1}: C—O (str.) in *acetates*. 3003 cm^{-1}: C—H (str.) in Me and/or CH_2. 1449 cm^{-1}: C—H (def.) in Me and/or CH_2. 1370 cm^{-1}: C—H (def.) in Me.

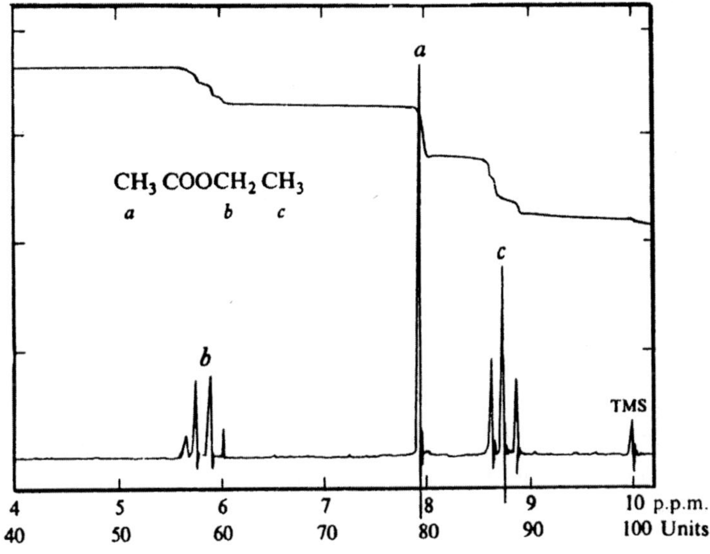

Fig. 4.16

The structure of the esters, like their parent acids, is best represented as a resonance hybrid:

$$R^1-C(=O)-OR^2 \longleftrightarrow R^1-C(-O^-)=\overset{+}{O}R^2$$

The infra-red absorption region of the carbonyl group (stretch) in saturated aliphatic esters is 1750–1735 cm^{-1}(s). This is a higher frequency than that of the carbonyl group in ketones and acids, and so esters may usually be identified.

Characteristic bands are also shown by the C—O (stretch) in formates, 1200-1180 cm^{-1}(s); acetates, 1250–1230 cm^{-1}(s); propionates, etc., 1200–1170 cm^{-1}(s).

The infra-red (film) ad NMR (in $CDCL_3$ at 60 MHz) spectra of ethyl acetate are shown in Figs. 4.15 and respectively.

General Reactions of Esters

1. Esters are hydrolysed by acids or alkalis:

$$R^1CO_2H + R^2OH \xleftarrow{H_3O^+} R^1CO_2R^2 \xrightarrow{OH^-} R^1CO_2^- + R^2OH$$

When hydrolysis is carried out with alkali, the salt of the acid is obtained, and since the alkali salts of the higher acids are soaps, alkaline hydrolysis is known as *saponification* (derived from Latin word meaning soap); saponification is far more rapid than acid hydrolysis.

Mechanisms of carboxyl esterification and hydrolysis of carboxylic esters. Mechanistic studies have shown that in alkaline or neutral hydrolysis, it is the neutral ester molecule which undergoes reaction, whereas in acid-catalysed hydrolysis, it is the conjugate acid of the ester which undergoes reaction.

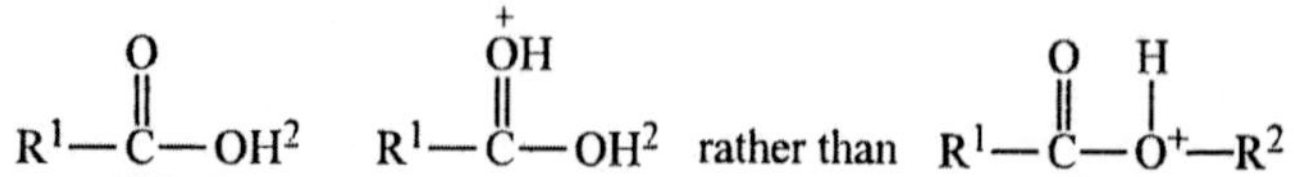

$$\underset{\text{neutral ester}}{R^1-\overset{\overset{\displaystyle O}{\|}}{C}-OH^2} \quad R^1-\overset{\overset{\displaystyle \overset{+}{O}H}{\|}}{C}-OH^2 \text{ rather than } \underset{\text{conjugate acid}}{R^1-\overset{\overset{\displaystyle O}{\|}}{C}-\overset{\overset{\displaystyle H}{|}}{O^+}-R^2}$$

The protonated carbonyl oxygen form of the cA is preferred, and this is based on the work of Fraenkel (1961), who examined the NMR spectrum of methyl formate in, *e.g.*, 100 per cent sulphuric acid and concluded from his results that the ester is protonated chiefly on the carbonyl oxygen. This formulation is also used for *acids*, partly from analogy with esters and partly from the fact that of the two possible conjugate acids, (I) and (II) , (I) may be written as a resonance hybrid to which two identical canonical forms contribute, whereas for (II) the two contributing structures are different and one has fewer bonds and carries unlike charges.

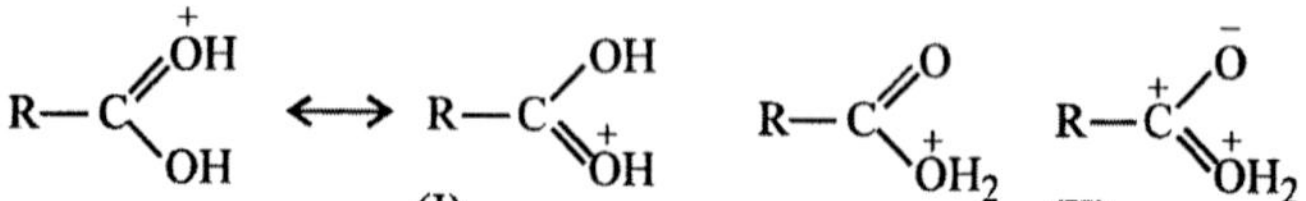

$$\underset{\text{(I)}}{R-C\begin{matrix}\nearrow\!\!\!\!/ \overset{+}{O}H \\ \searrow OH\end{matrix} \longleftrightarrow R-C\begin{matrix}\nearrow OH \\ \searrow\!\!\!\!\backslash \overset{+}{O}H\end{matrix}} \qquad \underset{\text{(II)}}{R-C\begin{matrix}\nearrow\!\!\!\!/ O \\ \searrow \overset{+}{O}H_2\end{matrix} \quad R-\overset{+}{C}\begin{matrix}\nearrow \bar{O} \\ \searrow\!\!\!\!\backslash \overset{+}{O}H_2\end{matrix}}$$

The hydrolysis of carboxylic esters may be formulated in two ways:

$$\underset{\text{acyl-oxygen heterolysis}}{R^1-\overset{\overset{\displaystyle O}{\|}}{C}\frown OR^2} \qquad \underset{\text{alkyl-oxygen heterolysis}}{R^1-\overset{\overset{\displaystyle O}{\|}}{C}-O-R^2}$$

Both modes of fission have been demonstrated experimentally, and kinetic studies have shown that two types of mechanism may operate for each mode, *unimolecular* and *bimolecular*. Hence, *eight* mechanisms are possible for ester hydrolysis but, so far, *six* have been observed Table 4.3. Also, since in principle, the *acid-catalysed* reactions (esterification and hydrolysis) are reversible, but not *alkaline* hydrolysis, then *four* mechanisms are possible for esterification. So far, *three* have been observed (Table 4.3). The eight possible mechanisms are indicated by the symbols shown in Table 4.3, where *A* represents the cA of the ester or acid (reaction in *acid* solution), *B* represents the unprotonated ester (reaction in *basic* or *neutral* solution), subscripts AC and AL, respectively, denote acyl- and alkyl-heterolysis, and numbers 1 and 2 represent the molecularity of the rate-determining step (Ingold, 1941):

Table 4.3

Type of mechanism	*Hydrolysis*	*Esterification*
$B_{AC}1$	—	—
$B_{AC}2$	Very common	—
$A_{AC}1$	Special cases	Special cases
$A_{AC}2$	Very common	Very common
$B_{AL}1$	Special cases	—
$B_{AL}2$	Rare	—
$A_{AL}1$	Very common for t-alcohols	Very common for t-alcohols
$A_{AL}2$	—	—

Most of the experimental work has been carried out on hydrolysis because this process can be studied in acidic, basic, or neutral media, whereas esterification can be studied only in acidic media. It might appear at first sight that esterification could also be studied in neutral media, but since carboxylic acids are acids, reaction in the absence of added acid (as catalyst) is still effectively in an acid medium. Because of this, mechanisms of esterification are partly based on the principle of microscopic reversibility.

Bimolecular basic hydrolysis with acyl-oxygen heterolysis (B_{AC} 2 mechanism). It has long been known that the alkaline hydrolysis of esters is a second-order reaction, *i.e.*, rate = *k* [ester] [OH^-]. Evidence for acyl-oxygen heterolysis has been obtained in several ways, *e.g.*, Polanyl *et al.*, (1934) showed that the alkaline hydrolysis of n-pentyl acetate in water enriched with ^{18}O gave n-pentyl alcohol containing *no* ^{18}O. Therefore acyl-oxygen heterolysis must have occurred:

$$\text{CH5CO—OC5H11} + \dot{O}H^- \longrightarrow CH_3CO\dot{O}H + C_5H_{11}O^- \longrightarrow$$

$$CH_3CO\dot{O}^- + C_5H_{11}OH$$

Another method uses optical activity as a diagnostic test for demonstrating acyl-oxygen heterolysis. If the alcohol group is optically active, and if the $B_{AC}2$ mechanism operates, then the ion RO^- will be liberated and will retain its optical activity since the bond R—O is never broken. Various examples of retention are known when the reaction is bimolecular.

A mechanism consistent with the above facts (and with the work of Bender (1951), who used esters labelled with ^{18}O in the carbonyl group) is:

$$HO^- + R^1\text{—}C(=O)\text{—}OR^2 \underset{\text{fast}}{\overset{\text{slow}}{\rightleftharpoons}} HO\text{—}C(O^-)(R^1)\text{—}OR^2 \underset{\text{slow}}{\overset{\text{fast}}{\rightleftharpoons}}$$

$$HO\text{—}C(=O)\text{—}R^1 + {}^-OR^2 \xrightarrow{\text{fast}} {}^-O\text{—}C(=O)\text{—}R^1 + R^2OH$$

Any R^1 group with – I effect should accelerate hydrolysis by creating a positive charge on the attached carbon atom, thereby facilitating attack by the hydroxide ion. On the other hand, because the r/d step is bimolecular, the larger R^1 is the larger can be expected to be the steric retardation, since the hybridisation of the carbon atom of the carbonyl group changes from sp^2 to sp^3 (with consequent increased crowding). These anticipated results have been observed in practice, *e.g.*,

– I effect

Ester	$MeCO_2Me$	$ClCH_2CO_2Me$	$CHCl_2CO_2Me$
Relative rates	1	761	16000

Steric effect

Ester	$MeCO_2Et$	$EtCO_2Et$	$isoPrCO_2Et$
Relative rates	1.00	0.47	0.10

Bimolecular acid-catalysed hydrolysis and esterification with acyl-oxygen heterolysis ($A_{AC}2$ mechanism). Acid-catalysed hydrolysis and acid-catalysed esterification are the reverse of each other, and so according to the principle of microscopic reversibility the mechanism of acid-catalysed hydrolysis will be that of acid-catalysed esterification, but in reverse.

The rate law for acid-catalysed hydrolysis has been shown to be second-order, first-order in both ester and hydrogen ion concentration, *i.e.*, rate = k [ester] [H_3O^+]. Also, evidence for acyl=oxygen heterolysis has been obtained in several ways, *e.g.*, Ingold *et al.* (1939) showed that the acid-catalysed hydrolysis of methyl hydrogen succinate in water enriched with ^{18}O gave methanol with no extra ^{18}O. Thus, acyl-oxygen heterolysis must have occurred:

$$HO_2CCH_2CH_2CO{-}OMe + H_2\dot{O} \xrightarrow{H^+} HO_2CCH_2CH_2CO\dot{O}H + MeOH$$

Furthermore, Robers *et al.* (1938) esterified benzoic acid with methanol enriched with 18O and obtained water *not* enriched with 18O. Thus, acyl-oxygen heterolysis must have occurred:

$$PhCO{-}OH + Me\dot{O}H \longrightarrow PhCO\dot{O}me + H_2O$$

A mechanism that fits these facts is:

$$R^1{-}\overset{\overset{O}{\|}}{C}{-}OH \underset{-H^+;\ fast}{\overset{+H^+;\ fast}{\rightleftharpoons}} R^1{-}\overset{\overset{\overset{+}{O}H}{\|}}{C}{-}OH \underset{-R^2OH;\ fast}{\overset{+R^2OH;\ slow}{\rightleftharpoons}} R^1{-}\underset{\underset{H-O^+-R^2}{|}}{\overset{\overset{OH}{|}}{C}}{-}OH \underset{fast}{\overset{fast}{\rightleftharpoons}}$$

$$R^1{-}\underset{\underset{OR^2}{|}}{\overset{\overset{\overset{+}{O}H_2}{|}}{C}}{-}OH \underset{+H_2O;\ slow}{\overset{-H_2O;\ fast}{\rightleftharpoons}} R^1{-}\underset{\underset{OR^2}{|}}{C}{=}\overset{+}{O}H \underset{+H^+;\ fast}{\overset{-H^+;\ fast}{\rightleftharpoons}} R^1{-}\underset{\underset{OR^2}{|}}{C}{=}O$$

It should be noted that the r/d step for esterification is the addition of alcohol, whereas that for hydrolysis is the addition of water. In both cases, the r/d step involves change of hybridisation of the carbon atom of the carbonyl group from sp_2 to sp_3, and consequently steric retardation can be expected to increase as R1 increase as size (*cf.* the BAC_2 mechanism), *e.g.* (esterification with methanol):

Acid	$MeCO_2H$	n-$PrCO_2H$	$MeCCO_2H$	Et_3CCO_2H
Relative rates	1.0	0.51	0.037	0.00016

Unimolecular acid-catalysed hydrolysis and esterification with acyl-oxygen heterolysis ($A_{AC}1$ mechanism). Let us consider the following *observed* results (*M* represents the mesityl group, 1,3,5-trimethylphenyl, in *mesitoic acid*, $Me_3C_6H_2CO_2H$):.

$PhCO_2H + MeOH \xrightarrow{HCl} PhCO_2Me$ *occurs readily*

$PhCO_2ME + OH^- \dashrightarrow PhCO_2^- + MeOH$ *occurs readily*

$$MCO_2H + MeOH \xrightarrow{HCl} \textit{No reaction}$$

$$MCO_2Me + OH^- \longrightarrow MCO_2^- + MeOH \textit{ very slow}$$

Hammett *et al.* (1937) showed that when methyl benzoate is dissolved in concentrated sulphuric acid and the solution then poured into ice/water, almost all of the ester is recovered. On the other hand, methyl mestioate, on the same treatment, gave a quantitative yield of mestioic acid. Then Newman (1941) showed that when a solution of mesitoic acid in concentrated sulphuric acid is poured into a cold methanol, methyl mesitoate is formed. This different behaviour of mesitoyl derivatives from benzoyl derivatives suggests that different mechanisms are operating in esterification and hydrolysis for these two sets of compounds. Now, Hammett (1937) had shown that when mesitoic acid is dissolved in concentrated sulphuric acid, the van't Hoff factor *i* is 4 (measured by depression of freezing point). Under the same conditions, benzoic acid gives $i = 2$. Hammett therefore proposed that the mesitoyl reactions proceed through an intermediate *acylium* (*oxocarbonium*) ion (*mesitoylium* ion in this case) formed by acyl-oxygen heterolysis:

$$\text{(i)}\ MCO_2H + H_2SO_4 \overset{\text{fast}}{\rightleftharpoons} M\overset{+}{C}(OH)_2 + HSO_4^-$$

$$\text{(ii)}\ M\overset{+}{C}(OH)_2 + H_2SO_4 \overset{\text{slow}}{\rightleftharpoons} M\overset{+}{C}\ O + H_3O+ + HSO_4^-$$

The overall equation may be written:

$$MCO_2H + 2H_2SO_4 \rightleftharpoons M\overset{+}{C}\ O + H_2O^+ + 2HSO_4^-$$

This accounts for $i = 4$ for mesitoic acid, and for $i = 2$ for benzoic acid if only reaction (*i*) is involved. Thus, the reactions of mesitoic acid and its ester may be formulated as:

$$\left.\begin{array}{l} MCO_2H \\ MCO_2H \end{array}\right] \xrightarrow[\text{slow}]{H_2SO_4} MC{=}O \xrightarrow{\text{fast}} \left[\begin{array}{l} \xrightarrow{MeOH} MCO_2Me \\ \xrightarrow{H_2O} MCO_2H \end{array}\right.$$

This type of reaction occurs with highly sterically hindered acids and esters, and so steric hindrance must be an important factor.

Further support for this mechanism has been provided by Olah *et al.* (1967), who showed that acids (and esters), when dissolved in FSO_3H—SbF_5—SO_2 (a very strongly acidic medium), at 60°C, lose a molecule of water at 0°C to form the acylium ion:

$$RC(=O)OH \xrightarrow[-60^\circ C]{H^+} RC(=\overset{+}{O}H)OH \longleftrightarrow RC(OH)=\overset{+}{O}H \xrightarrow[0^\circ C]{-H_2O} R\overset{+}{C}=O$$

Unimolecular basic hydrolysis with alkyl-oxygen heterolysis ($B_{AL}1$ mechanism). This mechanism has been shown to occur when the alkyl group of the alcohol is capable of forming a relatively stable carbonium ion, the solvent has high ionising power (thereby encouraging alkyl-oxygen heterolysis), and the hydroxide ion concentration is low (thereby suppressing the bimolecular mechanism). An example of the $B_{AL}1$ mechanism is the hydrolysis of optically active α-phenylethyl hydrogen phthalate in *faintly* alkaline solution. Kenyon *et al.* (1936) showed that the product was recemissed α-phenylethanol (phenylmethylmethanol); this is in keeping with the formation of an intermediate carbonium ion (sp^2 hybridised) [R = o–$HO_2CC_6H_4$—]:

$$RCO—O—CHMePh \underset{}{\overset{slow}{\rightleftharpoons}} RCO_2^- + PhMeCH^+ \xrightarrow[fast]{H_2O} PhMeCHOH_2^+ \xrightarrow[fast]{RCO_2^-}$$

$$(\pm)\text{-}PhMeCHOH + RCO_2H$$

Bimolecular basic hydrolysis with alkyl-oxygen heterolysis ($B_{AL}2$ mechanism). This mechanism is rare. One example is the reaction between methyl benzoate and sodium methoxide in methanol (Bunnett *et al.*, 1950). The products were dimethyl ether and sodium benzoate; these results are readily explained on the basis of the $B_{AL}2$ mechanism:

$$MeO^- \; Me—O—OCPh \longrightarrow Me_2O + PhCO_2^-$$

Unimolecular acid-catalysed hydrolysis and esterification with alkyl-oxygen heterolysis ($A_{AL}1$ mechanism). Definite evidence for this mechanism was obtained by Hughes, Ingold *et al.* (1939), who showed that the esterification of acetic acid with optically active octan-2-ol in the presence of sulphuric acid gave a large amount of racemised ester. Racemisation is to be expected if an intermediate carbonium ion is formed, and so the reaction may be formulated as $A_{AL}1$ (R = n-$C_6H_{11}CHMe$—):

$$ROH + H^+ \underset{fast}{\overset{fast}{\rightleftharpoons}} ROH_2^+ \underset{fast}{\overset{slow}{\rightleftharpoons}} H_2O + R^+$$

$$MeC(=\ddot{O})OH \;\; R^+ \underset{slow}{\overset{fast}{\rightleftharpoons}} MeC(=\overset{+}{O}R)O—H \underset{fast}{\overset{fast}{\rightleftharpoons}} MeC(OR)=O + H^+$$

Since all steps are reversible, the $A_{AL}1$ mechanism should, in principle, be possible for ester hydrolysis, Evidence for its occurrence has been obtained by, *e.g.*, Bunton *et al.* (1951), who carried out the acid-catalysed hydrolysis of t-butyl acetate in water enriched with ^{18}O and obtained t-butanol containing ^{18}O.

$$MeC(=O){-}O{-}Bu^t + H^+ \underset{fast}{\overset{fast}{\rightleftharpoons}} Me{-}C(=\overset{+}{O}H){-}O{-}Bu^t \underset{fast}{\overset{slow}{\rightleftharpoons}} MeC(OH)(=O) + \overset{+}{Bu^t}$$

$$\overset{+}{Bu^t} + H_2\dot{O} \underset{fast}{\overset{fast}{\rightleftharpoons}} Bu^t\dot{O}H_2^+ \underset{slow}{\overset{fast}{\rightleftharpoons}} Bu^t\dot{O}H + H^+$$

The $A_{AL}1$ mechanism is common for t-alcohols; these from relatively stable carbonium ions (*cf.* the $B_{AL}1$ mechanism).

2. Esters are converted into alcohols by the Bouveault-Blanc reduction, catalytic hydrogenation, and by metallic hydrides.

3. Esters react with ammonia to form amides. This reaction is example of *ammonolysis* (which means, literally, splitting by ammonia):

$$R^1COOR^2 + NH_3 \longrightarrow R^1CONH_2 + R^2OH\ (g.)$$

With hydrazine, esters form acid hydrazides:

$$R^1COOR^2 + H_2NNH_2 \longrightarrow R^1CONHNH_2 + R^2OH$$

Esters react with phosphorus pentachloride or thionyl chloride to form acyl chlorides, *e.g.*,

$$R^1CO_2R^2 + PCl_5 \longrightarrow R^1COCl + POCl_3 + R^2Cl$$

4. By means of *alcoholysis* (splitting by alcohol), an alcohol residue in an ester can be replaced by another alcohol residue. Alcoholysis is carried out by refluxing the ester with a large excess of alcohol, preferably in the presence of a small amount of acid or sodium alkoxide as catalyst.

$$CH_3CO_2C_2H_9 + C_2H_5OH \overset{C_2H_5ONa}{\rightleftharpoons} CH_3CO_2C_2H_5 + C_4H_9OH$$

Alcoholysis of esters is also known as *transesterification.*

The Mechanism of alcoholysis is uncertain; a possibility is (in basic media):

$$R^1C(=O)(OR^2) + EtO^- \rightleftharpoons R^1C(=O^-)(OR^2){-}OEt \rightleftharpoons R^1C(=O){-}OEt + R^2O^-$$

$R^2O^- + EtOH \rightleftharpoons R^2OH + EtO^-$

In *acidolysis*, the acid residue is displaced from its ester by another acid residue, *e.g.*,

$$CH_3CO_2C_2H_5 + C_5H_{11}CO_2H \rightleftharpoons C_5H_{11}CO_2C_2H_5 + CH_3CO_2H$$

Acidolysis is a useful reaction for converting the natural ester of a dibasic acid into its acid ester.

According to Parker *et al.* (1968), who used ^{18}O-labelled toluic acids and their esters, the mechanism of acidolysis is:

$$R^1-\overset{O}{\overset{\|}{C}}-OH + R^2-\overset{O}{\overset{\|}{C}}-OR^3 \overset{H^+}{\rightleftharpoons} R^1-C(=O)-O-C(-O(R_3))(R^2)(OH) \rightleftharpoons R^1-C(OR^3)(=O) + HO-C(=O)-R^2$$

5. When an ester – preferably the methyl or ethyl ester – is treated with sodium in an inert solvent, *e.g.*, ether, benzene or toluene, and subsequently with acid, an **acyloin** is formed (50–70 per cent yield). It is important that the reaction be carried out *in the absence of any free alcohol*. Acyloins are α,β-keto-alcohols, and according to Kharasch and his co-workers (1939), their formation takes place via a free-redical mechanism, *e.g.*, propionin from ethyl propionate:

$$2EtCO_2Et + 2e \xrightarrow{2Na} 2EtC(O^-)(OEt) \rightleftharpoons \begin{matrix} EtC(O^-)(OEt) \\ | \\ EtC(O^-)(OEt) \end{matrix} \rightleftharpoons 2EtO^- + \begin{matrix} EtC=O \\ | \\ EtC=O \end{matrix} \xrightarrow{2Na}$$

$$\begin{matrix} EtC-O^- \\ \| \\ EtC-O^- \end{matrix} \xrightarrow{H^+} \begin{matrix} EtC-OH \\ \| \\ EtC-OH \end{matrix} \xrightarrow{\text{isomerises}} \begin{matrix} EtC=O \\ | \\ EtCHOH \end{matrix}$$

Acyloins are reduced by the Clemmensen method to alkanes.

6. Carboxylic esters which contain α-hydrogen atoms react with sodamide in liquid ammonia solution to form the acid amide and a condensation product involving two molecules of the ester, *e.g.*, ethyl acetate gives acetamide and acetoacetic ester :

$$CH_3CO_2C_2H_5 + NaNH_2 \longrightarrow CH_3CONH_2 + C_2H_5ONa$$

$$CH_3CO_2C_2H_5 + NaNH_2 \longrightarrow NH_3 + NH_3 + [\ddot{C}H_2CO_2C_2H_5]^- Na^+$$

$$\xrightarrow{CH_3CO_2C_2H_5} CH_3COCH_2CO_2C_2H_5 + C2H_5OH$$

Of particular interest is the *carbonation* of esters, *i.e.*, the introduction of the carboxyl group.

$$R^1CH_2CO_2R^2 + NaNH_2 \xrightarrow{\text{liquid } NH_3}$$

$$NH_3 + [R^1CHCO_2R^2]^- Na^+ \xrightarrow[\text{ether}]{CO_2} R^1\underset{\displaystyle CO_2Na}{\underset{|}{C}}HCO_2R^2$$

The yields of these *malonic acid derivatives* are 54–60 per cent with acetates, and becomes progressively lower as the molecular weight of he acid increases.

7. The pyrolysis of esters, particularly acetates, gives alkenes.

Esters are used as solvents for cellulose, oils, gums, resins, etc., and as plasticisers. They are also used for making artificial flavours and essences, *e.g.*, isoamyl acetate – banana oil; amylbutyrate–apricot; isoamyl isovalerate–apple; methyl butyrate–banana oil; amyl butyrate–apicot; isoamyl isovalerate–apple; methyl butyrate–pineapple; etc.

Ortho-esters are compounds of the type $R^1C(OR^2)_3$. They are derived from the ortho-acids, $RC(OH)_3$, which have not yet been isolated, but which are possibly present in aqueous solution:

$$R{-}C{\overset{\displaystyle O}{\underset{\displaystyle OH}{\lessgtr}}} + H_2O \quad RC(OH)_3$$

The most important ortho-esters are the orthoformic esters, particularly ethyl orthoformate. This may be prepared by running ethanol and chloroform into sodium covered with ether:

$$2CHCl_3 + 6C_2H_5OH + 6Na \longrightarrow 2HC(OC_2H_5)_3 + 6NaCl + 3H_3\ (70\%)$$

A possible mechanism is via the formation of dichloromethylene:

$$CHCl_3 \xrightarrow{EtO^-} :CCl_2 \xrightarrow{EtOH} EtOCHCl_2 \xrightarrow{2EtO^-}$$

$$HC(OEt)_3 + 2Cl^-$$

Ethyl orthoformate may be used for preparing ketals and for preparing aldehydes by means of a Grignard reagent.

ESTERS OF THE INORGANIC ACIDS

Alkyl sulphates, $(RO)_2SO_2$. Only methanol and ethanol give a good yield of the alkyl sulphate by reaction with concentrated sulphuric acid; the higher alochols give mainly alkenes and ethers. On the other hand, all the alcohols give a fair yield of alkyl hydrogen sulphate when a mixture of alcohol and concentrated sulphuric acid is heated on a steam bath.

Methyl sulphate is prepared industrially:

(*i*) By heating methanol with concentrated sulphuric acid, and then distilling the methyl hydrogen sulphate under reduced pressure:

$$CH_3OH + H_2SO_4 \longrightarrow H_2O + CH_3OSO_3H \xrightarrow{2} (CH_3O)_2SO_2 + H_2SO_4$$

(*ii*) By treating methanol with sulphur trioxide at low temperatures:

$$2SO_3 + 2CH_3OH \longrightarrow (CH_3)_2SO_4 + H_2SO_4$$

Ethyl sulphate may be prepared by the same methods as methyl sulphate, but in addition, there is the industrial preparation by passing ethylene in excess into cold concentrated sulphuric acid:

$$2C_2H_4 + H_2SO_4 \longrightarrow (C_2H_5)_2SO_4$$

Sulphates of the higher alcohols are conveniently prepared by the oxidation of sulphites as follows:

$$2ROH_4 + SOCl_2 \longrightarrow 2HCl + (RO)_2SO \xrightarrow{KMnO_4} (RO)_2SO_2$$

Methyl sulphate, b.p. 188°C, and ethyl sulphate, b.p. 208°C, are heavy poisonous liquids. They are largely used as alkylating agents since the alkyl group will replace the hydrogen atom of the groups —OH, —NH— or —SH. The alkylation may be carried out by treating the compound with the alkyl sulphate in sodium hydroxide solution. Usually only one of the alkyl groups takes part in the reaction, *e.g.*, methylation of a primary amine:

$$RNH_2 + (CH_3)_2SO_4 + NaOH \longrightarrow RNHCH_3 + CH_3NaSO_4 + H_2O$$

The methyl and ethyl esters of the carboxylic acids maybe conveniently prepared by treating the sodium salt of the acid with respectively methyl or ethyl sulphate:

$$RCO_2Na + (C_2H_5)_2SO_4 \longrightarrow RCO_2C_2H_5 + C_2H_5NaSO_4$$

The sodium alkyl sulphates of the higher alcohols are used as detergents, *e.g.*, sodium lauryl sulphate:

Alkyl nitrates, $RONO_2$. The only important alkyl nitrate is ethyl nitrate, $C_2H_5ONO_2$. This may be prepared by heating ethyl iodide with silver nitrate in ethanolic solution:

$$C2H_5I + AgNO3 \longrightarrow C_2H_5ONO_2 + AgI$$

When concentrated nitric acid is added to ethanol, the reaction is usually violent; part of the ethanol is oxidised, and some of the nitric acid is reduced to nitrous acid. Apparently it is the presence of the nitrous acid which produces the violent oxidation of the alcohol by the nitric acid. This danger may be avoided by first boiling the nitric acid with urea, which destroys any nitrous acid present, and then adding this mixture to cool ethanol, any nitrous acid produced being immediately destroyed by the urea:

$$CO(NH_2)_2 + 2HNO_2 \longrightarrow CO_2 + 2N_2 + 3H_2O$$

Ethyl nitrate is a pleasant-smelling liquid, b.p. 87.5°C. When reduced with tin and hydrochloric acid, it forms hydroxylamine and ethanol:

$$C_2H_2ONO_2 \xrightarrow{Sn/HCl} C_2H_5OH + NH_2OH + H_2O$$

Alkyl nitrites, RONO. Alkyl nitrites are isomeric with the nitroalkanes. The only important alkyl nitrites are the ethyl and amyl nitrites; the latter is actually mainly isoamyl nitrite, since the amyl alcohol used is the isopentanol from fusel oil. Ethyl and amyl nitrites are prepared by adding concentrated hydrochloric acid or sulphuric acid to aqueous sodium nitrite and the alcohol:

$$R—O—H + NO^+ \longrightarrow R—O\begin{matrix} H \\ NO \end{matrix} \xrightarrow{-H^+} RONO$$

Ethyl nitrite, p.b. 17°C, and amyl nitrite, p.b. 99°C, are pleasant-smelling liquids, and are used as a means of preparing nitrous acid in anhydrous media; *e.g.*, an ethanolic solution of nitrous acid may be prepared by passing dry hydrogen chloride into amyl nitrite dissolved in ethanol.

ACID OR ACYL CHLORIDES

The general formula of the *acyl* groups is $R—C(=O)—$. Acid chlorides, which may be prepared by the replacement of the hydroxyl in the carboxyl group by chlorine, are also known as *acyl chlorides* because they contain the acyl group.

Nomenclature

According to the I.U.P.A.C. system the class suffix of the acyl chlorides is *-oyl*; the common names are formed by changing the suffix *-ic* of the *trivial* name of the acid into *-yl*, *e.g.*,

CH_3COCl	$(CH_3)_2CHCOCl$
ethanoyl chloride acetyl chloride	2-methylpropanoyl chloride isobutyryl chloride

If the carboxyl group is considered as a substituent, then according to the I.U.P.AC. system, the nomenclature of all substances containing acyl groups is in all cases based on the name 'carbonyl' for the CO group, *e.g.*,

$CH_3CH_2CH_2COCl$ propane-1-carbonyl chloride

When naming amides, the 'yl' is elided before the suffix amide, *e.g.*,

$CH_3CH_2CH_2CONH_2$ propane-1-carbonamide

General Methods of Preparation

1. The acid is heated with phosphorus trichloride or pentachloride, *e.g.*,

$$3RCOOH + PCl_3 \longrightarrow 3RCOCl + H_3PO_3 \ (g.)$$

$$RCOOH + PCl_5 \longrightarrow RCOCl + HCl + POCl_3 \ (v.g.)$$

The reaction with phosphorus trichloride is accompanied by the formation of small amounts of volatile phosphorus compounds, *e.g.*,

$$RCO_2H + PCl_3 \longrightarrow RCO_2PCl_2 + HCl$$

Thionyl chloride may be used instead of the phosphorus chlorides:

$$ROC_2H + SOCl_2 \longrightarrow RCOCl + SO_2 + HCl \ (v.g.)$$

The inorganic chloride is chosen according to the boiling point of the acyl chloride formed. Phosphorous acid decomposes at 200°C; the boiling point of phosphoryl chloride is 107°C, and that of thionyl chloride is 76°C. Since acetyl chloride boils at 52°C, any of the three inorganic halides may be used, but it is difficult to separate acetyl chloride from thionyl chloride (which is generally used in excess) by fractionation. n-Butyryl chloride boils at 102°C, and so phosphorus pentachoride cannot be used. Usually thionyl chloride is the most convenient.

The mechanism of the formation of acid chlorides is uncertain; it may be similar to that when an alcohol is the substrate.

Acyl halides undergo the following exchange reaction with acids (probably via the acid anhydride):

$$R^1COCl + R^2CO_2H \rightleftharpoons [R^1COOOCR^2 + HCl] \rightleftharpoons R^1CO_2H + R^2COCl$$

This therefore offers a means of preparing acyl halides, but the success depends on displacement of the equilibrium to the right. This is readily achieved by using oxalyl chloride (or bromide) as the acyl halide.

The half-acid chloride of oxalic acid produced is decomposed to carbon monoxide and carbon dioxide under the conditions of the experiment.

2. By distilling the salt of the acid with either phosphorus trichloride, phosphoryl chloride or sulphuryl chloride, *e.g.*,

$$3CH_3CO_2Na + PCl_3 \longrightarrow 3CH_3COCl + Na_3PO_3$$

$$2CH_3CO_2Na + POCl_3 \longrightarrow 2CH_3COCl + NaCl + NaPO_3$$

$$(CH_3CO_2)_2Ca + SO_2Cl_2 \longrightarrow 2CH_3COCl + CaSO_4$$

This method is used industrially since the salts are cheaper than the acid.

Acyl bromides may be prepared by the action of phosphorus tribromide or pentabromide (red phosphorus and bromine) on the acid, or by the action of excess of hydrogen bromide on the acyl chloride:

$$RCOCl + HBr \longrightarrow RCOBr + HCl$$

Acyl iodides may be prepared by the action of phosphorus tri-iodide on acid anhydrides. *Acyl fluorides* are also known, and may be prepared by the action of hydrogen fluoride on acid anhydrides:

$$(RCO)_2O + HF \longrightarrow RCOF + RCO_2H$$

General properties and reactions of the acyl chlorides. The lower acyl chlorides are colourless liquids with irritating odours; the higher members are colourless solids. The chlorine atom is very reactive and so the acid chlorides are important reagents.

The infra-red absorption region of the carbonyl group (stretch) in acyl chlorides is 1815–1785 cm^{-1}(s).

1. The acyl chlorides are readily hydrolysed by water, the lower members reacting vigorously:

$$RCOCl + H_2O \longrightarrow RCO_2H + HCl$$

Acyl chlorides usually react rapidly with compounds containing 'active' hydrogen atoms, *i.e.* hydrogen attached to oxygen, nitrogen or sulphur; *e.g.*, esters are formed with alcohols:

$$R^1COCl + R^2OH \longrightarrow R^1COOR^2 + HCl$$

Amides are formed with ammonia, and *N*-substituted amides with primary and secondary amines:

$$RCOCl + 2NH_3 \longrightarrow RCONH_2 + NH_4Cl$$

$$R^1COCl + R^2NH_2 \longrightarrow R^1CONHR^2 + HCl$$

Hydrazides are formed with hydrazine, and hydroxamic acids with hydroxylamine:

$$RCOCl + H_2NNH_2 \longrightarrow RCONHNH_2 + HCl$$

$$RCOCl + NH_2OH \longrightarrow RCONHOH + HCl$$

The mechanisms of these reactions are uncertain; a possibility is:

$$R^1{-}\overset{\overset{\displaystyle O}{\|}}{C}{-}Cl + R^2OH \underset{}{\overset{\text{slow}}{\rightleftharpoons}} R^1{-}\underset{\underset{\displaystyle H{-}O^+{-}R^2}{|}}{\overset{\overset{\displaystyle O^-}{|}}{C}}{-}Cl \xrightarrow{\text{fast}} R^1{-}\underset{\underset{\displaystyle OR^2}{|}}{\overset{\overset{\displaystyle O}{\|}}{C}} + HCl$$

2. Acyl chlorides may be reduced catalytically to aldehydes or to alcohols:

$$RCOCl \xrightarrow[\text{Pd}]{H_2} RCHO \xrightarrow{H_2} RCH_2OH$$

Acid chlorides are reduced to esters of an enediol by sodium amalgam in an inert solvent, *e.g.*, ether :

$$4RCOCl + 4Na \longrightarrow \begin{matrix} R & & R \\ & C{=}C & \\ RCOO & & OOCR \end{matrix} + 4NaCl$$

Under suitable conditions, small yields of α-diketones have been obtained. Acid chlorides are reduced to alcohols by lithium aluminium hydride or sodium borohydride, and to aldehydes by lithium hydride or lithium tri-t-butoxyaluminium hydride.

3. Acyl chlorides react with salts of the acids to form *acid anhydrides* (*q.v.*):

$$R^1COCl + R^2COONa \longrightarrow R^1CO{\cdot}O{\cdot}OCR^2 + NaCl$$

4. Acyl chlorides may be used in the Friedel–Crafts reaction to produce an aromatic ketone: *e.g.*, acetyl chloride reacts with benzene in the presence of anhydrous aluminium chloride to form acetophenone:

$$C_6H_6 + CH_3COCl \xrightarrow{AlCl_3} C_6H_5COCH_3 + HCl$$

5. Acyl chlorides react with Grignard reagents to produce ketones or tertiary alcohols, according to the conditions; *e.g.*, acetyl chloride forms butanone with ethylmagnesium iodide:

$$CH_3COCl + C_2H_5MgI \longrightarrow CH_3COC_2H_5 + MgClI$$

6. Acyl chlorides are readily halogenated in the α-position.

7. Acyl chlorides from esters (and other products) when heated with an ether in the presence of anhydrous zinc chloride as catalyst:

$$R^1COCl + R_2{}^2O \xrightarrow{ZnCl_2} R^1CO_2R^2 + R^2Cl$$

8. Acyl chlorides add on to the double bond of an alkene in the presence of a catalyst, *e.g.*, zinc chloride or aluminium chloride, to form a chloroketone which, on heating, eliminates a molecule of hydrogen chloride to form an unsaturated ketone:

$$(CH_3)_2C{=}CH_2 + CH_3COCl \xrightarrow{ZnCl_2} (CH_3)_2CClCH_2COCH_3$$

$$\xrightarrow{heat} (CH_3)_2C{=}CHCOCH_2 + HCl$$

Formyl fluoride, HCOF, has been isolated by Olah *et al.* (1961). Formyl chloride, HCOCl, has been obtained in solution by the action of hydrogen chloride on a solution of 1-formylimidazole in chloroform at – 10°C (Staab *et al.*, 1964).

A mixture of carbon monoxide and hydrogen chloride behaves as if it were formyl chloride in the Gattermann–Koch aldehyde synthesis.

Acetyl chloride is the most important acyl halide. It is a colourless fuming liquid, b.p. 52°C, and is soluble in ether and chloroform. It is largely used as an *acetylating* agent, *i.e.*, as a means of introducing the acetyl group, into compounds containing 'active' hydrogen atoms.

It is also used to detect the presence of hydroxyl groups in organic compounds, and to estimate their number.

The accepted abbreviation for the acetyl group is Ac; *e.g.*, AcOH is acetic acid; Ac_2O, acetic anhydride; AcCl, acetyl chloride; C_6H_5NHAc, acetanilide; etc.

ACID ANHYDRIDES

The acid anhydrides may be regarded theoretically as being derived from an acid by the removal of one molecule of water form two molecules of the acid:

$$2RCO_2H \xrightarrow{-H_2O} \overset{O}{\overset{\|}{R C}} O \overset{O}{\overset{\|}{C}} R$$

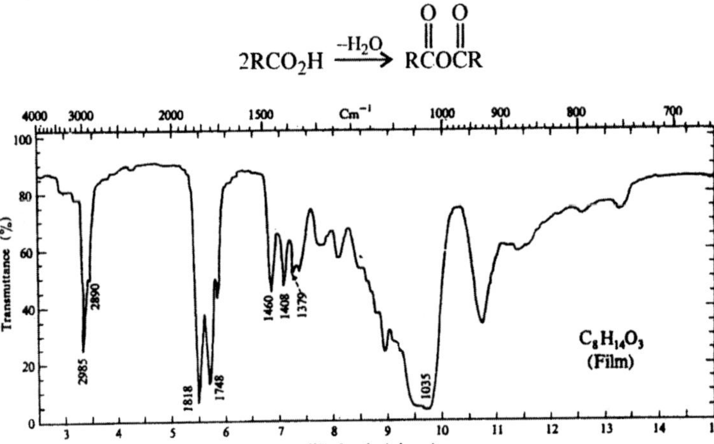

Fig. 4.17

1818 and 1748 cm^{-1}: C═O (str.) in acyclic anhydrides. 1035^{-1}: C—O (str.) in —C—C— (acyclic anhydrides). 2985 and 2890 cm^{-1}: C—H (str.) in Me and/or CH_2. 1460 and 1408 cm^{-1}: C—H (def.) in Me and/or CH_2. 1379 cm^{-1}: C—H (def.) in Me.

It is possible in practice to prepare many acid anhydrides by direct dehydration, but this method is usually confined to the anhydrides of the higher members of the acid series (see below).

The infra-red absorption regions of the carbonyl group (stretch) in acyclic anhydrides are 1840–1800 (s) and 1780–1740 cm^{-1} (s). There is also a band in the region 11801030 cm^{-1} (s) due to the C—O (stretch) in the grouping —C—O—C—.

Fig. 4.17 is the infra-red spectrum of n-butyric anhydride (as a film).

Nomenclature

The acid anhydrides are named as the anhydride of the acid groups present, the *trivial* name of the acid being used, *e.g.*,

$(CH_3CO)_2O$ acetic anhydride

$(CH_3CH_2CO)_2O$ propionic anhydride

Acetic anhydride may be conveniently prepared by distilling a mixture of anhydrous sodium acetate and acetyl chloride:

$$CH_3\overset{O}{\overset{\|}{C}}Cl + CH_3\overset{O}{\overset{\|}{C}}{-}O^- \rightleftharpoons CH_3\overset{\bar{O}}{\overset{|}{C}}(Cl)O\overset{O}{\overset{\|}{C}}CH_3 \longrightarrow CH_3\overset{O}{\overset{\|}{C}}O\overset{O}{\overset{\|}{C}}CH_3 + Cl^-$$

Aceti anhydride is prepared industrially in a number of ways, *e.g.*,

(*i*) By passing acetylene into glacial acetic acid in the presence of mercuric ions as catalyst, and distilling the resulting ethylidene diacetate:

$$C_2H_2 + 2CH_3CO_2H \xrightarrow{Hg^{2+}} CH_3CH(OCOCH_3)_2 \xrightarrow{heat}$$
$$(CH_3CO)_2O + CH_3CHO$$

(*ii*) By passing keten into glacial acetic acid:

$$CH_2{=}C{=}O + CH_3CO_2H \longrightarrow (CH_3CO)_2O$$

Anhydrides of the higher acids may be prepared by heating, and then fractionating, a mixture of the acid and acetic anhydride:

$$2RCO_2H + (CH_3CO)_2O \rightleftharpoons (RCO)_2O + 2CH_3CO_2H$$

This method is only satisfactory for anhydrides which have higher boiling points than acetic acid. A much better method of preparation is to the acid chloride with pyridine and benzene, and the acid, and then heat:

$$RCOCl + RCO_2H + C_5H_5N \longrightarrow (RCO)_2O + C_5H_5\overset{+}{N}HCl^- \ (g.\text{–}ex.)$$

Suschitzky *et al.* (1964) have shown that acid anhydrides which are unaffected by cold water may be readily prepared in high yield and purity by shaking an aqueous solution of a metal salt of the carboxylic acid with an acyl halide at room temperature in the presence of a t-amine, *e.g.*, pyridine; *e.g.*, heptanoic anhydride:

$$CH_3(CH_2)_5CO_2Na + CH_3(CH_2)_5COCl \xrightarrow{pyridine} [CH_3(CH_2)_5CO]_2O$$

Properties of Acetic Anhydride

Acetic anhydride is a colourless liquid, b.p. 139.5°C, with an irritating smell. It is neutral when pure, and is slightly soluble in water, but readily soluble in ether and benzene. It is hydrolysed slowly in water, but rapidly by alkali:

$(CH_3CO)_2O + H_2O \longrightarrow 2CH_3CO_2H$

In undergoes reactions similar to those of acetyl chloride, but with less vigour; only half of the acetic anhydride molecule is used in acetylation, the other half being converted into acetic acid:

$(CH_3CO)_2O + C_2H_5OH \longrightarrow CH_3CO_2C_2H_5 + CH_3CO_2H$

$(CH_3CO)_2O + NH_3 \longrightarrow CH_3CONH_2 + CH_3CO_2H$

Acetylation with aceti anhydride is usually best carried out in pyridine solution or in the presence of a small amount of sodium acetate or concentrated sulphuric acid as catalyst. Acetic anhydride reacts with dry hydrogen chloride to form acetyl chloride:

$(CH_3CO)_2O + HCl \longrightarrow CH_3COCl + CH_3CO_2H$

It is readily halogenated, and may be used in the Friedel–Crafts reaction. It reacts with aldehydes to form alkylidene diacetates; *e.g.*, with acetaldehyde it forms ethylidene diacetate:

$(CH_3CO)_2O + CH_3CHO \longrightarrow CH_3CH(OCOCH_3)2$

Acetic anhydride is reduced to ethanol by some metallic hydrides.

Acetic anhydride reacts with nitrogen pentoxide to form **acetyl nitrate**, b.p. 221°C/70 mm, which explodes violently if heated suddenly. It is hydrolysed by water to acetic and nitric acid:

$CH_3COONO_2 + H_2O \longrightarrow CH_3CO_2H + HNO_3$

Acetyl nitrate is very useful for preparing certain aromatic nitro-compounds, but is dangerous to handle.

Acylation of a hydroxy-compound to give an ester is generally a better procedure than direct esterification of the acid. It has already been pointed out that sterically hindered acids are difficult to esterify, but Parish *et al.* (1965) have shown that such acids may be readily converted into their esters by dissolving the aid in trifluoroacetic anhydride and adding the hydroxy-compound (yield: up to 97 per cent). According to the authors, the active agent is the mixed anhydride, $RCOOCOCF_3$, that is formed.

Acetyl peroxide, $CH_3COOOCOCH_3$, may be prepared by the action of barium peroxide on acetic anhydride. It is colourless, pungent-smelling liquid, p.b. 63°C/21 mm, and tends to explode on warming. It is powerful oxidising agent.

Propionic anhydride, p.b. 169°C, is manufactured, *e.g.*,

$$CH_3CH_2CO_2H + C_2H_4 + CO \xrightarrow[\text{pressure}]{\text{heat;}} (CH_3CH_2CO)_2O$$

It is mainly used for the preparation of cellulose esters.

Formic anhydride may be prepared (about 90 per cent purity) by the distillation of a mixture of diethoxymethyl formate and n-butyric formic anhydride (Schijf *et al.*, 1965).

$$HCO_2CH(OEt)_2 + PrCOOOCH \longrightarrow (HCO)_2O + PrCO_2CH(OEt)_2$$

Mixed anhydrides containing the formic acid group have been known for some time and may be prepared as follows, *e.g.*,

$$CH_3COCl + HCO_2Na \longrightarrow CH_3COOOCH + NaCl$$

This behaves as a *formylating* reagent.

ACID AMIDES

Acid amides are compounds in which the hydroxyl of the carboxyl group has been replaced by the amino-group, NH_2, to form the *amido-group*, —$CONH_2$.

There are three classes of amides: primary, $RCONH_2$; secondary, $(RCO)_2NH$; and tertiary, $(RCO)_2N$. Only the primary amides are important, and it an be seen from their formulae that all three classes may be regarded as the acyl derivatives of ammonia.

Nomenclature

According to the I.U.P.A.C. system of nomenclature, the suffix -*oic* of the parent acid is replaced by *amide,* but they are commonly named by replacing the suffix *-ic* of the trivial name of the parent acid by *amide*.

$HCONH_2$ methanamide; formamide

CH3$CONH_2$ ethanamide *or* methylcarbonamide; acetamide

General Methods of Preparation of the Amides

1. By heating the ammonium salt of the acid:

$$RCOONH_4 \longrightarrow RCONH_2 + H_2O \ (g.\text{–}v.g.)$$

Since the ammonium salts tend to dissociate on heating, the reaction is best carried out in the presence of some free acid, which represses the hydrolysis and the dissociation of the ammonium salt.

Amides may also be prepared by heating an acid with urea (Cherbuliez *et al.*, 1946):

$$RCO_2H + CO(NH_2)_2 \longrightarrow RCONH_2 + CO_2 + NH_3 \quad (g.\text{–}v.g.)$$

2. By *ammonolysis*, *i.e.*, the action of concentrated ammonia solution on acid chlorides, acid anhydrides or esters:

$$RCOCl + 2NH_3 \longrightarrow RCONH_2 + NH_4Cl \quad (v.g.)$$

$$(RCO)_2O + 2NH_3 \longrightarrow RCONH_2 + RCO_2NH_4 \quad (v.g.)$$

$$R^1COOR_2 + NH_3 \longrightarrow R^1CONH_2 + R_2OH \quad (g.)$$

The mechanisms of these reactions are uncertain, but a possibility is:

$$R—\overset{\overset{\displaystyle O}{\|}}{\underset{\underset{\displaystyle Z}{|}}{C}}—:NH_3 \rightleftharpoons R—\overset{\overset{\displaystyle O^-}{|}}{\underset{\underset{\displaystyle Z}{|}}{C}}—\overset{+}{N}H_3 \xrightarrow{NH_3} R—\overset{\overset{\displaystyle O^-}{|}}{\underset{\underset{\displaystyle Z}{|}}{C}}—NH_2 + NH_4^+$$

$$\longrightarrow R—\overset{\overset{\displaystyle O}{\|}}{C}—NH_3 + ZH + NH_3$$

N-substituted amides may be prepared by using primary or secondary amines instead of ammonia, *e.g.*,

$$R^1COCl + R^2NH_2 \longrightarrow R^1CONHR^2 + HCl$$

3. By the *graded* hydrolysis of alkyl cyanides. This may be carried out satisfactorily by dissolving the alkyl cyanide in concentrated sulphuric acid and then pouring the solution into cold water, or by shaking the alkyl cyanide with cold concentrated hydrochloric acid:

$$RC{\equiv}N + H_2O \xrightarrow{HCl} RCONH_2 \quad (g.)$$

Synder *et al.* (1954) have shown that the hydrolysis of cyanides with polyphosphoric acid gives very good yields of amide.

A possible mechanism for this hydrolysis is:

$$RC{\equiv}N: + \overset{+}{H} \longrightarrow RC{\equiv}\overset{+}{N}—H \longleftrightarrow R\overset{+}{C}{=}\ddot{N}—H \xrightarrow{H_2O} R\overset{\overset{\displaystyle \overset{+}{O}H_2}{|}}{C}{=}\ddot{N}\ H$$

$$\xrightarrow{-H^+} R\overset{\overset{\displaystyle OH}{|}}{C}{=}\ddot{N}H \rightleftharpoons R\overset{\overset{\displaystyle O}{\|}}{C}—NH_2$$

Sometimes the conversion of an alkyl cyanide into the amide may be effected by means of alkaline hydrogen peroxide. A possible mechanism is (Wiberg 1953):

$$HOOH + OH^- \longrightarrow H_2O + {}^-OOH$$

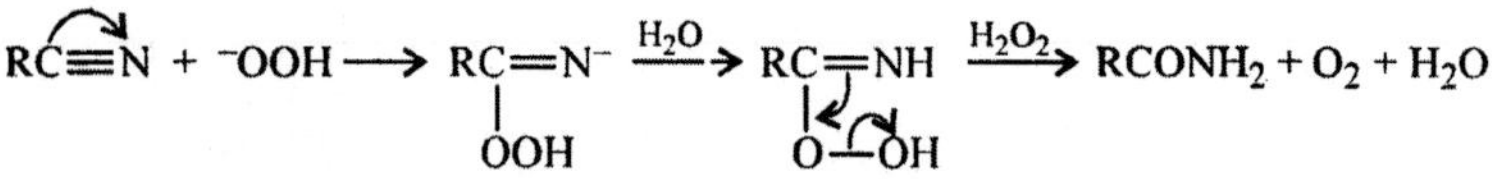

$$RC{\equiv}N + {}^-OOH \longrightarrow \underset{\displaystyle OOH}{RC{=}N^-} \xrightarrow{H_2O} \underset{\displaystyle O{-}OH}{RC{=}NH} \xrightarrow{H_2O_2} RCONH_2 + O_2 + H_2O$$

Other methods for preparing amides are the Ritter reaction, the action of ammonia on diazoketones, the Willgerodt reaction, and the Beckmann rearrangement.

General Properties and Reaction of the Amides

Except for formamide, which is a liquid, all the amides are colourless crystalline solids, and those of low molecular weight are soluble in water (with which they can from hydrogen bonds). The lower amides have much higher melting points and boiling points than are to be expected from their molecular weights; this indicates association (through hydrogen bonding):

$$H_2N{-}CR{=}O\cdots H{-}NH{-}CR{=}O\cdots H{-}NH{-}CR{=}O\cdots$$

This association has been demonstrated from infra-red spectroscopic studies on amides in solution and in the solid state (Richards *et al.*, 1947).

Infra-red spectroscopic evidence has also indicated that amides are resonance hybrids, and this is supported by N.M.R. studies, *e.g.*, the two methyl groups in dimethylformaide give two separate signals at room temperature. This can be explained on the basis that, because of the partial double bond

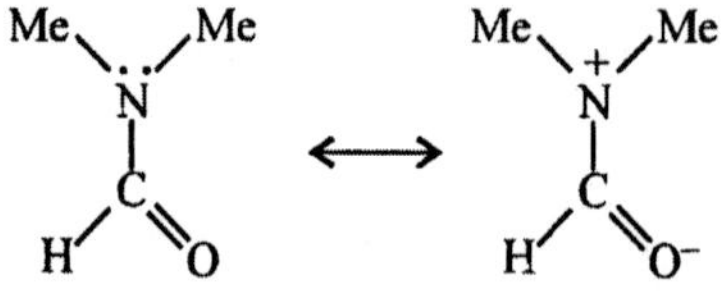

$$\underset{\displaystyle H\diagup\; C \;\diagdown\!\!\diagdown O}{\overset{\displaystyle Me\diagdown\ddot{N}\diagup Me}{|}} \longleftrightarrow \underset{\displaystyle H\diagup\; C \;\diagdown\!\!\diagdown O^-}{\overset{\displaystyle Me\diagdown\overset{+}{N}\diagup Me}{|}}$$

character of the C—N bond, the two methyl groups have *different* environments, one methyl group being *cis* and the other *trans* to the hydrogen atom. At room temperature the rate of rotation about the C—N bond is slow enough for the two methyl groups to be non-equivalent. As the temperature is raised, the signals broaden and finally collapse to a single line. At these higher temperatures, the molecule has absorbed sufficient energy to overcome the energy barrier to rotation to make the average environment of the two methyl groups the same *i.e.*, the two methyl groups are *now* equivalent.

The infra-red absorption regions of the carbonyl group (stretch) in primary amides are near 1690 (s) and 1650 cm^{-1}(s), and those of the free NH group (stretch) are near 3500 (s) and 3400 cm^{-1} (s). For a hydrogen-bonded NH group, the regions are near 3350 (m) and 3200 cm^{-1} (m).

Dimethylformamide, p.b. 152°C, is a very good solvent for polar and non-polar compounds, and is used as s formylating agent in conjunction with phosphoryl chloride.

It is manufactured by reaction between methyl formate and dimethylamine:

$$HCO_2Me + Me_2NH \longrightarrow HCONMe_2 + MeOH$$

1. Amides are hydrolysed slowly by water, rapidly by acids, and far more rapidly by alkalis:

$$RCONH_2 + H_2O \longrightarrow RCO_2H + NH_3$$

Possible mechanisms are:

Base hydrolysis

$$H\bar{O} + R\text{—}\overset{O}{\overset{\|}{C}}\text{—}NH_2 \rightleftharpoons HO\text{—}\underset{R}{\overset{\bar{O}}{C}}\text{—}NH_2 \rightleftharpoons HO\text{—}\underset{R}{\overset{O}{\overset{\|}{C}}} + \bar{N}H_2 \longrightarrow \bar{O}\text{—}\underset{R}{\overset{O}{\overset{\|}{C}}}\text{—}NH_3$$

Acid hydrolysis

$$\underset{R}{\overset{O}{\overset{\|}{C}}}\text{—}NH_2 + \overset{+}{H} \rightleftharpoons \underset{R}{\overset{O}{\overset{\|}{C}}}\text{—}\overset{+}{N}H_3 \overset{H_2O}{\rightleftharpoons} H_2\overset{+}{O}\text{—}\underset{R}{\overset{O^-}{C}}\text{—}\overset{+}{N}H_3 \rightleftharpoons$$

$$H_2\overset{+}{O}\text{—}\underset{R}{\overset{O}{\overset{\|}{C}}} + NH_3 \longrightarrow HO\text{—}\underset{R}{\overset{O}{\overset{\|}{C}}} + NH_4^+$$

2. Amides are very feebly basic and from unstable salts with strong inorganic acids, *e.g.*, $RCONH_2{\cdot}HCl$. The structure of these salts may be (I) or (II):

$$RC\begin{matrix}\nearrow\!\!\!\!/ \; O^+\text{—}H \\ \searrow NH_2\end{matrix} \longleftrightarrow RC\begin{matrix}\nearrow O\text{—}H \\ \searrow\!\!\!\!\backslash \; \overset{+}{N}H_2\end{matrix} \quad ; \quad RC\begin{matrix}\nearrow\!\!\!\!/ \; O \\ \searrow \overset{+}{N}H_3\end{matrix}$$

(I) (II)

Kutzennigg *et al.* (1962), from their proton magnetic resonance studies, have provided evidence that the salts of both acetamide and the *N,N*-disubstituted acetamides are protonated on the oxygen atom. This would be in agreement with the fact that amides are resonance hybrids, the O carrying a negative charge and the N a positive charge.

3. Amides are also feebly acidic; *e.g.*, they dissolve mercuric oxide to form covalent mercury compounds:

$$2RCONH_2 + HgO \longrightarrow (RCONH)_2Hg + H_2O$$

When amides are treated with sodium or sodamide in ethereal solution, the sodium *salt* is formed, the structure of which is probably a resonance hybrid (ambident cation):

$$RC(=O)NH^- \longleftrightarrow RC(O^-)=NH$$

Treatment of the *sodium* salt with iodide gives the *N*-alkyl derivative, whereas with the silver salt, the *O*-alkyl derivative (*imido ester*) is obtained. Imido esters are obtained by alkylation of amides with alkyl sulphates or oxonium salts, *e.g.*, $R_3O^+BF_4^-$

4. Amides are reduced by sodium and ethanol, catalytically, etc.

$$RCONH_2 \xrightarrow{Na/C_2H_5OH} RCH_2NH_2 + H_2O$$

5. When heated with phosphorus pentoxide, or better, with PCl_5, $POCl_3$, or $SOCl_2$, amides are dehydrated to alkyl cyanides:

$$RCONH_2 \xrightarrow{-H_2O} RCN$$

The mechanisms of these dehydrations are not certain, but according to Kirsanov (1954) and Lapidot *et al.* (1962), the mechanism with phosphorus pentachloride is:

$$RCONH_2 \xrightarrow{PCl_5} [RCONHPCl_4] \xrightarrow{-HCl} RC(=O)N=PCl_3 \longleftrightarrow$$

$$RC(=N-\overset{+}{P}Cl_3)O^- \longrightarrow RC\equiv N + POCl_3$$

Alkyl cyanides are also formed when the amides of the higher acids are heated to a high temperature:

$$2RCONH_2 \longrightarrow RCN + RCO_2H + NH_3$$

If this reaction is carried out in the presence of excess of ammonia, then all the amide is converted into alkyl cyanide:

$$RCO_2H + NH_3 \longrightarrow RCO_2NH_4 \xrightarrow{-H_2O} RCONH_2 \xrightarrow{-H_2O} RCN$$

6. When amides are treated with nitrous acid, nitrogen is evolved and the acid is formed:

$$RCONH_2 + HNO_2 \longrightarrow RCO_2H + N_2 + H_2O$$

This reaction is usually slow; a better method is use butyl nitrite:

$$RCONH_2 + BuONO \longrightarrow RCO_2H + N_2 + BuOH$$

7. **Hofmann preparation of primary amines** (1881). This is the conversion of an amide into a primary amine with one carbon atom less by means of bromine (or chlorine) and alkali. The overall equation for the reaction may be written:

$$RCONH_2 + Br_2 + 4KOH \longrightarrow RNH_2 + 2KBr + K_2CO_3 + 2H_2O$$

The mechanism of this reaction has been the subject of a great deal of work. A number of intermediates have been isolated: *N*-bromamides, RCONHBR; salts of these bromamides, $[RCONBr]^-K^+$; isocyanates, RNCO.

On this basis, a possible mechanism is the following, which is an example of the 1,2-shift:

$$RC(=O)\ddot{N}H_2 \xrightarrow{Br_2} RC(=O)\ddot{N}(Br)H \xrightarrow[-H^+]{KOH} [RC(=O)\ddot{N}{-}Br]^- K^+ \xrightarrow{-KBr}$$

$$[RC(=O)\ddot{N}:] \longrightarrow O{=}C{=}NR \xrightarrow{H_2O} RNH_2 + CO_2$$

(I)

The uncertainty of this mechanism lies mainly in the formation of the acyl nitrene. According to Wright (1968), who used ^{14}C- and ^{15}N-lableed compounds, the Hofmann, Curtius and Lossen rearrangements may not involve nitrenes as intermediates; he has proposed the following *concerted* mechanism:

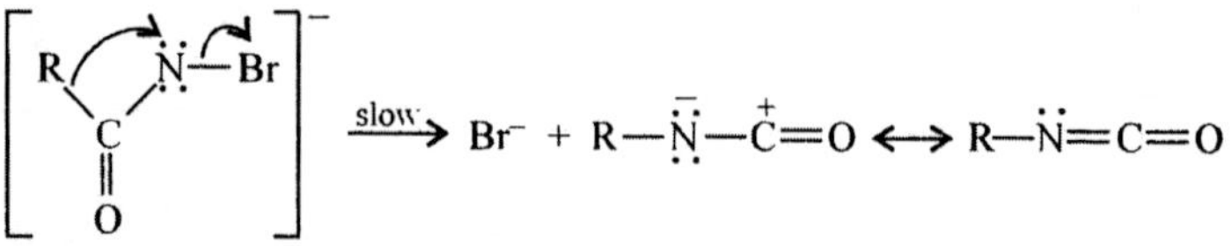

$$\left[\begin{matrix} R\curvearrowright\ddot{N}\!-\!Br \\ C \\ \| \\ O \end{matrix}\right]^{-} \xrightarrow{\text{slow}} Br^- + R-\bar{\ddot{N}}-\overset{+}{C}=O \longleftrightarrow R-\ddot{N}=C=O$$

Hauser *et al.* (1937) have shown that electron-releasing groups in the *para*-position of a migrating aryl group accelerate the reaction and electron-withdrawing groups retard the reaction. This supports the formation of a bridged ion (a phenonium ion), (I*a*), where the migrating group is aromatic, and by inference, a bridged ion (I*b*) when the migrating group is

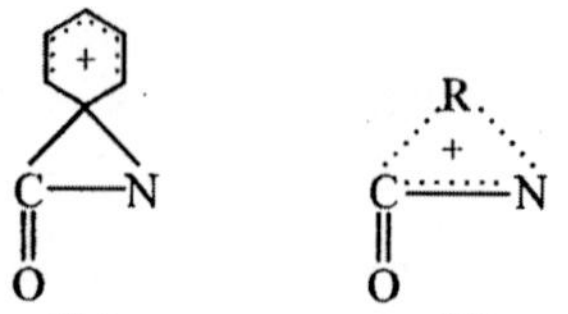

(I*a*) (I*b*)

alkyl. This is further supported by the fact that the rearrangement is intramolecular, *e.g.*, Wallis *et al.* (1931), using the amide of optically active α-benzylpropionic acid, found that there was retention of configuration in the product, α-bonzylethylamine; thus the migrating group is never free:

$$C_6H_5CH_2CHMeCONH_2 \longrightarrow C_6H_5CH_2CHMeNH_3$$

The Hofmann amine preparation can be carried out on all monocarboxylic acids, and the yields are 70–90 per cent for amides containing up to seven carbon atoms; amides with more than seven carbon atoms in the chain give mainly alkyl cyanide:

$$RCH_2CONH_2 \xrightarrow{Br_2KOH} RCN$$

The alkyl cyanide, however, may readily be reduced to the primary amine RCH_2NH_2.

It is possible to obtain the amine in good yield from the long-chain amides by modifying the procedure as follows. Bromine is added rapidly to a methanolic solution of the amide containing sodium methoxide:

$$RCOHN_2 + 2CH_3ONa + Br_2 \longrightarrow RNHCO_2CH_3 + 2NaBr + CH_3OH$$

The *N*-alkylurethan, on hydrolysis with alkali, gives the amine:

$$RNHCO_2CH_3 + 2NaOH \longrightarrow RNH_2 + Na_2CO_3 + CH_3OH$$

The Hofmann amine preparation offers an excellent method of preparing primary amines free from secondary and tertiary amines; it also affords a means of descending a series.

If the amide contains an electron-attracting group, the product is a bromide, *e.g.*, heptafluorobutyramide gives bromoheptafluoropropane (Haszeldine *et al.*, 1956):

$$C_3F_7CONH_2 \xrightarrow[NaOH]{Br_2} C_3F_7Br\ (85\%)$$

Another method for converting primary amides into amines is as follows (Beckwith *et al.*, 1965). On treatment with lead tetra-acetate, the amide gives an acylamine (40–81 per cent); this is readily hydrolysed to the amine:

$$RCONH_2 \xrightarrow[\text{benzene; reflux}]{Pb(OAc)_4} RNHAc$$

HYDROXAMIC ACIDS

The hydroxamic acids exhibit tautomerism; the keto form (I) is known as the *hydroxamic* form, and the enol form (II) as the *hydroximic* form:

$$\text{(I)}\ R-\overset{\overset{O}{\|}}{C}-NHOH \rightleftharpoons R-\overset{\overset{O}{|}}{C}=NOH\ \text{(II)}$$

Hydroxamic acids may be prepared by the action of hydroxylamine on esters or acid chlorides:

$$R-C\begin{matrix}\nearrow O \\ \searrow OC_2H_5\end{matrix} + NH_2OH \longrightarrow R-\overset{\overset{O}{/\!/}}{C}-NHOH + C_2H_5OH$$

$$RCOCl + NH_2OH \longrightarrow R-\overset{\overset{O}{/\!/}}{C}-NHOH + C_2H_5OH$$

Hydroxamic acids give a coloration with ferric chloride solution, a reaction which is characteristic of enols.

When treated with a strong inorganic acid, hydroxamic acids undergo the **Lossen rearrangement** (1875), which results in the formation of a primary amine.

The mechanism of the Lossen rearrangement is closely related to that of the Hofmann amine preparation, and so may be formulated:

$$RC(=O)NH{-}OH \xrightarrow{HCl} RC(=O)NH{-}OH_2^+ \xrightarrow{-H_2O} RC(=O)\overset{+}{N}H \xrightarrow{-H^+} RC(=O)\ddot{N}: \longrightarrow O{=}C{=}NR \xrightarrow{H_2O} RNH_2 + CO_2$$

This rearrangement (1,2-shift) has been shown to be intramolecular, and electron-releasing groups in a migrating aryl group accelerate the reaction, thereby supporting the formation of a phenonium ion.

The amides of the hydroxamic acids are known as **amidoximes**. These are tautomeric substances:

$$RC(NH_2){=}NOH \rightleftharpoons RC({=}NH)NHOH$$

Aliphatic amidoximes are best obtained by the action of hydroxylamine on an alkyl cyanide:

$$RCN + NH_2OH \longrightarrow RC(NH_2){=}NOH$$

IMIDIC ESTERS AND AMIDINES

Imidic esters, which are also known as *imino-ethers*, are best prepared by passing dry hydrogen chloride into a solution of an alkyl cyanide in anhydrous alcohol; the imidic ester hydrochloride is slowly deposited (Pinner, 1877):

$$R^1C{\equiv}N + R^2OH + HCl \longrightarrow R^1C({=}NH_2^+Cl^-)OR^2$$

Imidic ester hydrochlorides form ortho-esters when allowed to stand in an alcohol, and are readily hydrolysed to esters by water:

$$R^1C(OR^2)_3 + NH_4Cl \xleftarrow{2R^2OH} R^1C({=}NH_2^+Cl^-)OR^2 \xrightarrow{H_2O} R^1C({=}O)OR^2 + NH_4Cl$$

Imidic ester hydrochlorides are decomposed on heating:

$$R^1C({=}NH_2^+Cl^-)OR^2 \longrightarrow R^1CONH_2 + R^2Cl$$

When an ethereal solution of an imidic ester hydrochloride is treated with potassium carbonate, the free ester is obtained. When treated with an ethanolic solution of ammonia, the imidic ester hydrochloride forms the **amidine**:

$$R^1C(=NH_2^+Cl^-)OR^2 + 2NH_3 \longrightarrow R^1C(=NH)NH_2 + NH_4Cl + R^2OH$$

Formamidine acetate and acetamidine acetate may be prepared from ortho-esters (R = H or Me) as follows (Taylor *et al.*, 1960):

$$RC(OEt)_3 + NH_3 + CH_3CO_2H \longrightarrow N_2NCR{=}NH_2^+\ CH_3CO_2^-$$

These acetates are better precursors of their respective amidines than are the corresponding hydrochlorides.

Amidines are strong monoacid bases, forming salts with strong acids. Their basic strength may be explained by resonance. In the amidine, the resonating structures are different and one carries *unlike* charges.

$$R{-}C(=NH)NH_2 \longleftrightarrow R{-}C(=\bar{N}H)\bar{N}H_2$$

In the amidine *ion*, the gain of a proton results in *both* contributing structures being equivalent and carrying only a positive charge.

$$R{-}C(=\overset{+}{N}H_2)NH_2 \longleftrightarrow R{-}C(NH_2)=\overset{+}{N}H_2$$

Amidines are readily hydrolysed, hydrolysis being catalysed by acids or alkalis:

$$R{-}C(=NH)NH_2 \quad H_2O \longrightarrow R{-}C(=O)NH_2 \quad NH_3 \xrightarrow{H_2O} RCO_2NH_4$$

ACID HYDRAZIDES AND ACID AZIDES

Acid hydrazides may be prepared by the action of hydrazine on esters or acyl chlorides:

$$RCOOC_2H_5 + N_2H_4 \longrightarrow RCONHNH_2 + C_2H_5OH$$

$$RCOCl + N_2H_4 \longrightarrow RCONHNH_2 + HCl$$

Girard's reagent 'T' is trimethylaminoacetohydrazide, which may be prepared by interaction between trimethylamine, ethyl chloroacetate and hydrazine:

$$(CH_3)_3N + ClCH_2CO_2C_2H_5 + N_2H_4 \longrightarrow [(CH_3)_3NCH_2CONHNH_2]^+ Cl^- + C_2H_5OH$$

Girard's reagent 'P' is formed in a similar manner, except that pyridine is used instead of trimethylamine.

The acid hydrazides resemble the acid amides, but differ in the following ways:

(*i*) They are less readily hydrolysed than the amides:

$$RCONHNH_2 + H_2O \longrightarrow RCO_2H + N_2H_4$$

(*ii*) They are reducing agents–hydrazine is a powerful reducing agent.

(*iii*) They from acid azides when treated with nitrous acid; no nitrogen is evolved:

$$RCONHNH_2 + HNO_2 \longrightarrow RCON_3 + 2H_2O$$

Acid azides may also be prepared by the reaction between an acyl chloride and sodium azide:

$$RCOCl + NaN_3 \longrightarrow RCON_3 + NACl$$

The structure of the acid azides is best represented as a resonance hybrid:

$$R-\overset{\overset{O}{\|}}{C}-\ddot{N}=\overset{+}{N}=\overset{-}{\underset{..}{N}}: \longleftrightarrow R-\overset{\overset{O}{\|}}{C}-\underset{..}{\overset{..}{N}}^{-}-\overset{+}{N}\equiv N: \longleftrightarrow R-\overset{\overset{O^-}{|}}{C}=\ddot{N}-\overset{+}{N}\equiv N:$$

When boiled with an alcohol, acid azides undergo rearrangement to form *N*-alkyl-substituted urethans:

$$R^1CON_3 + R^2OH \longrightarrow RNHCO_2R^2 + N_2 \ (v.g.\text{-}ex.)$$

The mechanism of this rearrangement – the **Curtius rearrangement** (1894)– is similar to that of the Hofmann and Lossen rearrangements, and so may be formulated:

$$R-\overset{\overset{O}{\|}}{C}-\overset{-}{\ddot{N}}-\overset{+}{N}\equiv N \xrightarrow{-N_2} \left[R-\overset{\overset{O}{\|}}{C}-\ddot{N}\right] \longrightarrow R-N=C=O \xrightarrow{ROH} RNHCO_2R$$

This mechanism is supported by the fact that when the terminal nitrogen atom is ^{15}N in 3,5-dinitrobenzazide, all of this tracer is found in the nitrogen eliminated in the reaction (Bothner–By *et al.*, 1951).

The Curtius reaction may be used to step down a series (*cf.* the Hofmann amine preparation):

$$RCO_2H \xrightarrow{C_2H_5OH} RCO_2C_2H_5 \xrightarrow{N_2H_4} RCONHNH_2 \xrightarrow{HNO_2}$$

$$RCON_3 \xrightarrow{CH_3OH} RNHCO_2CH_3 \xrightarrow{NaOH} RNH_3$$

The Curtius reactions offers a very good method for preparing primary amines free from secondary and tertiary amines, and may also be used for preparing isocyanates and urethans; *e.g.*, when heated in benzene or chloroform solution, an acid azide rearranges to the alkyl isocyanate which can be isolated. If the isocyanate is warmed with an alcohol, an *N*-substituted urethan is formed:

$$R^1CON_3 \xrightarrow[\text{solution}]{\text{benzene}} R^1NCO \xrightarrow{R^2OH} R^1NHCO_2R^2$$

HALOGEN DERIVATIVES OF THE MONOCARBOXYLIC ACIDS

The halogen derivatives of the monocarboxylic acids are compounds in which one or more hydrogen atoms in the *carbon chain* have been replaced by halogen.

Nomenclature

The following examples illustrate the methods:

CH_2ClCO_2H	monochloroacetic acid
$CH_3CHClCO_2H$	α-chloropropionic acid or 2-chloropropanoic acid
$(CH_3)_2CClCHBrCO_2H$	α-bromo-β-chloroisovaleric acid or 2-bromo-3-chloro-3-methylbutanoic acid

α-Halogeno-acids. Although the monocarboxylic acids themselves are not readily halogenated, their acid chlorides and anhydrides may be halogenated very easily. Bromination takes place only at the α-position, but chlorination, although it occurs mainly at the α-position, may also take place further in the chain; *e.g.*, chlorination of propionic acid results in the formation of the α- and β-chloro-derivatives (see also below):

$$CH_3CH_2CO_2H \xrightarrow{Cl_2} CH_3CHClCO_2H + ClCH_2CH_2CO_2H$$

Fluorination of the monocarboxylic acids has not yet been studied in great detail, but the work done so far seems to indicate that fluorine enters the chain somewhat indiscriminately; *e.g.*, fluorination of butyric acid in carbon tetrachloride solution gives the β-and γ-fluoro-derivatives.

Iodo-derivatives are prepared from the chloro- or bromo-compound by the action of potassium iodide in methanolic or acetaone solution.

The usual method for preparing α-chloro- or bromo-acids is by the **Hell-Volhard-Zelinsky reaction (H.V.Z. reaction),** which is carried out by treating the acid with chlorine or bromine in the presence of a small amount of red phosphorus.

The mechanism of the H.V.Z. reaction is uncertain, but there is some evidence to show that it may proceed through an enol intermediate (Zimmerman *et al.*, 1959):

$$RCH_2CO_2H \xrightarrow{P + Br_2} RCH_2C(=O)Br \rightleftharpoons RCH{=}C(OH)Br \xrightarrow{Br_2}$$

$$HBr + RCHBrCObr \xrightleftharpoons{RCH_2CO_2H} RCHBrCO_2H + RCH_2COBr, \text{etc.}$$

Enolisation of the acid bromide is far more rapid than that of the acid.

The second α-hydrogen atom may be replaced by chlorine or bromine by using excess of the halogen, but whereas bromination ceases when both α-hydrogen atoms have been replaced, chlorination proceeds further in the chain (see above).

Since the H.V.Z. reaction with bromine is specific for α-hydrogen atoms, it can be used to detect the presence of α-hydrogen in an acid; *e.g.*, trimethylacetic acid does not undergo the H.V.Z. reaction with bromine.

The H.V.Z. reaction with bromine is applicable to di- and polycarboxylic acids, all the α-positions being substituted if sufficient bromine is used.

Another convenient method of preparing a-bromo-acids is by brominating an alkylmalonic acid and heating the bromo-acid, whereupon it is decarboxylated to the monobasic acid:

$$RCH(CO_2H)_2 \xrightarrow{Br_2} RCBr(CO_2H)_2 \xrightarrow{heat} RCHBrCO_2H + CO_2$$

Sulphuryl chloride, in the presence of small amount of iodine, chlorinates monocarboxylic acids in the α-position, but in the presence of organic peroxides (to give a free-radical reaction), the α-, β-, γ-, etc. positions are substituted. On the other hand, Louw (1966) has shown that cupric chloride (plus LiCl) chlorinates acid chlorides, anhydrides, and carboxylic acid (containing the acid chloride or anhydride as 'catalyst') at the α-position only (yields > 95 per cent). The reaction is carried out in a polar inert solvent, *e.g.*, sulpholane, and only the α-monochloro derivative is obtained regardless of the number of α-hydrogen atoms in the compound.

β-Halogeno-acids may be prepared by treating in α,β-unsaturated acid with halogen acid; *e.g.*, acrylic acid forms β-bromopropionic acid when treated with hydrogen bromide in acetic acid solution:

$$CH_2{=}CHCO_2H + HBr \longrightarrow BrCH_2CH_2CO_2H$$

β-Halogeno-acids may also be prepared by treating an α,β-unsaturated aldehyde with halogen acid, and oxidising the β-chloroaldehyde produced, *e.g.*, acraldehyde gives b-chloropropionic acid when treated with concentrated hydrochloric acid, and the product then oxidised with concentrated nitric acid:

$$CH_2{=}CHCHO + HCl \longrightarrow ClCH_2CH_2CHO \xrightarrow{[O]} ClCH_2CH_2CO_2H \ (60\text{-}65\%)$$

The addition of all the halogen acids to α,β-unsaturated carbonyl compounds takes place very readily, and in a direction contrary to Markownikoff's rule.

When an alkene cyanohydrin is refluxed with 40 per cent hydrobromic acid, the β-bromoacid is obtained; *e.g.*, ethylene cyanohydrin gives β-bromopropionic acid:

$$CH_2(OH)CH_2CN + 2HBr + H_2O \longrightarrow BrCH_2CH_2CO_2H + NH_4Br \ (82\text{-}83\%)$$

Alternatively, when an alkene halohydrin is oxidised with concentrated nitric acid, the halogeno-acid is produced; *e.g.*, trimethylene chlorohydrin gives β-chloropropionic acid:

$$ClCH_2CH_2CH_2OH \longrightarrow ClCH_2CH_2CO_2H \ (78\text{-}79\%)$$

γ-Halogeno-acids may be prepared by the addition of halogen acid to a β,γ-unsaturated acid; the addition occurs contrary to Markownikoff's rule (*cf.* above); *e.g.*, pent-3-enoic acid gives γ-bromo-n-valeric acid when treated with hydrobromic acid:

$$CH_3CH{=}CHCH_2CO_2H + HBr \longrightarrow CH_3CHBrCH_2CH_2CO_2H$$

γ-Halogeno-acids may also be prepared by heating a γ-hydroxyacid with oncentrated halogen acid solution *e.g.*, γ-hydroxybutyric acid gives γ-chlorobutyric acid when treated with concentrated hydrochloric acid:

$$HOCH_3CH_2CH_2CO_2H + HCl \longrightarrow ClCH_3CH_2CH_2CO_2H + H_2O$$

The methods described above apply to particular types of halogeno-acids. However, the following method may be used for preparing any halogeno-acid; *e.g.*, a δ-halogeno-acid:

$$MeO_2C(CH_2)_4CO_2H \xrightarrow[KOH]{AgNO_3} MeO_2(CH_2)_4CO_2Ag \xrightarrow[CCl_4]{Br_2}$$

$$MeO_2C(CH_2)_3CH_2Br \ (65\text{-}68\%)$$

Properties of Halogeno Acids

The α-halogeno-acids undergo most of the reactions of the alkyl halides, but the halogen atom in the acid is far more reactive than that in the alkyl halide; the enhanced reactivity is due to the – I effect of the adjacent carbonyl group. The reactions of the carboxyl group are unchanged. β-, γ- and δ-halogeno-acids undergo some of the reactions of the alkyl halides, but tend to eliminate a molecule of hydrogen halide to form an unsaturated acid or a lactone. They do not form Grignard reagents, but the halogeno-acids esters react with Grignard reagents to form halogen-substituted tertiary alcohols. The halogeno-acids are reduced to the corresponding acids by sodium amalgam, but the reduction of acids of the type $R^1R^2CClCO_2H$ with lithium aluminium hydride gives chlorohydrins (I), alcohols (II), aldehydes (III) and glycols (IV).

$$\underset{(I)}{R^1R^2CClCH_2OH} \qquad \underset{(II)}{R^1R^2CHCH_2OH} \qquad \underset{(III)}{R^1R^2CHCHO}$$

$$\underset{(IV)}{R^1R^2C(OH)CH_2OH}$$

The amount of each depends on the nature of R^1 and R^2, *e.g.*, very little glycol is obtained unless at least one of these groups is phenyl (Eliel *et al.*, 1956).

The behaviour of an halogeno-acid with alkali depends on the position of the halogen atom relative to the carboxyl group.

(*a*) α-Halogeno-acids are converted into the corresponding α-hydroxyacid:

$$RCHXCO_2H + H_2O \xrightarrow[\text{(ii) } H^+]{\text{(i) NaOH}} RCH(OH)CO_2H$$

On the other hand, if the α-halogeno-acid ester is heated with a tertiary amine, the α,β-unsaturated acid ester is formed by the elimination of a molecule of hydrogen halide; *e.g.*, ethyl α-bromobutyrate gives ethyl crotonate when heated with dimethylaniline, $C_6H_5N(CH_3)_2$:

$$CH_3CH_2CHBrCO_2C_2H_5 \xrightarrow{-HBr} CH_3CH{=}CHCO_2C_2H_5$$

(*b*) β-Halogeno-acids are converted into the corresponding β-hydroxyacid, which, on continued reflux with alkali, eliminates a molecule of water to form the α,β-unsaturated acid:

$$RCHBrCH_2CO_2H \xrightarrow{NaOH} RCH(OH)CH_2CO_2^- \xrightarrow[-H_2O]{NaOH} RCH{=}CHCO_2^-$$

(*c*) γ- and δ-Halogeno-acids are converted into lactones; *e.g.*, γ-chlorobutyric acid gives γ-butyrolactone:

$$ClCH_2CH_2CH_2CO_2H \xrightarrow{OH^-} ClCH_2CH_2CH_2CO_2^- \longrightarrow Cl^- + \underset{\text{(}CH_2CH_2CH_2CO\text{–O ring)}}{\gamma\text{-butyrolactone}}$$

This is an example of neighbouring group participation.

(*d*) ε-Halogeno-acids, etc., give the corresponding hydroxyacid; *e.g.*; ε-bromocaproic acid gives ε-hydroxycaproic acid:

$$BrCH_2CH_2CH_2CH_2CH_2CO_2H \xrightarrow[\text{(ii) } H^+]{\text{(i) NaOH}} HOCH_2CH_2CH_2CH_2CH_2CO_2H$$

The most characteristic reaction of the α-halogeno-acid esters is the *Reformatsky reaction.*

The halogeno-acids are all stronger acids than the parent acid, and for a group of isomeric acids, the further the halogen is removed from the carboxyl group the weaker is the acid.

The increase in acid strength may be explained by high electronegativity of the halogen atom which exerts a strong inductive effects, thereby facilitating the release of proton from the carboxyl group.

$X \leftarrow CH_2 \leftarrow C(=O) \leftarrow O \leftarrow H$ $\qquad$ $CH_3 \rightarrow C(=O) \rightarrow O \rightarrow H$

$X_2 \rightarrow CH \leftarrow\leftarrow C(=O) \leftarrow\leftarrow O \leftarrow\leftarrow H$ $\qquad$ $(CH_3)_2 \rightarrow CH \rightarrow\rightarrow C(=O) \rightarrow\rightarrow O \rightarrow\rightarrow H$

$X_3 \rightarrow C \leftarrow\leftarrow\leftarrow C(=O) \leftarrow\leftarrow\leftarrow O \leftarrow\leftarrow\leftarrow H$ $\qquad$ $(CH_3)_3 \rightarrow C \rightarrow\rightarrow\rightarrow C(=O) \rightarrow\rightarrow\rightarrow O \rightarrow\rightarrow\rightarrow H$

The larger the number of halogen atoms on the α-carbon atom, the stronger is the inductive effect, and consequently the stronger is the acid; also, the further removed the halogen atom is from the carboxyl group, the weaker is the inductive effect at the carboxyl group, and consequently the weaker is the acid. In the same way, since an alkyl group is electron-repelling, increasing their number on the α-carbon atom should *decrease* the strength of the acid since release of the proton will be hindered. Since – I groups in the α-position, *e.g.*, X, NO_2, OH, increase the strength of the acid, and + I groups, *e.g.*, Me, weaken the acid, the measurement of the pK_a's of various substituted acetic acids affords a means of measuring the *relative* inductive effects of different groups.

Relative inductive effects (or electronegativities) may also be measured by NMR spectroscopy. The results, however, must be accepted with reserve, since the applied field itself may affect the electron distribution in the molecule.

To study the strengths of the different members of the carboxylic acids quantitatively, we must use the equations:

(*i*) HA (solvated) $\rightleftharpoons$ H^+ (solvated) + A^- (solvated)

(*ii*) $-RT \ln K_a = \Delta G^\theta$ (ionisation) $= \Delta H^\theta$ (ionisation $- T\Delta S^\theta$) (ionisation)

This in theory, may be done by using the following thermochemical cycle:

HA (solvated) —1 (ionisation)→ **H^+ (solvated) + A^- (solvated)**

2 (desolvation) ↓ $\qquad$ 6 (solvation) ↑ $\qquad$ 7 (solvation) ↑

H^+ (vapour) $\qquad$ **A^- (vapour)**

4 ($-e$) ↑ $\qquad$ 5 ($+e$) ↑

HA (vapour) —3 (dissociation)→ **H· (vapour + A· (vapour)**

$$\Delta H_1 = \Delta H_2 + \Delta H_3 + \Delta H_4 + \Delta H_5 + \Delta H_6 + \Delta H_7$$

It then follows that:

$$\Delta S_1 = \Delta S_2 + \Delta S_3 + \Delta S_4 + \Delta S_5 + \Delta S_6 + \Delta S_7$$

Unfortunately, very few of these values are known for the various organic acids. However, it is possible to evaluate ΔH_1 from experimental values of K_a over a range of temperatures, and since ΔG is also, obtained, ΔS_1 can thus be calculated. Some values of ΔH_1 (kJ) and ΔS_1 (JK^{-1}), respectively, are in water: formic aicd, – 0·08368, – 73·64; acetic, – 0·3766, – 92·47; propionic, – 0·6694, – 95·40; chloroacetic, – 4·853, – 71·13.

It can be seen that ΔH_1 is quite small, but ΔS_1 can vary a great deal. Thus the difference in acid strengths must be due to entropy effects. An interesting point to note is the smaller entropy loss for chloroacetic acid makes this a stronger acid than acetic. ΔS_1 for trichloroacetic acid is – 8·368 JK^{-1}, and this acid is a very strong acid. These entropy changes may be explained as follows. Acetic acid is far less polar than the acetate ion, and so the latter is far more solvated than the former. Since solvation produces some sort of arrangement of the solvent molecules around the anion, this loss of randomness of solvent molecules means a decrease in entropy (*i.e.*, ΔS is negative). Because trichloroacetic acid is highly polar, it is highly solvated both in the un-ionised form and in the anion, and so the loss in entropy in this case is very much smaller. It should also be noted that when ionisation occurs, the liberated proton is also solvated. This solvation causes the 'freezing' of four molecules of water (to form $H^+(H_2O)_4$), and so this contributes to the decrease in entropy when the acid ionises. However, since the solvated proton is the same for all acids, only the change in entropy of the anion need be considered.

Chloroacetic acid may be prepared by the H.V.Z. reaction; the reaction can be carried out in direct sunlight, and in the absence of phosphorus. Chloroacetic acid is prepared industrially:

(*i*) by heating trichloroethylene with 74 per cent sulphuric acid:

$$CHCl{=}CCl_2 + 2H_2O \xrightarrow[140°C]{H_2SO_4} ClCH_2CO_2H + 2HCl$$

(ii) By the chlorination of acetic acid in the presence of acetic anhydride-sulphuric acid mixture.

Chloroacetic acid is a deliquescent solid, m.p. 61°C, soluble in water and ethanol. It finds many uses in organic syntheses, and is used in the industrial preparation of indigotin (*q.v.*). It is readily esterified with ethanol in the presence of sulphuric acid.

Ethyl chloroacetate is converted into chloroacetamide when shaken with aqueous ammonia:

$$ClCH_2COOC_2H_5 + NH_3 \longrightarrow ClCH_2CONH_2 + C_2H_5OH \text{ (78-84\%)}$$

Chloroacetamide is dehydrated to chloroacetonitrile (p.b. 123°C) by phosphorus pentoxide:

$$ClCH_2CONH_2 \xrightarrow[-H_2O]{P205} ClCH_2CN \text{ (61-70\%)}$$

Dichloroacetic acid may be prepared in the laboratory and industrially by adding calcium carbonate to a warm aqueous solution of chloral hydrate, then adding an aqueous solution of sodium cyanide, and finally heating the mixture:

$$2CCl_3CH(OH)_2 + 2CaCO_3 \xrightarrow{NaCN} (CHCl_2CO_2)_2Ca + 2CO_2 +$$

$$CACl_2 + 2H_2O \xrightarrow{HCl} CHCl_2CO_2H \text{ (88-92\%)}$$

The action of the sodium cyanide is not understood.

Dichloroacetic acid, p.b. 194°C, gives glyoxylic acid (*q.v.*) on hydrolysis:

$$CHCl_2CO_2H + H_2O \xrightarrow[(ii)\,H^+]{(i)\,NaOH} OCHCO_2H + 2HCl$$

Vigorous hydrolysis with concentrated alkali gives oxalate and glycollate, due to the glyoxylic acid undergoing he Cannizzaro reaction:

$$2OCHCO_2H + 3NaOH \longrightarrow (CO_2Na)_2 + CH_2(OH)CO_2Na + 2H_2O$$

Trichloroacetic acid is best prepared by oxidising chloral hydrate with concentrated nitric acid:

$$CCL_3CH(OH)_2 + [O] \xrightarrow{HNO_3} CCl_3CO_2H + H_2O$$

Trichloroacetic acid is a deliquescent solid, m.p. 58°C, and is one of the strongest organic acids.

The presence of three chlorine atcms on a carbon atom adjacent to a carbonyl group causes the C—C bond to break very easily. Thus, when trichloroacetic acid is boiled with dilute sodium hydroxide, or even water, chloroform is obtained:

$$CCL_3CO_2H \longrightarrow CHCl_3 + CO_2$$

The ready fission of the C—C bond may be attributed to the strong inductive effect of the chlorine atoms, and a possible mechanism is:

$$Cl_3C\text{—}CO_2^- \longrightarrow CO_2 + CCl_3^- \xrightarrow{H_2O} CHCl_3$$

With *concentrated* alkali, formates are produced, and these probably result via the formation of dichloromethylene.

Trifluoroacetic acid is conveniently prepared by the oxidation of *p*-trifluoromethyltoluidine with chromic acid:

$$CF_3C_6H_4NH_2 \xrightarrow{[O]} CF_3CO_2H$$

Trifluoroacetic acid is a liquid fumes in the air, and is one of the strongest organic acids known. It does not form fluoroform when heated with alkali, and is reduced to trifluroacetaldehyde by lithium a aluminium hydride. Peroxytrifluoroacetic acid, CF_3CO_3H, is a useful oxidising agent (Emmons, 1954; *cf.* peroxyacetic acid).

Fluoroacetic acid may be prepared by heating a mixture of carbon monoxide, hydrogen fluoride and formaldehyde under pressure.

$$CO + HF + HCHO \longrightarrow FCH_2CO_2H$$

The simplest chloro-acid would be chloroformic acid, $ClCO_2H$, but it is unknown; chlorination of formic acid results in the formation of hydrogen chloride and carbon dioxide:

$$HCO_2H \xrightarrow{Cl_2} 2HCl + CO_2$$

On the other hand, esters of chloroformic acid are known.

EXERCISES

1. Convert n-$C_5H_{11}CO_2H$ into n-$C_4H_9CO_2H$ by three methods.
2. Which of the following would be most and least readily hydrolysed with NaOH, and why?

   (i) $MeCO_2Me$; (ii) Me_2CHCO_2Me; (iii) $MeCO_2Bu^t$.
3. Using MeI, $\overset{*}{Me}I$ NaCN, $Na\overset{*}{C}\overset{*}{N}$ ($\overset{*}{C} = {}^{14}C$) and any other chemicals, suggest a synthesis for each of the following:

   (i) $\overset{*}{C}H_3CO_2H$;

(ii) $CH_3\overset{*}{C}O_2H$

(iii) $\overset{*}{C}H_3\overset{*}{C}O_2H$;

(iv) $\overset{*}{C}H_3\overset{*}{C}O\overset{*}{C}H_3$;

(v) $\overset{*}{C}H_3\overset{*}{C}HO$;

4. Complete the equations and comment:

(i) $CHCl_2COCl \xrightarrow{LAH} ?$

(ii) $HOCH_2CH_2CN \xrightarrow{HBr-H_2O} ?$

(iii) $B_{AC}2$ mech. $Me_3CCH_2OAc + OH^- \longrightarrow ?$

(iv) $B_{AL}1$ mech. $Me_3CCH_2OAc + OH^- \longrightarrow ?$

(v) $ClCH_2CO_2H + NaI \xrightarrow{acetone} ?$

(vi) $Me_2CBrCONH_2 \xrightarrow[NaOH]{Br_2} ?$

5. Calculate $\Delta G^{\ominus}$ for the equilibrium:

AcOH + EtOH AcOEt + H_2O

moles 1/3 1/3 2/3 2/3

What information does the value of $\Delta G^{\emptyset}$ give on the time required for this equilibrium to be reached?

6. Draw the energy profile for the $A_{AC}2$ esterification mechanism and comment.

7. A solution of Ph_3CCO_2H in conc. H_2SO_4 gives $MeOCPh_3$ when poured into methanol. Discuss.

8. Convert:

(i) n-BuOH into n-$PrNH_2$;

(ii) Et_2CHCO_2H into Et_2CHCHO.

9. Can $MeCH_2COCl$, $MeCHClCO_2H$ and $ClCH_2CH_2CO_2H$ be differentiated by NMR spectroscopy? Give reasons for your answer.

10. Draw the diagrammatic NMR spectra of:

(i) $MeCO_2Et$; (ii) $EtCO_2Me$.

11. Give the fragmentation paths that could produce the following ions:

(i) m/e	31	43	59	71	74	102
Me n-butyrate (%)	6	100	25	55	69	2
(ii) m/e	16	43	44	59	71	87
n-Butanamide (%)	3	40	94	100	6	3

12. Complete the following equations and comment where necessary:

(i) $\begin{matrix} CH_2CO \\ | \quad\quad O \\ CH_2CO \end{matrix} + MeOH \xrightarrow{heat} ?$

(ii) $CH_2(CO_2H)_2 + 2M\ e_2C = CH_2 \xrightarrow[Et_2O]{H_2SO_4} ?$

(iii) $HO_2C(CH_2)_4CO_2Me \xrightarrow{AgNO_3} ? \xrightarrow[CCl_4]{Br_2} ?$

(iv) $\begin{matrix} CH_2CO \\ | \quad\quad NH \\ CH_2CO \end{matrix} \xrightarrow[KOH]{Br_2} ?$

(v) $PhCH(CO_2H)(CH_2)_2CO_2H \xrightarrow[heat]{Ac_2O} ?$

(vi) $CH_2 = CH(CH_2)nCO_2R1 \xrightarrow[R^3OH]{O_3} ?$

(vii) Cyclopentanone $\xrightarrow{H_2SO_4}$

(viii) $But\ OH + NH_2CONH_2 \xrightarrow{H_2SO_4} ?$

(ix) $KNCS + NH_2NH_2 \xrightarrow[H_2O]{H_2SO_4} ?$

(x) $CH_3CH = CHCH_2CH_3 \xrightarrow{NBS} ?$

(xi) $HOCH_2(CH_2)_n\ CH_2OH \longleftarrow HO_2C(CH_2)_nCO_2Et$

$\longrightarrow HO_2C(CH_2)_nCH_2OH$

13. Synthesise from readily accessible materials:

(i) $EtCH(CN)CO_2Et$;

(ii) $EtN(CO_2\ Et)_2$;

(iii) $EtCH(CO_2H)CH_2\ CO_2H$

14. Suggest a mechanism for the reaction:

$$NH_2CONH_2{\cdot}HNO_3 \xrightarrow{H_2SO_4} NH_2CONHNO_2 + H_2O$$

15. Convert succinic acid into cyclopentanone.

16. Acetic anhydride, when refluxed with potassium premanganate, gives compound A, $C_4H_4O_3$. The i.r. specturm of A shows bands at 1860 c m^{-1} and 1784 cm^{-1}. What is A? Suggest a mechanism for its formation.

17. The electroylsis of the sodium salt of b, b-dimethylglutaric acid gives A, C_5H_{10}, which on ozonolysis, gives spectrum showed three signals, τ 9.1, 7.8, 7.6. What are A and B? Give your reasons.

18. The N.M.R. spectrum of the compound $C_8H_{14}O_4$ shows three signals: triplet, τ 8.74; quartet, τ 5.82; singlet, τ 7.4. Suggest a structure for the compound.

19. Write out all the possible isomeric dicarboxylic acids of molecular formula $C_6H_{10}O_4$ and name them by the 1. U.P.A.C. system.